Design of Digital Phase Shifters for Multipurpose Communication Systems

RIVER PUBLISHERS SERIES IN COMMUNICATIONS

Series Editors:

ABBAS JAMALIPOUR
The University of Sydney
Australia

MARINA RUGGIERI
University of Rome Tor Vergata
Italy

JUNSHAN ZHANG
Arizona State University
USA

Indexing: All books published in this series are submitted to the Web of Science Book Citation Index (BkCI), to SCOPUS, to CrossRef and to Google Scholar for evaluation and indexing.

The "River Publishers Series in Communications" is a series of comprehensive academic and professional books which focus on communication and network systems. Topics range from the theory and use of systems involving all terminals, computers, and information processors to wired and wireless networks and network layouts, protocols, architectures, and implementations. Also covered are developments stemming from new market demands in systems, products, and technologies such as personal communications services, multimedia systems, enterprise networks, and optical communications.

The series includes research monographs, edited volumes, handbooks and textbooks, providing professionals, researchers, educators, and advanced students in the field with an invaluable insight into the latest research and developments.

For a list of other books in this series, visit www.riverpublishers.com

Design of Digital Phase Shifters for Multipurpose Communication Systems

Binboga Siddik Yarman

Istanbul University-Cerrahpasa
Turkey
and
University of Lincoln
UK

Routledge
Taylor & Francis Group
LONDON AND NEW YORK

Published 2019 by River Publishers
River Publishers
Alsbjergvej 10, 9260 Gistrup, Denmark
www.riverpublishers.com

Distributed exclusively by Routledge
4 Park Square, Milton Park, Abingdon, Oxon OX14 4RN
605 Third Avenue, New York, NY 10017, USA

First issued in paperback 2023

Design of Digital Phase Shifters for Multipurpose Communication Systems / by Binboga Siddik Yarman.

Routledge is an imprint of the Taylor & Francis Group, an informa business

Publisher's Note
The publisher has gone to great lengths to ensure the quality of this reprint but points out that some imperfections in the original copies may be apparent.

While every effort is made to provide dependable information, the publisher, authors, and editors cannot be held responsible for any errors or omissions.

ISBN 13: 978-87-7022-969-2 (pbk)
ISBN 13: 978-87-7022-094-1 (hbk)
ISBN 13: 978-1-003-33785-0 (ebk)

Contents

Preface

This book aims to cover a new emerging need in designing digital phase shifter for modern communication systems.

With the advancement of new-generation mobile communication systems, directed beams of antenna arrays save substantial amount of power as well as improve the communication quality.

In this regard, beam-forming circuits such as digital phase shifters constitute essential parts of the antenna array systems.

Therefore, this book is devoted to the design of digital phase shifters for various communications systems.

In the past, phase shifters design requirements used to demand narrow bandwidth without physical size constraints.

Nowadays, phase shifters must be compact suitable for Very Large Scale Integrated Circuits (VLSI) or Microwave Monolithic Integrated Circuit (MMIC) implementation with wide frequency band.

Since the early 1980s, we have been designing digital phase shifters for various applications. We started with loaded lines phase shifters and then employed loaded branch line couplers to achieve wider frequency bands. In order to reduce the physical size, we used to employ three-element LC ladder networks in T or PI configurations.

In order to achieve broad frequency band with substantial amount of phase shift, usage of LC lattice structures is inevitable.

Lately, we designed phase shifters using both low-pass and high-pass LC lattice structures suitable for monolithic implementation. In the course of design, we utilized PIN diodes and FETs as switching elements.

In this book, first, we cover the major concepts of phase shifters as building blocks of antenna array systems (Chapter 1). Then, we introduce scattering parameters for two-port networks to analyze phase shifter circuits (Chapter 2).

In Chapter 3, we introduce theory of transmission lines.

In Chapter 4, loaded line phase shifters are presented.

Chapter 5 is devoted to analyze three-element LC low-pass and high-pass sections as phase shifting units.

In Chapter 6, a novel 180° Low-Pass-based T-Section Digital Phase Shifters Topology (LPT-DPS) is introduced.

In Chapter 7, a novel 180° Low-Pass-based PI-Section Digital Phase Shifter Topology (LPπ-DPS) is presented.

In Chapter 8, a novel High-Pass-based T-Section Digital Phase Shifter Topology is outlined.

All the phase shifter configurations introduced up to Chapter 8 yield narrow-frequency bandwidth. They perhaps result in 10–20% bandwidth around the center frequency of the passband.

In Chapter 9, we introduce a broadband and large phase range digital phase shifter topology based on symmetrical lattice structures. In this chapter, we also discuss MMIC implementation of wideband digital phase shifter topology. Complete layouts for 45°, 90°, and 180° phase shifters are presented as they are designed using TSMC – 180 nm library on Cadence. These structures are also suitable for new-generation communication systems.

In Chapter 10, we present a T-Section-based digital phase shifter topology, which provides phase shift range over 0°–360° phase. This topology is simpler than that of lattice-based phase shifter. It utilizes only three solid-state switches. In return, it offers smaller bandwidth.

In Chapter 11, we introduce a PI-Section digital phase shifter topology, which covers the phase range of 0°–360°. In a similar manner to that of Chapter 11, 360° PI-Section-based digital phase shifter topology is simpler than that of lattice-based digital phase shifter topology. However, it offers narrow frequency band.

Finally, in Chapter 12, we outlined a "180° High-Pass-based Digital Phase Shifter Topology". This topology can be directly be derived from the topology presented in Chapter 11 by performing a "two-pair-wire" type of operation in one digital phase state and a "high-pass LC ladder" operation in the other switching state.

Each chapter of this book is organized stand alone in such a way that the reader requires no specific background acquired from the other chapters.

For each phase shifter topology introduced in this book, the reader is furnished with explicit design equations to construct the circuit under consideration. Furthermore, design equations are programmed under MathLab to assess the electrical performance of the phase shifters with ideal (or lossless components) and lossy components.

It is hoped that the interested reader can immediately identify the "optimum phase shifter topology" for the need under consideration with its estimated electric performance.

If you face any problem as you go alone with your readings or with your phase shifter designs, please feel free to be in touch with me.

I hope, you enjoy reading the book.

Binboga Siddik Yarman
Vanikoy, Istanbul, Turkey

Readers of the Book

This book is prepared for a variety of readers. Personally, I used the content of the first eight chapters for my senior students in the Microwave Engineering course and in their graduation projects at Istanbul University, Technical University of Istanbul, Middle East Technical University, Anadolu University of Turkey, Ruhr University of Germany, Tokyo Institute of Technology of Japan, Wuhan Technological University of China. Furthermore, I also utilized the full content of the book for my graduate courses. In fact, all the phase shifter topologies presented in the book invented as either government projects or PhD dissertations of my doctoral students at the above-mentioned universities.

The topic "digital phase shifter designs" is the integral part of the commercial and military phase array systems. Therefore, major communication companies and research institutes, which involve designing antenna array systems, must possess this book to understand the ingredients of phase shifter designs.

Acknowledgments

This book is the consequence of my long professional journey started at Radio Corporation of America (RCA), General Sarnoff Research Center, Princeton, N.J, USA back in the early 1980s, which has not yet ended.

I was acquainted with digital phase shifters as a young research-project engineer at Microwave Technology Center of Sarnoff Laboratories. At that time, my job was to invent new circuit topologies and design methods to construct matching networks, amplifiers, and phase shifters that can operate up to high end of millimeter waves. Our projects were successfully completed. We published many state-of-the-art patents, special reports, and papers on the subject. During that era, my dear friend Dr. Arye Rosen was tremendously helpful and supportive to my family and me. In this regard, I am grateful to Dr. Rosen and his lovely wife Daniela for creating us a cozy atmosphere to run our life in New Jersey, USA.

This book was partially prepared at Frostburg State University of Maryland, USA and University of Lincoln of Lincolnshire, UK during my sabbatical year of 2017–2018. In this regard, I would like to express my gratitude to my hosts Professors Hilkat-Oguz Soysal of Frostburg and Dr. Saket Srivastava of Lincoln who furnished outstanding academic environment with me to finalize the book.

Like many other researchers, I have stolen from the family time to do research and write all my papers, patents, and books. To our convoy, my grandchildren Deha and Ilim-Su of Perse College of Cambridge, UK, have joined. Therefore, I dedicate this book to my expanding family starting from my warm-hearted wife Prof. Dr. Sema Yarman, my dear son Dr. Can Evren Yarman; including my daughter in law Belgin and my grandchildren, my brothers and sisters Professors Tolga-Ozan, Erbil-Ayse, Huseyin-Fatos, and Faruk, and finally my past away holly mothers Cemile - Zehra and holly father Vecdi Yarmans'. Let them remain in peace in their glorious space.

March 31, 2019
Vanikoy, Istanbul, Turkey

List of Figures

List of Tables

List of Abbreviations

3S-DPS	Symmetric Single and Simple Digital Phase Shifter Structure
AAS	Antenna Array Systems
AOF	Actual Operating Frequency
APF	All Pass Function
BB	Broadband
BFB	Broad Frequency Band
BIT-IN	Inserted phase bit in the n-cascaded phase shifter unit. It is designated by logical state "1"
BIT-OUT	Extracted phase bit in the n-cascaded phase shifter unit. It is designated by logical state "0"
BPE	Bit Phase Error
BPL	Bandpass Ladder
BPLS	Bandpass Ladder Structure
BPR	Broad Phase Range
BR	Bounded Real
BSY	Binboga Siddik Yarman
CMOS Device	Complementary Metal Oxide Semiconductor Device
DPS	Digital Phase Shifter
EBPE	Effective Bit Phase Error
ESBA	Electronically Steered Beam Antenna
HP	Highpass
HPL	Highpass Ladder
HPSP	Highpass Symmetric PI
HPS-PI Section	Highpass Symmetric PI Section
HPS-T Section	Highpass Symmetric T Section
HPST	Highpass Symmetric T
HPT-DPS	Highpass Based T Section Digital Phase Shifter
IM-DPS	Intrinsically Mismatch Digital Phase Shifter
ISLL-DPS	Inductively Series Loaded Line Digital Phase Shifter
Lagging LSLS	Lagging Symmetric Lattice Structure
LC-LPT-DPS	LC Lowpass Based T Section Digital Phase Shifter

Leading LSLS	Leading Symmetric Lattice Structure
LL-DPS	Loaded Line Digital Phase Shifter
LOP	Lossless One Port
LP	Lowpass
LPI-DPS	Lowpass Based PI Section Digital Phase Shifter
LPL	Lowpass Ladder
LPSP	Lowpass Symmetric PI
LPST	Lowpass Symmetric T
LPST-DPS	Lowpass Based Symmetric T Section Digital Phase Shifter
LPT-DPS	Lowpass Based T Section Digital Phase Shifter
LSLS	Lossless Symmetric Lattice Structure
LTL	Loaded Transmission Line
LTP	Lossless Two Ports
MatLab	Copy Righted Trade name of Software package of Mat Works Inc.
MMIC	Microwave Monolithic Integrated Circuits
MOSFET Device	Metal Oxide Semiconductor Field Effect Device
NMOS transistors	n-type Metal Oxide Semiconductor Field Effect Transistor
PI-DPS	PI Section Digital Phase Shifter
PIN Diode	p-Insulation-n diode
PLL-DPS	Parallel Loaded Line Digital Phase Shifter
PR	Positive Real
QF	Quality Factor
ROP	Reactance One Port
RTP	Reactance Two Port
RX	Receiver
SLPI	Symmetric Lowpass PI Section
SLS	Symmetric Lattice Section
S-Par	Scattering Parameters
SSS-DPS	Symmetric Single and Simple Digital Phase Shifter Structure
STS	Symmetric T Section
T-DPS	T Section Digital Phase Shifter
TEM	Transverse Electromagnetic
TL	Transmission Line
TPG	Transducer Power Gain
TSMC	Taiwanese Semiconductor Manufacturing Company

TSP	Transfer Scattering Parameter
TX	Transmitter
Type-I Lattice Topology	Symmetric LC-Lagging All Pass Lattice Topology
Type-II Lattice Topology	Symmetric LC-Leading All Pass Lattice Topology
VLSI	Very Large Scale Integrated Circuits
VSWR	Voltage Standing Wave Ratio
WB	Wide Band

1

Fundamentals of Digital Phase Shifters

Summary

In this chapter, the concepts of phase array antenna systems and digital phase shifters are introduced. Fundamental performance criteria to design digital phase shifters (such as gain, insertion loss, phase bits) are given.

1.1 Introduction

Phase shifters are the major building blocks of "phase array antenna" systems.

They are mostly employed in the back of transmitter antennas to electronically steer the antenna beams, as shown in Figure 1.1.

A phased array is a computer-controlled antenna, which produces a beam of radio waves that can be steered electronically to point in different directions, without mechanically turning the antenna.

They are widely used in military radars. In these days, they are employed in wireless communication systems for point-to-point communication, to reduce the radio-frequency (RF) power.

The phased array consists of an array of multiple identical antenna elements *(A)*, excited by a transmitter *(TX)*. The feed current for each antenna passes through a two-port device is called a "phase shifter", which is controlled by a computer control system *(C)*.

Phase shifters are lossless two-ports. They change the transmission phase from input to output. The phase shift is measured by the phase $\varphi_{21}(\omega)$ of the transfer scattering parameter, which is described by

$$S_{21}(j\omega) = |S_{21}(j\omega)| \, e^{j\varphi_{21}(\omega)} \tag{1.1}$$

The scattering parameters (or in short S-parameters) of two-ports are detailed in Chapter 2. Therefore, we have left the details of S-parameters to next the chapter.

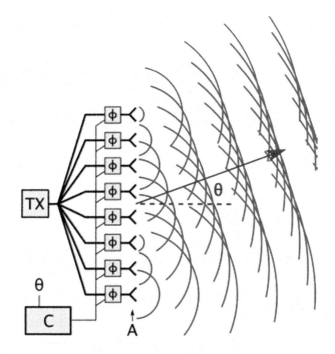

Figure 1.1   A typical schematics of a phase array antenna structure of a transmitter.

In Figure 1.1, the red lines show the constant phase-wave fronts emitted by each antenna. The waves of the individual antennas are spherical, and they add up or combine in the air, generating a plane wave and a beam of radio waves traveling in a specific direction.

By progressively delaying the phase of each antenna going up the line, the phase shifters form the beam of radio waves to be emitted at an angle of "θ" to the plane of the antennas. By changing the phase shifts, the computer control system can instantly steer the beam to any direction over a wide angle in front of the antenna.

From the design point of view, phase shifters, which are two-ports, must possess the following major features:

(a) Ideally, they must be lossless. In practice, however, they show some component losses. The phase shifter loss is either measured by its gain G, which is described as

$$G\left(\omega\right) = 10log_{10}|S_{21}\left(j\omega\right)|^2 \text{ in dB} \tag{1.2}$$

or by its insertion loss, which is described as

$$IL(\omega) = 10log_{21}\left[\frac{1}{|S_{21}(j\omega)|^2}\right] \text{ in dB} \qquad (1.3)$$

(b) They must present equal gain or insertion loss characteristics as a function of frequency for all phase states. Actually, this property is not easy to accomplish. However, at a specified operation frequency f or equivalently angular frequency $\omega = 2\pi f$, one can design the phase shifter in such a way that in all phase state gain or insertion loss is equalized.

(c) If the phase shifter two-port is lossless and consist of ideal inductors and capacitors, it is reciprocal. Therefore, it provides the same phase shift in two directions. This property makes phase shifters to be utilized at both receiver and the transmitter ends.

The above characteristics describe the electrical performance of phase shifters.

As far as actual operation is concerned, the usable bandwidth of a phase shifter and its power-handling capability are quite important, especially, when it is connected to a high-power transmitter. In this book, however, our focus is on the circuit design aspects of the phase shifters rather than on the power issues. Therefore, the power issues are omitted here.

The transmission phase $\varnothing$ of a phase shifters can be controlled electrically, magnetically, or mechanically. All the phase shifters described in this book are assumed electronically controlled, lossless, and reciprocal two-ports.

The total phase variation of a phase shifter need only be 360° to control an electronically steerable antenna (ESA) over a prescribed bandwidth. Most practical applications demand narrow bandwidth. With the advancement of the new design ideas and production technologies such as Microwave Monolithic Integrated Circuit (MMIC) production technology, one is able to construct low-loss, wide-phase-range, wide-frequency-band, and compact size phase shifters, which can steer complete 360°. Especially, in Chapter 9, the recently invented lattice type phase shifter structure provides excellent phase performance over 360°.

1.2 Concept of Digital Phase Shift

In the previous section, we have introduced the concept of transmission phase and its related device "phase shifter", which is described as a lossless,

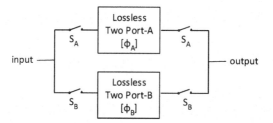

Figure 1.2   Generic schematics of a digital phase shifter.

reciprocal two-port. The phase shifter may be analog or digital. An analog phase shifter can continuously shift the transmission phase as it is designed. A digital phase shifter is an electronic circuit, which includes several switches in it. Depending on the position of switches, it varies the transmission phase as in analog manner.

A "Digital Phase Shifter (DPS)", is a two-state device, which changes transmission phase from $\emptyset_A$ to $\emptyset_B$, as shown in Figure 1.2.

Referring to Figure 1.2, in a generic digital phase shifter,

(a) when the switch S_A is closed and S_B is opened, the phase change from input to output is $\varphi_{21} = \emptyset_A$.

(b) when the switch S_B is closed and S_A is opened, the phase change from input to output is $\varphi_{21} = \emptyset_B$.

(c) the net phase change $\Delta\theta$ between the states A and B is given by

$$\Delta\theta = \emptyset_A - \emptyset_B \tag{1.4}$$

1.3 Digital Phase Bits

Most phase shifters are digitally controlled because they are immune to noise on their voltage control lines.

Digital phase shifters provide a discrete set of phase states that are controlled by two-state "phase bits." The highest order bit is 180°, the next highest is 90°, then 45°, etc., as 360° is divided into smaller and smaller binary steps. A three-bit phase shifter would have a 45° least significant bit (LSB), while a six-bit phase shifter would have a 5.6° least significant bit.

In fact, for an $n - bit$ digital phase shifter, the least significant phase bit $\emptyset_0$ is specified by

$$\emptyset_0 = \frac{360°}{2^n} \tag{1.5}$$

Table 1.1 Phase bits of an eight-bit digital phase shifter

	Phase-Bit in degree $\varphi_k = 2^k \times \emptyset_0 = 2^k \left[\frac{360^0}{2^n}\right]$
$k = 0$ *(Least Significant Bit)*	$\varphi_0 = 1 \times \emptyset_0 = 1.4063$
$k = 1$	$\varphi_1 = 2^1 \times \emptyset_0 = 2.8125$
$k = 2$	$\varphi_2 = 2^2 \times \emptyset_0 = 5.6250$
$k = 3$	$\varphi_3 = 2^3 \times \emptyset_0 = 11.250$
$k = 4$	$\varphi_4 = 2^4 \times \emptyset_0 = 22.500$
$k = 5$	$\varphi_5 = 2^5 \times \emptyset_0 = 45.00$
$k = 6$	$\varphi_6 = 2^6 \times \emptyset_0 = 90.00$
$k = n - 1 = 7$ *(Most Significant Bit)*	$\varphi_7 = 2^7 \times \emptyset_0 = 180.0$

Then, the phase bits are defined as

$$\varphi_k = 2^k \emptyset_0 = 2^k \left[\frac{360°}{2^n}\right] ; \quad k = 0, 12, 3, 4, .., (n-1) \qquad (1.6)$$

For example, an eight-phase bit digital phase shifter (DPS) must consist of the phase bits as listed in Table 1.1.

1.4 n-Bit Phase Shifter

An n-Bit digital phase shifter block can be constructed by cascading n-unit of phase shifting cells starting from the most significant phase bit of 180° at the left end, down to least significant bit of $360°/2^n$ at the right end as shown in Figure 1.3.

The least significant phase shift bit (LSB) is $360°/2^n$. The most significant phase shift bit (MSB) is 180°. At this point, we should note that k^{th} bit phase shift bit is defined as

$$\Delta\theta_k = 2^k \frac{360}{2^n} = \left[2^{(k-n)}\right][360^0] = \frac{360^0}{2^{n-k}} = \emptyset_{Ak} - \emptyset_{Bk};$$
$$k = 0, 1, 2, \ldots, (n-1) \qquad (1.7)$$

Now, let us run an example to understand (1.7) for a given phase shifter block shown in Figure 1.3.

Example 1.1: In this example, we wish to generate a truth table for a three phase-bit digital phase shifter block shown in Figure 1.3. In this block, we have cascade connection of 180, 90, and 45° phase shifters. Binary phase

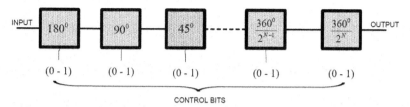

Figure 1.3   Cascaded bits of a digital phase shifter block.

states are controlled by means of control bits C_k. $C_k = 1$ refers to State-A of kth phase-bit. Similarly, $C_k = 0$ refers to State-B of the same bit. In other words, when the control bit $C_k = 1$, it refers to transmission phase $\varphi_{21A(k)} = \varnothing_{Ak}$. When $C_k = 0$, transmission phase $\varphi_{21B(k)} = \varnothing_{Bk}$.

The net phase shifts between the states are computed using (1.7) as $\Delta\theta_0 = 45°$, $\Delta\theta_1 = 90°$, and $\Delta\theta_1 = 180°$. Let us assume that transmission phase between the states are equally distributed. That is, $\Delta\theta_k = \varnothing_{Ak} - \varnothing_{Bk} > 0$ and $\varnothing_{Ak} = -\varnothing_{Bk} > 0$. In this case, for $k = 0$, $\varnothing_{A0} = -\varnothing_{B0} = 22.5° > 0$, which yields $\Delta\theta_0 = 45°$. Similarly, $\varnothing_{A1} = -\varnothing_{B1} = 45° > 0$, with $\Delta\theta_1 = 90°$; and $\varnothing_{A1} = -\varnothing_{B1} = 90° > 0$ with $\Delta\theta_1 = 180°$. Eventually, for perfectly match cascaded phase-shifter block, transmission phase φ_{21} is determined as

$$\varphi_{21} = \sum_{k=0}^{n-1} \varnothing_{A(B)k} \tag{1.8}$$

Hence, we end up with the following truth table. A close examination of Table 1.2 reveals that the phase block under consideration is able to steer a total of eight phase directions, with transmission phase such that

$$\varphi_{21} = \{-157.5, \ -112.5, \ -67.5, \ -22.5, \ +22.5, \ +67.5, \ +112.5, \ +157.5 \}$$

If the phase block under consideration is placed on an air-born platform, negative constant phase wave fronts are meaningful. However, if it is placed in the back of an antenna placed on the earth, then negative phase planes is not meaningful.

We may as well consider a three-bit phase shifter block where all $\theta_{Bk} = 0$, meaning that, in State-B, all phase shifting cells are shorted. In order to exhibit this situation, let us run the following example.

Table 1.2 Phase control truth table of a four-bit phase shifter block

Position of Phase Bits	Control Bits $C_k; k=0,1,2$			$\Delta\theta_2 = 180°$		$\Delta\theta_1 = 90°$		$\Delta\theta_0 = 45°$		φ_{21}
	C_2	C_1	C_0	$Ø_{A2}$	$Ø_{B2}$	$Ø_{A1}$	$Ø_{B1}$	$Ø_{A0}$	$Ø_{B0}$	$\sum_{k=0}^{n-1} Ø_{A(B)k}$
0	0	0	0	0	−90	0	−45	0	−22.5	−157.5
1	0	0	1	0	−90	0	−45	0	+22.5	−112.5
2	0	1	0	0	−90	+45	0	0	−22.5	−67.5
3	0	1	1	0	−90	+45	0	+22.5	0	−22.5
4	1	0	0	+90	0	0	−45	0	−22.5	+22.5
5	1	0	1	+90	0	0	−45	+22.5	0	67.5
6	1	1	0	+90	0	+45	0	0	−22.5	+112.5
7	1	1	1	+90	0	+45	0	+22.5	0	+157.5

Table 1.3 Phase control truth table of a four-bit phase shifter block

Position of Phase Bits	Control Bits $C_k; k = 0,1,2$			$\Delta\theta_2 = 180°$		$\Delta\theta_1 = 90°$		$\Delta\theta_0 = 45°$		φ_{21}
	C_2	C_1	C_0	$\varnothing_{A2}$	$\varnothing_{B2}$	$\varnothing_{A1}$	$\varnothing_{B1}$	$\varnothing_{A0}$	$\varnothing_{B0}$	$\sum_{k=0}^{n-1}\varnothing_{A(B)k}$
0	0	0	0	0	0	0	0	0	0	0
1	0	0	1	0	0	0	0	0	+45	+45
2	0	1	0	0	0	+90	0	0	0	+90
3	0	1	1	0	0	90	0	45	0	+135
4	1	0	0	+180	0	0	0	0	0	+180
5	1	0	1	+180	0	0	0	45	0	+225
6	1	1	0	+180	0	+90	0	0	0	+270
7	1	1	1	+180	0	+90	0	+45	0	+315

Example 1.2: For a three-bit phase-shifter block, all $\{\theta_{Bk} = 0; k = 0, 1, 2\}$ and $\Delta\theta_2 = \varnothing_{A2} = 180°, \Delta\theta_1 = \varnothing_{A1} = 90°$ and $\Delta\theta_0 = \varnothing_{A0} = 45°$. Complete the phase truth table.

Solution

Table 1.2 can be altered using the input data specified as $\{\theta_{Bk} = 0; k = 0, 1, 2\}$ and $\Delta\theta_2 = \varnothing_{A2} = 180°, \Delta\theta_1 = \varnothing_{A1} = 90°$ and $\Delta\theta_0 = \varnothing_{A0} = 45°$. Hence, we end up with Table 1.2.

Examination of Table 1.3 indicates that the phase shifter block sweeps the phase directions specified by transmission phase

$$\varphi_{21} = \{0, +45, +90, +135, +180, +225, +270, +315\}$$

For an earth-based antenna system, transmission phases $\varphi_{21} = \{0, +45, +90, +135, +180\}$ are meaningful.

1.5 Phase Error

One of the important characteristic of phase shifters is the phase error.

In Figure 1.4a, typical phase characteristic of a lossy high pass T-Section digital phase shifter is depicted. The phase shifter is designed for $\Delta\theta(\omega_0) = \theta_A - \theta_B = 45°$ at the normalized angular frequency $\omega_0 = 1$.

Corresponding gain characteristics are depicted in Figure 1.4b.

Due to small mismatch of the phase shifting circuit, gain characteristics for both State A (blue line) and State B (red line) are in the order of -1 dB and pretty much equalized in the neighborhood of $\omega_0 = 1$ over 10% bandwidth.

The phase shift is drifted from its original value introducing a small bit-phase error of $\varepsilon_\theta(\omega_0) = 0.09$.

In this regard, we can define the following phase error for the phase shifter under consideration.

$$\varepsilon_\theta(\omega) = \Delta\theta(\omega) - \theta_0 \qquad (1.9)$$

where θ_0 is the target phase shift between the States A and B.

In Figure 1.4c, error characteristics of the lossy high-pass-based T-Section digital phase shifter of $45°$ are shown.

There may be several methods to describe the quantified bit-phase error. For example, we can introduce the concept of effective bit-phase error

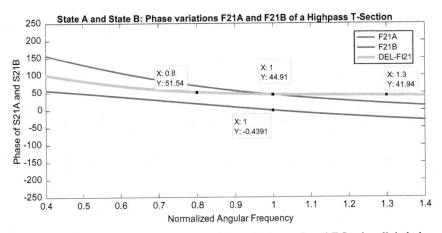

Figure 1.4a   Phase characteristic of a typical lossy high-pass-based T-Section digital phase shifter.

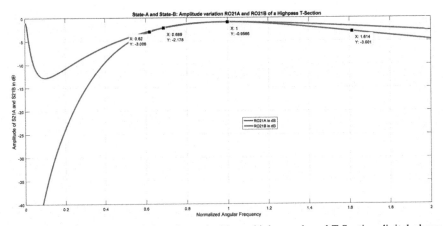

Figure 1.4b   Gain characteristics of a typical lossy high-pass-based T-Section digital phase shifter.

$\varepsilon_{\theta-eff}$ over a prescribed frequency band of $\Delta\omega$, which is defined as

$$\varepsilon_{\theta-eff} = \frac{1}{\Delta\omega} \int_{\omega_1}^{\omega_2} \varepsilon_\theta^2(\omega)d\omega \tag{1.10a}$$

where ω_1 and ω_2 are the low-end and high-end of the passband, respectively.

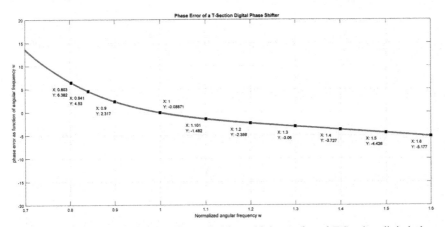

Figure 1.4c Error characteristics of a typical lossy high-pass based T-Section digital phase shifter.

If $\varepsilon_{\theta-eff}$ is evaluated over N discrete sample points, then it is approximated as

$$\varepsilon_{\theta-eff} = \frac{1}{N} \sum_{j=1}^{N} \varepsilon_\theta^2 (\omega_j) \qquad (1.10b)$$

We can define a relative bit-phase error at a specified frequency ω_j such that

$$\delta_\theta (\omega_j) = \frac{\varepsilon_\theta (\omega_j)}{\theta_0} \qquad (1.10c)$$

In this regard, we can define the maximum of $\delta_\theta (\omega_j)$ over band of interest, or we can define the useful bandwidth of a phase shifter for a pre-fixed relative error. The useful bandwidth of a phase shifter may be defined to satisfy two major criteria simultaneously such that

(a) ω_{1a} and ω_{2a} are determined to maximize the bandwidth $(\Delta\omega)_a = \omega_{2a} - \omega_{1a}$ satisfying the inequality

$$\varepsilon_\theta (\omega_j) \leq \varepsilon_{\theta a}$$

or

$$\delta_\theta (\omega_j) \leq \delta_{\theta a}$$

(b) ω_{1b} and ω_{2b} are determined to maximize the bandwidth $(\Delta\omega)_b = \omega_{2b} - \omega_{1b}$ for -3dB half-power gain.

For example, regarding the criteria of (a), if one selects $\varepsilon_{\theta a} = 4.5°$ (i.e. 10% of $\theta_0 = 45°$), then Figure 1.4c yields $\omega_{2a} = 1.4$ and $\omega_{1a} = 0.841$. Regarding the half-power criteria, Figure 1.4b reveals that
$\omega_{2b} = 1.614$ and $\omega_{1a} = 0.62$. Hence, the useful bandwidth must be selected as $(\Delta\omega)_a = 1.4 - 0.841 = 0.5590$.

1.6 Practical Issues

Digital phase shifters are more versatile than analog phase shifters. Their major advantages may be listed as below:

- Analog phase shifters may provide better loss and perhaps less production cost over the digital phase shifters. On the other hand, digital phase shifters offer the following advantages;
- A digital phase shifter is a noise-immune device on control lines;
- Digital phase shifters may provide more uniform electrical performance, unit-to-unit;
- Depending on the selected design type, a digital phase shifter may offer flat phase response over a wide frequency band;
- Digital phase shifters are less susceptible to phase pulling when embedded in networks that are not perfectly impedance-matched;
- If produced using discrete components, digital phase shifters may be easier to assemble;
- Digital phase shifters potentially offer higher power handling and linearity.

1.7 Types of Digital Phase Shifters

One can generate many types of passive digital phase shifters based on switching concepts introduced in Figure 1.3. In this figure, lossless two-ports may be selected as LC filters. For example, in one state, the LC filter could be a symmetric low-pass-T, and in the other state, it could be a symmetric high-pass-T, as shown in Figure 1.5.

Symmetric T-Sections may be replaced by symmetric PI-Sections.

Lossless two-ports may be leading and lagging symmetric lattice structures, as shown in Figure 1.6.

It may be practical to construct the lossless two-ports by means of loaded transmission lines.

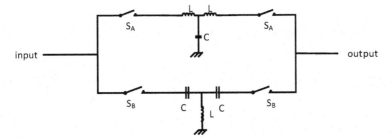

Figure 1.5 Generic schematics of a digital phase shifter constructed with low-pass/high-pass T-sections.

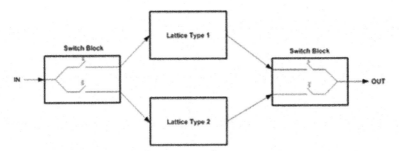

Figure 1.6 Generic schematics of a digital phase shifter constructed with symmetrical lattice sections.

One may construct the lossless two-ports using rat-race type hybrids or branch-line couplers, etc.

Digital phase shifters can be obtained by symmetrically loading these structures using reactive impedances as they are switched from one state to other.

In this book, we focus on the circuit design aspects of the digital phase shifters. In this regard, we drive the scattering parameters of various types of lossless two-ports such as low-pass/high-pass-based symmetric T/PI-sections, symmetric lattice sections, and loaded transmission lines. Their component values are determined to achieve desired phase shifts between the states.

Chapter 4 covers many types of loaded line phase shifters.

Chapter 5 includes phase shifting properties of three-element symmetric LC ladders using the scattering parameter approach.

Chapters 6–8, and 10–12 deal with digital phase shifters constructed on the 3-LC ladders.

Chapter 9 is devoted to the design of wide-phase-range, wideband digital phase shifters employing symmetric lattice two-ports.

All the work presented in this book is the outcome of our research teams from the United States and the published works may as well be traced in the references [1–7].

Further reading on phase array antennas and phase shifters can be found in the following references.[1]

Reading List on Antenna Array Systems and Phase Shifters

1. Milligan, Thomas A. (2005). Modern Antenna Design, 2nd Ed. John Wiley & Sons. ISBN 0471720607.
2. Balanis, Constantine A. (2015). Antenna Theory: Analysis and Design, 4th Ed. John Wiley & Sons. pp. 302–303. ISBN 1119178983.
3. Stutzman, Warren L.; Thiele, Gary A. (2012). Antenna Theory and Design. John Wiley & Sons. p. 315. ISBN 0470576642.
4. Lida, Takashi (2000). Satellite Communications: System and Its Design Technology. IOS Press. ISBN 4274903796.
5. Laplante, Phillip A. (1999). Comprehensive Dictionary of Electrical Engineering. Springer Science and Business Media. ISBN 3540648356.
6. Visser, Hubregt J. (2006). Array and Phased Array Antenna Basics. John Wiley & Sons. pp. xi. ISBN 0470871180.
7. Golio, Mike; Golio, Janet (2007). RF and Microwave Passive and Active Technologies. CRC Press. p. 10.1. ISBN 142000672X.
8. Mazda, Xerxes; Mazda, F. F. (1999). The Focal Illustrated Dictionary of Telecommunications. Taylor & Francis. p. 476. ISBN 0240515447.
9. Pandey, Anil (2019). Practical Microstrip and Printed Antenna Design. USA: Artech House. p. 480. ISBN 9781630816681.
10. This article incorporates public domain material from the General Services Administration document "Federal Standard 1037C" (in support of MIL-STD-188). Definition of Phased Array Archived 2004-10-21 at the Wayback Machine. Accessed 27 April 2006.
11. "Archived copy" (PDF). Archived (PDF) from the original on 2008-07-06. Retrieved 2009-04-22. Braun's Nobel Prize lecture. The phased array section is on pages 239–240.

---

[1]It is noted that the webpage called "Microwave 101" provides useful introductory material to understand phase array antennas and phase shifters with relevant references. (see https://www.microwaves101.com)

12. "Die Strassburger Versuche über gerichtete drahtlose Telegraphie" (The Strassburg experiments on directed wireless telegraphy), Elektrotechnische und Polytechnische Rundschau (Electrical technology and polytechnic review [a weekly]), (1 November 1905). This article is summarized (in German) in: Adolf Prasch, ed., Die Fortschritte auf dem Gebiete der Drahtlosen Telegraphie [Progress in the field of wireless telegraphy] (Stuttgart, Germany: Ferdinand Enke, 1906), vol. 4, pages 184–185.

13. http://www.100jahreradar.de/index.html?/gdr_5_deutschefunkmesstech nikim2wk. html Archived 2007-09-29 at the Wayback Machine Mamut1 first early warning PESA Radar

14. "A Fully Integrated 24GHz 8-Path Phased-Array Receiver in Silicon"(PDF). Archived (PDF) from the original on 2018-05-11.

15. "A 24GHz Phased-Array Transmitter in 0.18 μm CMOS" (PDF). Archived (PDF) from the original on 2018-05-11.

16. "A 77GHz 4-Element Phased Array Receiver with On-Chip Dipole Antennas in Silicon" (PDF). Archived (PDF) from the original on 2018-05-11.

17. "A 77GHz Phased-Array Transmitter with Local LO-Path Phase-Shifting in Silicon" (PDF). Archived (PDF) from the original on 2015-09-09.

18. World's Most Complex Silicon Phased Array Chip Developed at UC San Diego Archived 2007-12-25 at the Wayback Machine in UCSD News (reviewed 2 November 2007)

19. See Joseph Spradley, "A Volumetric Electrically Scanned Two-Dimensional Microwave Antenna Array," IRE National Convention Record, Part I – Antennas and Propagation; Microwaves, New York: The Institute of Radio Engineers, 1958, 204–212.

20. "AEGIS Weapon System MK-7". Jane's Information Group. 2001-04-25. Archived from the original on 1 July 2006. Retrieved 10 August 2006.

21. Scott, Richard (April 2006). "Singapore Moves to Realise Its Formidable Ambitions". Jane's Navy International. **111**(4): 42–49.

22. Corum, Jonathan (April 30, 2015). "Messenger's Collision Course With Mercury". New York Times. Archived from the original on 10 May 2015. Retrieved 10 May 2015.

23. Wallis, robert; Sheng Cheng. "Phased – Array Antenna System for the MESSENGER Deep Space Mission" (PDF). Johns Hopkins. Archived from the original (PDF) on 18 May 2015. Retrieved 11 May 2015.

24. National Oceanic and Atmospheric Administration. PAR Backgrounder-Archived 2006-05-09 at the Wayback Machine. Accessed 6 April 2006.
25. Otsuka, Shigenori; Tuerhong, Gulanbaier; Kikuchi, Ryota; Kitano, Yoshikazu; Taniguchi, Yusuke; Ruiz, Juan Jose; Satoh, Shinsuke; Ushio, Tomoo; Miyoshi, Takemasa (February 2016). "Precipitation Nowcasting with Three-Dimensional Space–Time Extrapolation of Dense and Frequent Phased-Array Weather Radar Observations". Weather and Forecasting. **31**(1): 329–340. Bibcode:2016WtFor.31.329O. doi:10.1175/WA F-D-15-0063.1.
26. P. D. Trinh, S. Yegnanarayanan, F. Coppinger and B. Jalali Silicon-on-Insulator (SOI) Phased-Array Wavelength Multi/Demultiplexer with Extremely Low-Polarization Sensitivity Archived 2005-12-08 at the Wayback Machine, IEEE Photonics Technology Letters, Vol. 9, No. 7, July 1997
27. "Electronic Two-Dimensional Beam Steering for Integrated Optical Phased Arrays" (PDF). Archived (PDF) from the original on 2017-08-09.
28. "An 8×8 Heterodyne Lens-less OPA Camera" (PDF). Archived (PDF from the original on 2017-07-13.
29. "A One-Dimensional Heterodyne Lens-Free OPA Camera" (PDF). Archi ved (PDF) from the original on 2017-07-22.
30. "Virgin, Qualcomm Invest in OneWeb Satellite Internet Plan". Space-News.com. 2015-01-15. Retrieved 2019-03-05.
31. Elon Musk, Mike Suffradini (7 July 2015). ISSRDC 2015 – A Conversation with Elon Musk (2015.7.7) (video). Event occurs at 46:45–50:40. Retrieved 2015-12-30.
32. "Mojix Star System" (PDF). Archived (PDF) from the original on 16 May 2011. Retrieved 24 October 2014.
33. "Airborne Ultrasound Tactile Display". Archived from the original on 18 March 2009. SIGGRAPH 2008, Airborne Ultrasound Tactile Display
34. "Archived copy". Archived from the original on 2009-08-31. Retrieved 2009-08-22. SIGGRAPH 2009, Touchable holography.
35. Active Electronically Steered Arrays – A Maturing Technology (ausair-power.net)
36. "YIG-sphere-based phase shifter for X-band phased array applications". Scholarworks. Archived from the original on 2014-05-27.
37. "Ferroelectric Phase Shifters". Microwaves 101. Archived from the original on 2012-09-13.

38. "Total Ownership Cost Reduction Case Study: AEGIS Radar Phase Shifters" (PDF). Naval Postgraduate School. Archived (PDF) from the original on 2016-03-03.

References

[1] B. Yarman, "T Section Digital Phase Shifter Apparatus". US Patent: 4.603.310, 29 July 1986.
[2] Binboga S. Yarman, "π-section digital phase shifter apparatus," U.S. Patent: 4604593, Washington DC, USA, August 5, 1986.
[3] B. S. Yarman, "New Circuit Configurations for Designing 0–180° Digital Phase Shifters," *IEE Proceeding H,* Vol. 134, pp. 253–260, 1987.
[4] B. S. Yarman, "Design of Digital Phase Shifters Suitable for Monolithic Implementations," *Bull. of Technical University of Istanbul,* pp. 185–205, 1985.
[5] B. S. Yarman, "Novel Cicuit Configurations to Desgn Loss Balanced 0–360° Digital Phase Shifters," *AEU,* Vol. 45, No. 2, pp. 96–104, 1991.
[6] B. Yarman, A. Rosen and P. Stabile, "Lowloss EHF Digital Phase Shifters Suitable for Monolithic Implementation," In *IEEE ISCAS,* Montreal, 1984.
[7] B. S. Yarman, "PI Section Digital Phase Shifter Apparatus". USA Patent: 460.604.593, 5 August 1986.

2

Scattering Parameters
for Lossless Two-Ports

2.1 Introduction

For many engineering applications, two-ports may be constructed to optimize
the system performance under consideration. In this case, descriptive
parameters of two-ports must be chosen in such a way that they are easily
generated on the computer to reach to the pre-set targets or goals, which are
linked to system performance.

Impedance, admittance, or hybrid parameters are somewhat idealized
since they are measured under open- or short-circuit conditions. Furthermore, they may be unbounded such as the impedance of an open circuit or
admittance of a short circuit On the other hand, scattering parameters are
bounded and very practical to describe linear active and passive systems,
such as phase-shifting circuits, since they are defined or measured under
the operational conditions. As far as computer-aided design or synthesis
problems are concerned, they present excellent numerical stability in number
crunching process.

For many engineering applications, there is a demand to construct lossless
two-ports for various kinds of problems such as filters, phase shifters,
power transfer networks, or equalizers. Therefore, in the following, first, a
formal definition of scattering parameters is given to describe, specifically,
linear active and passive two-ports. However, these definitions can easily
be extended to describe n-ports. Then, some important properties of these
parameters are summarized in classical complex variable $p = \sigma + j\omega$.

2.2 Formal Definition of Scattering Parameters

Linear two-ports may be described in terms of two wave quantities called incident and reflected waves. Literally speaking, these wave quantities are defined by means of port voltages and currents.

Referring to Figure 2.1, let us define the incident waves for port 1 and port 2 as functions of time "t" as follows

$$a_1(t) = \frac{1}{2}\left[\frac{v_1(t)}{\sqrt{R_0}} + \sqrt{R_0}i_1(t)\right] \qquad (2.1a)$$

$$a_1(t) = \frac{1}{2\sqrt{R_0}}[v_1(t) + R_0 i_1(t)] \qquad (2.1b)$$

$$a_2(t) = \frac{1}{2}\left[\frac{v_2(t)}{\sqrt{R_0}} + \sqrt{R_0}i_2(t)\right] \qquad (2.1c)$$

$$a_2(t) = \frac{1}{2\sqrt{R_0}}[v_2(t) + R_0 i_2(t)] \qquad (2.1d)$$

where $v_1(t)$ and $v_2(t)$ and $i_1(t)$ and $i_2(t)$ designate the port voltages and currents with selected polarities and directions, as shown in Figure 2.1. R_0 is the normalization resistance, which is essential to define the wave quantities. It may as well be utilized to terminate the ports under consideration. It is also known as the port normalization number.

It should be noted that the above relations are preserved under the linear transformations such as Fourier or Laplace transformation

Let the Laplace transformation of $a_1(t)$ and $a_2(t)$ be $a_1(p)$ and $a_2(p)$, respectively.

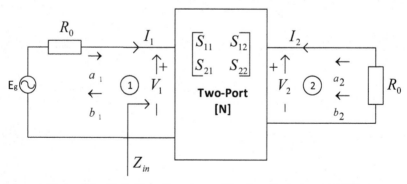

Figure 2.1 Definition of wave quantities for a linear two-port [N].

Then, they are given by

$$a_j(p) = \int_{-\infty}^{+\infty} a_j(t)e^{-pt}dt; j = 1, 2 \tag{2.2a}$$

where $p = \sigma + j\omega$ is the conventional complex frequency of the Laplace domain.

Then, denoting the time and the Laplace domain pairs as

$$a_j(t) \leftrightarrow a_j(p),$$

$v_j(t) \leftrightarrow V_j(p)$ and $i_j(t) \leftrightarrow I_j(p)$ the above equations can be directly written in p domain.

$$a_1(p) = \frac{1}{2}\left[\frac{V_1(p)}{\sqrt{R_0}} + \sqrt{R_0}I_1(p)\right] \tag{2.2b}$$

$$a_2(p) = \frac{1}{2}\left[\frac{V_2(p)}{\sqrt{R_0}} + \sqrt{R_0}I_2(p)\right] \tag{2.2c}$$

Specifically, at port-1, the incident wave quantity can be given in terms of the excitation $e_g(t) = v_1(t) + R_0 i_1(t)$ or in its Laplace domain pair $E_g(g) = V_1(p) + R_0 I_1(p)$. Thus, it is found that

$$a_1(t) = \frac{e_g(t)}{2\sqrt{R_0}} \tag{2.2c}$$

or in p-domain

$$a_1(p) = \frac{E_g(p)}{2\sqrt{R_0}} \tag{2.2d}$$

For many applications, the actual frequency response of a systems is determined by setting $p = j\omega$. In this case, the complex port quantities $a_j(j\omega)$, $V_j(j\omega)$, and $I_j(j\omega)$ are measured with properly designed equipments called network analyzers. Here, in this representation, f is being the actual operating frequency measured in Hertz and $\omega = 2\pi f$ represents the actual angular frequency in $radian/sec$.

For passive reciprocal two-ports, which consist of lumped circuit elements such as capacitors, inductors, resistances, and ideal transformers, the voltage and current quantities are expressed by means of ratio of polynomials in the complex variable "p", known as rational functions. Obviously, for passive systems, these functions must be bounded for all bounded excitations.

Similarly, reflected wave quantities are defined as time and Laplace domain pairs $b_j(t) \leftrightarrow b_j(p)$ as follows

$$b_1(t) = \frac{1}{2}\left[\frac{v_1(t)}{\sqrt{R_0}} - \sqrt{R_0}i_1(t)\right] \tag{2.2e}$$

$$b_2(t) = \frac{1}{2}\left[\frac{v_2(t)}{\sqrt{R_0}} - \sqrt{R_0}i_2(t)\right] \tag{2.2f}$$

In p-domain

$$b_1(p) = \frac{1}{2}\left[\frac{V_1(p)}{\sqrt{R_0}} - \sqrt{R_0}I_1(p)\right] \tag{2.2g}$$

$$b_2(p) = \frac{1}{2}\left[\frac{V_2(p)}{\sqrt{R_0}} - \sqrt{R_0}I_2(p)\right] \tag{2.2h}$$

At this point, it is appropriate to mention that based on the maximum power theorem, a complex exponential (or equivalently a sinusoidal) excitation $e_g(t) = E_m e^{j\omega t}$ (or equivalently $e_g(t) = E_m cos(\omega t)$ which is the real part of $\{E_m e^{j\omega t}\}$, with internal resistance R_0, delivers its maximum power to the input port, when it sees a driving point input impedance $Z_{in}(j\omega) = R_0$. In this case, the maximum average power P_A over a period $T = \frac{1}{f}$ s is given by

$$P_A = real\left\{\frac{1}{T}\int_0^T v_1(t)i_1^*(t)\,dt\right\} \tag{2.2i}$$

where

$$v_1(t) = \frac{e_g(t)}{2}$$

$$i_1(t) = \frac{e_g(t)}{(Z_{in} + R_0)} = \frac{e_g(t)}{2R_0}$$

and "*" designates the complex conjugate of a complex quantity. Thus, (2.2i) yields

$$P_A = \frac{E_m^2}{4R_0} \tag{2.2j}$$

On the other hand, amplitude square of the incident wave is given by

$$a_1(t)a_1^*(t) = |a_1|^2 = \frac{E_m^2}{4R_0}$$

which also a measure of the maximum deliverable power of the input excitation. Therefore, we say that $P_A = |a_1|^2$ is the available power of the generator.

As a matter of fact, definition of the incident wave is made in such a way that for $p = j\omega$, incident or available power of the excitation is equal to the amplitude square of the incident wave $a_1(j\omega)$ or simply, $P_A = |a_1(j\omega)|^2$.

Based on the above introduction, in the Laplace domain $p = \sigma + j\omega$, for any linear two-port, reflected waves $b_1(p)$ *and* $b_2(p)$ can be expressed in terms of the incident waves $a_1(p)$ *and* $a_2(p)$ such that

$$b_1 = S_{11}a_1 + S_{12}a_2$$
$$b_2 = S_{21}a_1 + S_{22}a_2 \qquad (2.3a)$$

or in the matrix form

$$\left[\begin{array}{c} b_1(p) \\ b_2(p) \end{array} \right] = \left[\begin{array}{cc} S_{11}(p) & S_{12}(p) \\ S_{21}(p) & S_{22}(p) \end{array} \right] \left[\begin{array}{c} a_1(p) \\ a_2(p) \end{array} \right] \qquad (2.3b)$$

where $S_{ij}(p)$; $i,j = 1,2$ are called the scattering parameters; and the scattering matrix S(p) is given by

$$S(p) = \left[\begin{array}{cc} S_{11}(p) & S_{12}(p) \\ S_{21}(p) & S_{22}(p) \end{array} \right] \qquad (2.3c)$$

Specifically, $S_{11}(p)$ and $S_{22}(p)$ are called input and the output reflectances, under R_0 terminations, and $S_{21}(p)$ and $S_{12}(p)$ are called forward and backward transfer scattering parameters from port-1 to port-2 and from port-2 to port-1, respectively.

It is important to note that for any linear active and passive two-ports, for a pre-selected normalization port numbers R_0, for any excitation applied to either port-1 and port-2, and under arbitrary port terminations, (2.1–2.3) hold for all measured sets of $\{ v_i(t) \leftrightarrow V_i(p) f$ *and* $i_i(t) \leftrightarrow I_i(p); i = 1,2\}$. In other words, scattering parameters must be invariant under any excitation and terminations.

Therefore, we can say that real normalized (or normalized with respect to R_0 ohms) scattering parameters, which are measured or derived over the entire frequency band $(-\infty < \omega = 2\pi f < +\infty)$, completely describe the linear two-ports. Hence, proper generation of the scattering parameters is essential.

In the following section, we introduce a straightforward technique to generate or measure the scattering parameters for linear networks.

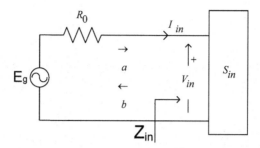

Figure 2.2   Definition of wave quantities for a one-port network.

Definition 1: Reflectance of a port

Referring to Figure 2.2, for a one-port, we can always define an incident $a(t) \leftrightarrow a(p)$ and reflected $b(t) \leftrightarrow b(p)$ waves in terms of the one port's input voltage $v_{in}(t) \leftrightarrow V_{in}(p)$ and current $i_{in}(p) \leftrightarrow I_{in}(p)$ pair. In the Laplace domain "p", the incident wave $a(p)$ is defined by

$$a(p) = \frac{1}{2}\left[\frac{V_{in}(p)}{\sqrt{R_0}} + \sqrt{R_0}I_{in}(p)\right] = \frac{1}{2\sqrt{R_0}}[V_{in}(p) + R_0 I_{in}] \quad (2.4a)$$

Similarly, the reflected wave $b(p)$ is given by

$$b(p) = \frac{1}{2}\left[\frac{V_{in}(p)}{\sqrt{R_0}} - \sqrt{R_0}I_{in}(p)\right] = \frac{1}{2\sqrt{R_0}}[V_{in}(p) - R_0 I_{in}] = \frac{V_G(p)}{2\sqrt{R_0}}$$
$$(2.4b)$$

Then, the input reflectance (or the input reflection coefficient) $S_{in}(p)$ of the one-port is defined by

$$S_{in}(p) \triangleq \frac{b(p)}{a(b)} = \frac{V_{in} - R_0 I_{in}}{V_{in} + R_0 I_{in}} = \frac{Z_{in}(p) - R_0}{Z_{in}(p) + R_0} \quad (2.4c)$$

where

$$Z_{in}(p) = \frac{V_{in}(p)}{I_{in}(p)} \quad (2.4d)$$

is called the input impedance of the one-port under consideration.

Definition 2: Normalized Impedance

The ratio

$$z_{in} = \frac{Z_{in}}{R_0} \quad (2.4e)$$

is called the normalized impedance with respect to normalization resistance R_0.

Then, using (2.4c) in (2.4e), the input reflectance is given by

$$S_{in}(p) = \frac{z_{in} - 1}{z_{in} + 1} \tag{2.4f}$$

Now, let us investigate power relations for the "one-port" under consideration.

As discussed above, available power or the incident power of the generator is given by

$$P_A = |a(j\omega)|^2 \tag{2.4g}$$

Similarly, it may be appropriate to define the reflected power from the one-port as

$$P_R = |b(j\omega)|^2 \tag{2.4h}$$

In this case, at a given angular frequency ω, the net power P_{in} delivered to the "one-port" must satisfy the relation

$$P_{in} = P_A - P_R$$

or

$$P_{in} = P_A \left[1 - \frac{P_R}{P_A} \right] = |a|^2 \left[1 - \left| \frac{b}{a} \right|^2 \right] = |a|^2 \left[1 - |S_{in}(j\omega)|^2 \right] \tag{2.4i}$$

or

Let the power transfer ratio $T(\omega)$ be defined as the ratio of the input power to the available power. Then,

$$T = \frac{P_{in}}{P_A} = 1 - |S_{in}(j\omega)|^2 \tag{2.4j}$$

(2.4j) can be expressed in terms of the normalized input impedance as

$$z_{in}(j\omega) = r_{in}(\omega) + jx_{in}(\omega)$$

$$T = 1 - \left| \frac{z_{in} - 1}{z_{in} + 1} \right|^2$$

$$T = 1 - \frac{(r_{in} - 1)^2 + x_{in}^2}{(r_{in} + 1)^2 + x_{in}^2}$$

$$T = \frac{4r_{in}}{(r_{in} + 1)^2 + x_{in}^2} \tag{2.4k}$$

It should be noted that the above relation is valid, if the definition given for the reflected power $P_R = |b(j\omega)|^2$ results in the actual power P_{ac} delivered to one-port equal to $P_{in} = P_A - P_R$. In other words, one has to show that the actual power delivered to one-port is $P_{ac} = real\{V_{in}I_{in}^*\} = P_{in}$.

In fact, this is true as shown by the following derivations.

Definition 3: Normalized voltage

The ratio

$$v_{in} = \frac{V_{in}}{\sqrt{R_0}} \tag{2.5a}$$

is called the normalized voltage.

Definition 4: Normalized current

The ratio

$$i_{in} = \frac{I_{in}}{\sqrt{R_0}} \tag{2.5b}$$

is called the normalized current.

Based on the above definition, $e_g = \frac{E_g}{\sqrt{R_0}}$ defines the normalized voltage of the excitation.

Considering definitions of incident and reflected waves, the normalized voltage and current expressions can be given as

$$v_{in}(j\omega) = a(j\omega) + b(j\omega) \tag{2.5c}$$
$$i_{in}(j\omega) = a(j\omega) - b(j\omega) \tag{2.5d}$$

Thus, the actual power delivered to the one-port is given by

$$P_{ac} = real\{(a+b)(a+b)^*\} = real\{aa^* - bb^* - ab^* + a^*b\}$$

Notice that the quantity $a^*b - ab^*$ is purely imaginary.
Therefore,

$$P_{ac} = |a|^2 - |b|^2 \tag{2.5e}$$

which is described as the input power P_{in}. Thus, we conclude that definition given for the reflected power $P_R = |b|^2$ is appropriate.

Now, let us investigate some important properties of the input reflection coefficient S_{in}.

Property 1: Positive Real Function

If one-port consists of passive lumped elements such as inductors, capacitors, resistors, and transformers, then the input impedance $z_{in}(p)$ is a positive real

(PR) and rational function in the complex variable $p = \sigma + j\omega$ and it is expressed as

$$z_{in}(p) = \frac{N(p)}{D(p)} \tag{2.6}$$

where both numerator $N(p)$ and denominator $D(p)$ polynomials are Hurwitz[1], perhaps with simple roots on the imaginary axis $j\omega$ of the complex $p -$ plane. Thus, the input admittance

$$y_{in}(p) = \frac{1}{z_{in}(p)}$$

is also a PR, rational function in p. In this case, the corresponding input reflectance $S_{in} = \frac{z_{in}-1}{z_{in}+1}$ must be a real rational function in p such that

$$S_{in}(p) = \frac{h(p)}{g(p)} \tag{2.7a}$$

where

$$h(p) = N(p) - D(p) = h_1 p^n + h_2 p^{n-1} + \cdots + h_n p + h_{n+1}$$

$$g(p) = N(p) + D(p) = g_1 p^n + g_2 p^{n-1} + \cdots + g_n p + g_{n+1}$$

In (2.7), degree "n" of polynomials $h(p)$ and $g(p)$ refers to the total number of reactive elements (i.e. capacitor and inductors) in one-port under consideration.

Let us go through the following example to find the input reflectance of a simple one-port consist of parallel combination of a capacitor $C = 1$ F and resistor $R = 1\Omega$ as shown in Figure 2.3.

Example 2.1: Determine the input reflectance of the one-port shown in Figure 2.3.

Solution

The input impedance of the one-port is given as

$$Z_{in} = \frac{1}{p+1} = \frac{N(p)}{D(p)}$$

---

[1]A Hurwitz polynomial must have all its roots in the closed left half plane (LHP) in the complex domain $p = \sigma + j\omega$. Or equivalently, a Hurwitz polynomial must be free of open right half plane (RHP) roots. These definitions imply that a root on the $j\omega$ is allowable.

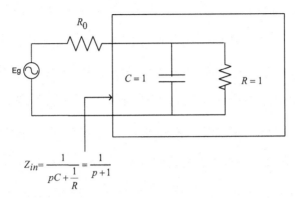

Figure 2.3 R//C load as a one-port.

with

$$N(p) = 1, \ D(p) = p + 1$$

Then,

$$h(p) = N(p) - D(p) = 1 - p - 1 = -p$$

and

$$g(p) = N(p) + D(p) = 1 + p + 1 = 2 + p$$

Thus,

$$S_{in}(p) = -\frac{p}{p+2}$$

Property 2: Subtraction of Hurwitz polynomials

Subtractions of Hurwitz polynomials result in an arbitrary polynomial with real coefficients.

Therefore, the numerator polynomial $h(p)$ is an ordinary polynomial with real coefficients.

Property 3: Addition of Hurwitz polynomials

Addition of two Hurwitz polynomials results in Strictly Hurwitz polynomial which has all its roots in the closed left half plane (LHP)[2].

According to Property 3, the denominator polynomial $g(p)$ must be strictly Hurwitz, which makes the input reflection coefficient $S_{in}(p)$ regular on the $j\omega$ axis as well as in open right half plane (RHP) of the complex domain $p = \sigma + j\omega$. In other words, $S_{in}(p)$ is bounded over the entire frequency band as well as in open RHP.

---

[2]By definition, in strictly Hurwitz polynomials, $p = j\omega$ roots are not permissible.

It is also important to note that on the $j\omega$ axis, positive real functions must always yield non-negative real parts. Therefore, if $z_{in}(j\omega) = r_{in}(\omega) + jx_{in}(\omega)$ is a PR function, then $r_{in}(\omega)$ must be non-negative over the entire frequency axis, which bounds the amplitude square of the input reflectance by 1 as detailed below.

$$|S_{in}(j\omega)|^2 = \left|\frac{z_{in} - 1}{z_{in} + 1}\right| = \frac{(r_{in} - 1)^2 + x_{in}^2}{(r_{in} + 1)^2 + x_{in}^2} \qquad (2.7b)$$

or

$$|S_{in}(j\omega)|^2 = \left|\frac{h(j\omega)}{g(j\omega)}\right|^2 \leq 1 \; for \; all \; \omega \qquad (2.7c)$$

or equivalently

$$|h(j\omega)| \leq |g(j\omega)| \qquad (2.7d)$$

By analytic continuation[3], in Equation (6.27), complex variable $p = j\omega$ can be replaced by its full version $p = \sigma + j\omega$ yielding $S_{in}(p) = \frac{h(p)}{g(p)} \leq 1$ *for all* $p = \sigma \geq 0$ since $g(p)$ is strictly Hurwitz (free of closed RHP zeros). Now let us verify (2.7) for the one-port shown in Figure 2.3.

Example 2.2: Show that (2.7) holds for the one-port depicted in Figure 2.3.

Solution

By Example (6.1), $h(j\omega)$ and $g(j\omega)$ are given as $h(j\omega) = -j\omega$ and $g(j\omega) = 2 + j\omega$ or $|h(j\omega)|^2 = \omega^2$ and $|g(j\omega)|^2 = 4 + \omega^2$. Obviously, $\omega^2 < 4 + \omega^2$ for all ω. Thus, (2.7) holds.

Property 4: Lossless One-Ports

If there is no reflection from the one-port network ($P_R = |b(j\omega)|^2 = 0$), then the incident power is dissipated on the one port yielding $T = \frac{P_{in}}{P_A} = 1 - |S_{in}|^2 = 1$ or equivalently $S_{in} = 0$. In this case, $z_{in} = 1$ or the actual input impedance Z_{in} must be equal to the normalization resistance R_0. This situation describes a perfect match between the excitation with internal resistance R_0 and one-port network.

On the other hand, if one-port network is purely reactive ($r_{in} = 0$), then $|S_{in}|^2 = 1$. In this case, one-port reflects the incident power back to the generator. Obviously, this power must be dissipated on the internal impedance

---

[3]That is to say by expanding $p = j\omega$ to the right half plane by setting $p = \sigma + j\omega$ with $\sigma \geq 0$.

of the generator. In practice, this fact may burn the excitation source. There-fore, one has to take necessary measures to avoid this undesirable situation, perhaps by introducing attenuation over a circulator to dissipate the reflected power.

It is remarkable to note that the above properties can be derived from (2.4–2.6) in a straightforward manner as follows.

For a passive one-port network, the actual power delivered to its port must be non-negative. Therefore, $P_{ac} = |a|^2 - |b|^2 \geq 0$. If one-port is lossless (i.e. purely reactive yielding $z_{in}(j\omega) = jx_{in}(\omega)$, then there will be no power dissipated on it. In this case, $P_{ac} = 0$ and the equality holds.

By analytic continuation (i.e. replacing $j\omega$ by $p = \sigma + j\omega$; $\sigma \geq 0$), we can state that

$$a(p)\,a(-p) - b(p)\,b(-p) \geq 0 \; for \; all \; p = \sigma \geq 0$$

Replacing $b(p)$ by $S_{in}\,a(p)$, it is found that

$$a(p)\,[1 - S_{in}(p)\,S_{in}(-p)]\,a(-p) \geq 0 \, for \; all \; p = \sigma \geq 0$$

or equivalently,

$$1 - S_{in}(p)\,S_{in}(-p) \geq 0 \; for \; all \; p = \sigma \geq 0.$$

For lossless one-ports,
$$S_{in}(p)\,S_{in}(-p) = 1$$

or in general for a passive one port, the inequality

$$S_{in}(p)\,S_{in}(-p) \leq 1$$

must be satisfied by all $p = \sigma \geq 0$.

Because of the aforementioned properties, the rational form of

$$S_{in}(p) = \frac{h(p)}{g(p)}$$

is called the bounded-real (BR) function.

Example 2.3: Referring to Figure 2.4, let us compute the reflectance for a one-port consisting of a single inductor.

Figure 2.1. A one-port consisting of a single inductor (a purely imaginary input impedance)

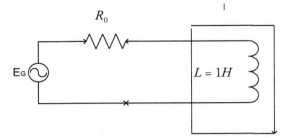

Figure 2.4 R//C load as a one-port.

In this case, $z_{in} = p$. On the $j\omega$ axis,

$$|S_{in}|^2 = \left|\frac{b(j\omega)}{a(j\omega)}\right|^2 = \left|\frac{j\omega - 1}{j\omega + 1}\right|^2 = \left|\frac{1 + \omega^2}{1 + \omega^2}\right| = 1$$

This result indicates that the reflected power is equal to incident power. In other words, one-port dissipates no power if it is purely reactive as expected.

Property 5: Bounded Realness of the Input Reflectance

In (2.7), amplitude square functions $|h(j\omega)|^2$ and $|g(j\omega)|^2$ of ω define non-negative even polynomials $H(\omega^2)$ and $G(\omega^2)$ such that

$$H(\omega^2) = h(j\omega)h(j\omega) = H_1\omega^{2n} + H_2\omega^{2(n-1)} + \cdots + H_n\omega^2 + H_{n+1}$$
$$G(\omega^2) = g(j\omega)g(j\omega) = G_1\omega^{2n} + G_2\omega^{2(n-1)} + \cdots + G_n\omega^2 + G_{n+1}$$
$$H(\omega^2) \le G(\omega^2) \tag{2.7e}$$

At this point, we can comfortably state that one can always find a non-negative even polynomial

$$F(\omega^2) = f(j\omega)f(j\omega) = F_1\omega^{2n} + F_2\omega^{2(n-1)} + \cdots + F_n\omega^2 + F_{n+1}$$

such that the last inequality can be transformed to an equality as follows

$$G(\omega^2) = H(\omega^2) + F(\omega^2) \tag{2.7f}$$

In fact, the physical existence of $F(\omega^2)$ can easily be verified by a close examination of the power transfer ratio (PTR) given by (2.4k)

Obviously, the power transfer ratio $T\left(\omega^2\right)$ of (2.4k) describes an even, non-negative, bounded, and real function over the entire frequency axis ω as follows

$$T = 1-|S_{in}|^2 = 1 - \frac{H\left(\omega^2\right)}{G\left(\omega^2\right)} = \frac{G\left(\omega^2\right)-H\left(\omega^2\right)}{G\left(\omega^2\right)} = \frac{F\left(\omega^2\right)}{G\left(\omega^2\right)} \leq 1; \ \forall\omega$$

(2.7g)

Thus, $F\left(\omega^2\right)$ appears as the numerator polynomial of $T\left(\omega^2\right)$.

Clearly, zero $F\left(\omega^2\right)$ are the zeros of the power transfer ratio.

Here, it is interesting to note that if the passive one-port consists of lumped elements and transformers, then one can immediately generate the power transfer ratio function $T\left(\omega^2\right)$ in terms of network descriptive functions step by step as follows.

In the first step, normalized input impedance $z_{in}\left(p\right)$ is generated as a positive real function as $z_{in}\left(p\right) = \frac{N(p)}{D(p)}$ with $N\left(p\right) = a_1 p^{na} + a_2 p^{na-1} + \cdots + a_{na} + a_{na+1}$ and $D\left(p\right) = b_1 p^{nb} + b_2 p^{nb-1} + \cdots + b_{nb} + b_{nb+1}$ such that $0 \leq |na - nb| \leq 1$ Here, the real coefficients $\{a_i, b_j\}$ are explicitly computed by means of the lumped element values of the one-port under consideration.

In the second step, input reflection coefficient is derived as $S_{in}\left(p\right) = \frac{h(p)}{g(p)}$ where $h\left(p\right) = N\left(p\right) - D(p)$ and $g\left(p\right) = N\left(p\right) + D(p)$. In this step, all the coefficients of $h(p)$ and $g(p)$ are computed in terms of the coefficients of N(p) and D(p). Obviously, degree "n" is determined as $max\ of\ (na, nb)$.

In the third step, even functions $H\left(p^2\right) = h\left(p\right) h\left(-p\right)$ and $G\left(p^2\right) = g\left(p\right) g\left(-p\right)$ are derived.

In the fourth step, $F\left(p^2\right)$ of Equation (6.30) is generated as $F\left(p^2\right) = G\left(p^2\right) - H\left(p^2\right)$

Finally, in the fifth step, replacing p^2 by $-\omega^2$, $T\left(\omega^2\right)$ is generated as in (2.7g).

Let us run a simple exercise to generate the power ratio for a one-port.

Example 2.4: Derive the power transfer ratio function for the one-port depicted in Figure 2.4.

Solution

In Example 2.1, we have already obtained $S_{in}\left(p\right) = \frac{h(p)}{g(p)} = \frac{-p}{p+2}$ yielding

Then,

$$H\left(p^2\right) = h\left(p\right) h\left(-p\right) = -p^2$$

and

$$G\left(p^2\right) = (p+2)(-p+2) = -p^2 + 4$$

Thus,

$$F\left(p^2\right) = G\left(p^2\right) - H\left(p^2\right) = 4$$

yielding

$$T\left(\omega^2\right) = \frac{4}{\omega^2 + 4}$$

Property 6: Darlington's Theorem

In his remarkable PhD dissertation, it was shown by Sydney Darlington that any positive real immittance function[4] could be realized as a lossless two-port in resistive termination R. However, this resistive termination can always be leveled to the normalizing resistance R_0 by using a transformer, which is also lossless. This property is known as the Darlington theorem that constitutes the heart of the classical filter theory[5].

Now, let us exercise the Darlington theorem in the following example.

Example 2.5: Show that $z\left(p\right) = \frac{2p^2+8p+1}{p+4}$ is a positive real impedance function.

Find the zeros of the even part of $z(p)$.

Synthesize $z(p)$ as a lossless two-port in unit termination.

Solution

In order $z\left(p\right)$ to be a positive real function, the followings must be satisfied.

When $p = \sigma$, $z\left(\sigma\right)$ must be real. In fact, this is the case.

When $\sigma \geq 0$, $z\left(\sigma\right)$ must be non-negative. As a matter of fact

$$z\left(\sigma\right) = \frac{2\sigma^2 + 8\sigma + 1}{\sigma + 1} \geq 0 \; for \; all \; \sigma \geq 0$$

Now let us find the even part of $z(p)$, which is designated by $r\left(p^2\right)$

$$r\left(p^2\right) = \frac{1}{2}\left[z\left(p\right) + z(-p)\right] = \frac{4}{-p^2 + 16}$$

---

[4]Immittance function is a generic name such that it either refers to an impedance or an admittance function.

[5]Darlington Theorem, PhD dissertation, 1939, University of New-Hampshire.

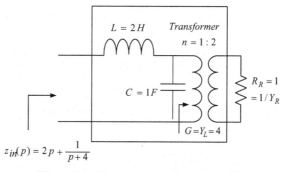

Figure 2.5   Darlington synthesis of $z_{in}(p)$.

A close examination the above results reveals that that $r\left(p^2\right)$ is free of finite zeros and its zeros are all at infinity of order 2. Zeros at infinity can easily be extracted by continuous fraction expansion of $z(p)$, which is also known as long division. Thus,

$$z\left(\mathrm{p}\right) = 2\mathrm{p} + \frac{1}{\mathrm{p} + 4}$$

which yields a series inductance $L = 2\ Henry$ and a parallel capacitor $C = 1\ Farad$, and finally, terminated in a conductance of 4 Siemens. Eventually, the resistive termination can be shifted to 1 ohm by means of a transformer, as shown in Figure 2.5.

We can extend the above properties given for the one-port reflectance $S_{in}\left(p\right)$ to cover the complete scattering parameters of lossless two-ports. However, in the following section, first let us consider the generation procedures of the scattering parameters for two-ports and then study the properties of the lossless two-ports.

2.3 Generation of Scattering Parameters for Linear Two-Ports

Referring to Figure 2.6, let $\{S_{i,j}(p);\ i,j = 1,2\}$ be the scattering parameters of a linear passive two-port. Let E_{Gk} be an arbitrary excitation with internal complex impedance $Z_{Gk}\left(p\right)$ applied to either port-1 or port-2. Let $Z_{Lk}\left(p\right)$ be any complex impedance that terminates the other end of the two-port. Since the scattering parameters are invariant under any terminations, port reflected waves $b_{1k}\left(p\right)\ and\ b_{2k}\left(p\right)$ are always related to port incident waves $a_{1k}\left(p\right)\ and\ a_{2k}\left(p\right)$ linearly by means of scattering parameters.

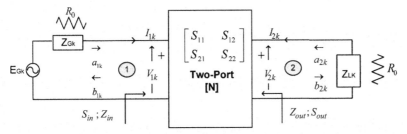

Figure 2.6 Two-port under arbitrary terminations.

$$b_{1k}(p) = S_{11}(p)\, a_{1k}(p) + S_{12}(p)\, a_{2k}(p)$$
$$b_{2k}(p) = S_{21}(p)\, a_{1k}(p) + S_{22}(p)\, a_{2k}(p)$$

for any "k" (i.e. for any source and load terminations).

Obviously, under arbitrary port terminations $Z_{Gk}(p)$ and $Z_{Lk}(p)$, one can always measure port voltages and currents or related incident and reflected wave quantities for a selected common port normalization number R_0 satisfying the equation set given by Equation (6.31). Then, we can always compute input and output reflectances yielding $S_{in} = \frac{b_{1k}}{a_{1k}} = \frac{Z_{in}-R_0}{Z_{in}+R_0}$ and $S_{out} = \frac{b_{2k}}{a_{2k}} = \frac{Z_{out}-R_0}{Z_{out}+R_0}$, respectively.

In this representation, Z_{in} is the driving point input impedance when the two-port is terminated in Z_{Lk}. Similarly, Z_{out} is the output impedance when the input port is terminated in Z_{Gk}.

On the other hand, (2.8) implies that S_{11} *and* S_{22} can directly be measured for any excitation applied to port-1 (with an arbitrary internal impedance Z_{Gk}) when a_{2k} is zero.

This can easily be achieved by choosing $Z_{Lk} = R_0$ meaning that $V_{2k} = -Z_{Lk}I_{2k} = -R_0 I_{Lk}$ or $a_{2k} = \frac{1}{2\sqrt{R_0}}[V_{2k} + R_0 I_{2k}] = 0$. In this case, the ratio $\frac{b_{1k}}{a_{1k}}$ measures S_{11} and S_{21} is measured as the ratio of $\frac{b_{2k}}{a_{1k}}$. Similarly, when port-1 is terminated in R_0 forcing $a_{1k} = 0$ and port-2 is derived by an arbitrary excitation, then, S_{22} *and* S_{12} are measured as $S_{22} = \frac{b_{2k}}{a_{2k}}$ *and* $S_{12} = \frac{b_{1k}}{a_{2k}}$.

The above measurements can be summarized as follows

$$S_{11} = \frac{b_{1k}}{a_{1k}} \; and \; S_{21} = \frac{b_{2k}}{a_{1k}} \; when \; a_{2k} = 0 \; which \; requires \; Z_{Lk} = R_0$$

$$S_{22} = \frac{b_{2k}}{a_{2k}} \; and \; S_{12} = \frac{b_{1k}}{a_{2k}} \; when \; a_{1k} = 0 \; which \; requires \; Z_{Gk} = R_0$$

$$(2.8a)$$

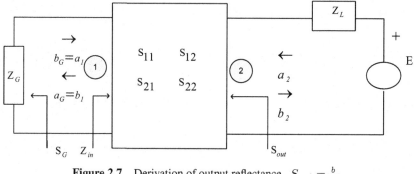

Figure 2.7   Derivation of output reflectance $S_{out} = \frac{b}{a}$

Once the scattering parameters are measured, input (S_{in}) and output (S_{out}) reflectances of the two-port can be easily be obtained in terms of these parameters.

Referring to Figure 2.7, let us first determine the output reflection coefficient

$S_{out} = \frac{b_2}{a_2}$ when port 1 in terminated in arbitrary impedance Z_G.

Considering (2.8), by definition,

$$S_{out} = \frac{b_2}{a_2} = S_{21}\left(\frac{a_1}{a_2}\right) + S_{22}$$

On the other hand,

$$\frac{b_1}{a_1} = S_{11} + S_{12}\left(\frac{a_1}{a_1}\right)$$

Now, let us consider the input termination Z_G as a passive one-port with incident wave a_G and reflected wave b_G. By definition, the reflectance S_G is given by $S_G = \frac{b_G}{a_G}$ and in terms of the impedance Z_G, $S_G = \frac{Z_G - R_0}{Z_G + R_0}$ as shown in the previous section. However, a close examination of Figure 2.7 reveals that, at port-1, $a_G = b_1$ and $b_G = a_1$

Hence,

$$\frac{b_1}{a_1} = \frac{1}{S_G}$$

Then, one obtains that $\frac{1}{S_G} = S_{11} + S_{12}\left(\frac{a_2}{a_1}\right)$ or $\frac{a_1}{a_2} = \frac{S_{12}S_G}{1 - S_{11}S_G}$. Employing this result in S_{out}, we have

$$S_{out} = S_{22} + \frac{S_{12}S_{21}}{1 - S_{11}S_G}S_G \tag{2.9a}$$

Similarly,

$$S_{in} = S_{11} + \frac{S_{12}S_{21}}{1 - S_{22}S_L}S_L \tag{2.9b}$$

2.4 Transducer Power Gain in Forward and Backward Directions

At this point, it is interesting to look into the ratio specified by

$$|S_{21}(j\omega)|^2 = \left|\frac{b_2(j\omega)}{a_1(j\omega)}\right|^2$$

when $a_2 = 0$

By definition, $b_2 = \frac{1}{2\sqrt{R_0}} = [V_2 - R_0I_2]$ and the voltage V_L across the termination R_0 is $V_L = -R_0I_2$. Then,

$$|b_2|^2 = \frac{|V_L|^2}{R_0}$$

which is the power P_L delivered to termination R_0 at port-2[6].

On the other hand, we know that $|a_1|^2$ is the available power P_A. Thus, $|S_{21}|^2$ measures the rate of power transfer from port-1 to port-2 (forward power transfer ratio), as the ratio of power delivered to the termination R_0 to the available power of the generator with internal resistance R_0.

In fact, this is the formal definition of transducer power gain $(TPG)_F$ under resistive terminations R_0 at both ends yielding

$$(TPG)_F = \frac{P_L}{P_A} = |S_{21}|^2 \tag{2.9c}$$

where subscript "*F*" refers to power transfer in forward direction (or from port-1 to port-2).

Similarly, referring to Figure 2.5, transducer power gain $(TPG)_B$ in backward direction is given by

$$(TPG)_B = \left|\frac{b_1}{a_2}\right|^2 = |S_{12}|^2 \tag{2.9d}$$

when $a_1 = 0$.

---

[6]It is interesting to note that $|b_2|^2$ is the reflected power of port 2, which is directly dissipated on the termination R_0 when $a_2 = 0$.

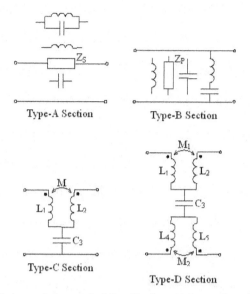

Figure 2.8   Simple lumped element building blocks, which constitute lossless two-ports.

2.5 Properties of the Scattering Parameters of Lossless Two-Ports

A lumped element lossless two-port, which consists of ideal inductors, capacitors, coupled coils, and transformers, can be constructed by cascading the sections of type A, B, C, and D, as shown in Figure 2.8.

Let $S_{i,j}(p)$; $i, j = 1, 2$ designate the scattering parameters of the lumped-lossless two-port under consideration. Let R_0 be the normalizing resistance for both input and output ports. Since the two-port is lossless, there will be no power dissipation on it. In other words, the total power delivered to its ports must be zero. Let us elaborate this statement by mathematical equations as follows.

Referring to Figure 2.9, let $P_1 = |a_1|^2 - |b_1|^2$ be the net power delivered to port-1. Similarly, let $P_2 = |a_2|^2 - |b_2|^2$ be the net power delivered to port-2.

Thus, the total power delivered to $[N]$ is given by

$$P_T = P_1 + P_2 = \begin{bmatrix} a_1^* & a_2^* \end{bmatrix} \begin{bmatrix} a_1 \\ a_2 \end{bmatrix} - \begin{bmatrix} b_1^* & b_2^* \end{bmatrix} \begin{bmatrix} b_1 \\ b_2 \end{bmatrix} \qquad (2.10a)$$

must be zero.

Recall that by definition $b = \begin{bmatrix} b_1 \\ b_2 \end{bmatrix} = S \begin{bmatrix} a_1 \\ a_2 \end{bmatrix} = Sa.$

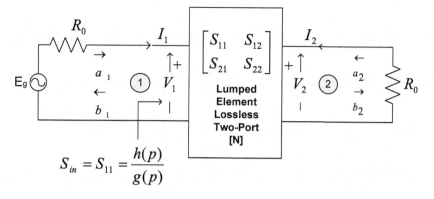

Figure 2.9 Scattering parameters for lossless two-ports.

Then, (2.10a) becomes

$$P_T = a^\dagger \left[I - S^\dagger S \right] a \qquad (2.10b)$$

Since the two-port $[N]$ is lossless, P_T must be zero.
Thus,

$$S^\dagger S = I \qquad (2.10c)$$

where the sign "$\dagger$" is called the dagger, which indicates the complex conjugate of the transposed matrix and $I = \begin{bmatrix} 1 & 0 \\ 0 & 1 \end{bmatrix}$ is the identity matrix.

It should be noted that the matrix $S^\dagger S$ must be symmetrical since the identity matrix is symmetrical. Therefore,

$$S^\dagger S = SS^\dagger = I \qquad (2.10d)$$

The expression $SS^\dagger = I$ yields four equations but one of the zero equalities is redundant. Therefore, the open form of (2.10d) is written as

$$S_{11}(p) S_{11}(-p) + S_{12}(p) S_{12}(-p) = 1$$
$$S_{22}(p) S_{22}(-p) + S_{21}(p) S_{21}(-p) = 1$$
$$S_{11}(-p) S_{21}(p) + S_{12}(-p) S_{22}(p) = 0 \qquad (2.10e)$$

or simply,

$$S_{11}(p) S_{11}(-p) = 1 - S_{12}(p) S_{12}(-p)$$
$$S_{22}(p) S_{22}(-p) = 1 - S_{21}(p) S_{21}(-p)$$

$$S_{22}(p) = -\frac{S_{21}(p)}{S_{12}(-p)} S_{11}(-p) \tag{2.10f}$$

Similarly, the open form of $S^\dagger S = I$ yields

By replacing $p = j\omega$, it is interesting to note that comparing Equations (2.10e) and (2.10f) reveals that

$$|S_{11}(j\omega)| = |S_{22}(j\omega)|$$

$$\left|\frac{S_{12}(j\omega)}{S_{21}(-j\omega)}\right| = \left|\frac{S_{21}(j\omega)}{S_{12}(-j\omega)}\right| \tag{2.10g}$$

For lumped element two-ports, reflection coefficients $S_{11}(p)$ and $S_{22}(p)$ can be regarded as one-port reflection coefficients $S_{in}(p) = S_{11}(p)$, when port-2 is terminated in R_0 and $S_{out}(p) = S_{22}(p)$, when port-1 is terminated in R_0. Therefore, they must be bounded real (BR) rational functions. In other words, they must be both regular in the closed RHP $i.e.(\sigma \geq 0)$.

If $S_{11}(p) = \frac{h(p)}{g(p)}$, then (2.10f) reveals that

$$S_{12}(p) S_{12}(-p) = 1 - \frac{h(p)}{g(p)} \frac{h(-p)}{g(p)}$$

or

$$S_{12}(p) S_{12}(-p) = \frac{g(p) g(-p) - h(p)h(-p)}{g(p) g(-p)} = \frac{F(p^2)}{G(p^2)} \tag{2.10h}$$

Then, (2.10h) suggests that

$$S_{12}(p) = \frac{f_{12}(p)}{g(p)} \tag{2.10i}$$

where $f_{12}(p)$ is constructed on propery selected roots of

$$F(p^2) = g(p) g(-p) h(p)h(-p)$$

Clearly, the solution is not unique.

Similarly,

$$S_{21}(p) = \frac{f_{21}(p)}{g(p)} \tag{2.10j}$$

where $f_{21}(p)$ is also obtained by the factorization of $F(p^2)$.

In this case,

$$S_{22}(p) = -\frac{f_{12}(p)}{f_{21}(-p)}\frac{h(-p)}{g(p)} \tag{2.10k}$$

Hence, for lossless two-ports, one must have

$$\frac{f_{12}(p)}{f_{21}(-p)} = \frac{f_{21}(p)}{f_{12}(-p)} \tag{2.10l}$$

Bounded realness of the scattering parameters requires that the ratios given by (2.10k) should be free of open RHP poles. This could be achieved with possible cancellations in

$$\eta_a = \frac{f_{12}(p)}{f_{21}(-p)}$$

and

$$\eta_b = \frac{f_{21}(p)}{f_{21}(-p)}$$

Furthermore, these ratios must have unity amplitude on the $j\omega$ axis $|S_{11}(j\omega)| = |S_{22}(j\omega)|$. Henceforth, we can say that $f_{12}(p) = \mu f_{21}(-p)$.

In this case, let us set $f(p) = f_{12}(p)$, then $f_{21}(p) = \mu f(-p)$, where μ is a uni-modular constant (i.e. $\mu = \mp 1$).

Moreover, if the lossless two-port is reciprocal, then

$$S_{12}(p) = S_{21}(p) \ or \ f_{12}(p) = f_{21}(p) \tag{2.10m}$$

which automatically satisfies (2.10k). In this case, $f(p)$ cannot be a full polynomial. Rather, it must be even or odd. If this is not true, lossless two-port cannot be reciprocal.

If $f(p)$ is even, i.e. $f(p) = f(-p)$, then, $f_{21}(p) = \mu f_{21}(p) = \mu f_{12}(-p) \triangleq \mu f(-p) = \mu f(p)$

Thus, reciprocity condition given by (2.10m) is satisfied with $\mu = +1$. Therefore, for reciprocal networks, if $f(p)$ is and even polynomial then, the uni-modular constant must be $\mu = +1$.

On the other hand, if $f(p)$ is an odd polynomial, then reciprocity demands negative uni-modular constant such that

$$f_{12}(p) = f(p) \triangleq -f(-p)$$

then, by (2.10m)

$$f_{21}(p) = \mu f_{12}(-p) \triangleq \mu f(-p) = -\mu f(p), \text{ which requires } \mu = -1$$

Thus, scattering matrix of a lossless reciprocal two-port is given by

$$S = \frac{1}{g} \begin{bmatrix} h & f \\ \mu f_* & -\frac{f}{\mu f_*} h_* \end{bmatrix} = \frac{1}{g} \begin{bmatrix} h & f \\ f & -\mu h_* \end{bmatrix} \qquad (2.10n)$$

where "*" designates the para-conjugate of a function $x(p)$ such that $x(p) = x(-p)$.

In summary, for a reciprocal lossless two-port, $f(p)$ is either even or odd polynomial and the uni-modular sign μ is given by

$$\mu = \frac{f(p)}{f(-p)} = \left\{ \begin{array}{l} +1 \; when \; f(p) \; is \; even \\ -1 \; when \; f(p) \; is \; odd \end{array} \right\}$$

The above results are summarized as follows[7].

Referring to Figure 2.9, for a lumped element-reciprocal lossless two-port, if the BR input reflection coefficient has the following form

$$S_{11}(p) = \frac{h(p)}{g(p)} \qquad (2.11a)$$

with real polynomials $h(p) \, and \, g(p)$ of degree "n", then rest of the scattering parameters are given as

$$S_{12}(p) = S_{21}(p) = \frac{f(p)}{g(p)} \qquad (2.11b)$$

and

$$S_{22}(p) = -\frac{f(p)}{f(-p)} \frac{h(p)}{g(p)} \qquad (2.11c)$$

provided that $f(p)$ is either even or odd polynomial satisfying

$$F(p^2) = g(p) g(-p) - h(p) h(-p) \qquad (2.11d)$$

In above forms, all the scattering parameters must be bounded-real rational functions (BR) in the complex variable $p = \sigma + j\omega$ From Equation (2.11b), construction of $f(p)$ requires a little bit of care. Therefore, let us investigate the possible situations within the following discussions.

---

[7]A comprehensive discussion on the subject can be found in "Wideband Circuit Design" by H. J Carlin and P. P Civalleri, CRC Press, 1997, pp. 231–235.

For a given lossless two-port, one may wish to obtain the full scattering matrix perhaps, starting from the driving point input impedance $z_{in}(p)$, which directly specifies the input reflection coefficient

$$S_{11}(p) = \frac{h(p)}{g(p)} = \frac{z_{in} - 1}{z_{in} + 1}$$

and then, by spectral factorization of

$$F(p^2) = f(p) f(-p) = g(p) g(-p) - h(p) h(-p)$$

$f(p)$ is generated on the selected roots of $F(p^2)$, which in turn yields

$$S_{21}(p) = \frac{f(p)}{g(p)}$$

Eventually, $S_{22}(p)$ is generated as

$$S_{22}(p) = -\frac{f(p)}{f(-p)} \frac{h(-p)}{g(p)} = -\mu \frac{h(-p)}{g(p)}$$

as in (2.11). However, this process requires some care as discussed in the following.

In the literature, the above forms of the scattering parameters are known as the Belevitch canonical form[8].

As it was shown in the previous section, under the standard terminations of Figure 2.9, $|s_{21}(j\omega)|^2 = \frac{|b_2|^2}{|a_1|^2}$ measures the transducer power gain in forward direction (from port-1 to port-2).

In this equation, $|a_1|^2$ is the measure of the available power P_A of port-1 and $|b_2|^2$ specifies the power P_L delivered to load R_0[9].

Let us now investigate some properties of the lossless two-ports.

Property 7: As was proven by Darlington, any positive real function Z_{in}, which is specified as a rational function $Z_{in}(p) = \frac{N(p)}{D(p)}$ or corresponding input reflection coefficient $S_{in}(p) = S_{11}(p) = \frac{Z_{in}-1}{Z_{in}+1} = \frac{h(p)}{g(p)}$, can be realized as a reciprocal lossless two-port in unit termination.

In this case, proof of the Darlington's theorem becomes straightforward; since $h(p)$ and $g(p)$ are specified, the complete scattering parameters of the

---

[8]V. Belevitch, Classical Network Theory, S Francisco, Holden – Day, 1968, p. 277

[9]In the literature, it is common to designate Transducer Power Gain as $TPG = \frac{P_L}{P_A}$.

lossless two-port can be determined by factorization of the even polynomial $F\left(p^2\right) = f\left(p\right)f\left(-p\right) = g\left(p\right)g\left(-p\right) - h\left(p\right)h(-p)$. Then, $f\left(p\right)$ is constructed on the selected roots of $F\left(p^2\right)$. At this point, it is very important to note that if the factorization process does not yield odd or even polynomial for $f\left(p\right)$, then we can always augment $S_{21}\left(p\right) = \frac{f(p)}{g(p)}$ with unity products in the form of $\frac{\gamma_r+p}{\gamma_r+p}$ with γ_r real and positive such that

$$S_{21}\left(p\right) = \frac{\widehat{f}(p)}{\widehat{g}(p)} = \frac{f\left(p\right)\left(\gamma_r + p\right)}{g\left(p\right)\left(\gamma_r + p\right)}$$

yielding odd or even $\widehat{f}\left(p\right)$. The same augmentation is also applied to $S_{11}\left(p\right)$ and $S_{22}(p)$ to end up with the common denominator polynomial

$$\widehat{g}\left(p\right) = g\left(p\right)\left(\gamma_r + p\right)$$

in all the scattering parameters.

Example 2.6: Let $S_{11}\left(p\right) = \frac{1-p}{\sqrt{2}+\sqrt{2p}}$. Generate the scattering parameters for the corresponding lossless reciprocal lumped element two-port.

Solution

Here,

$$h(p) = 1-p$$

and

$$g(p) = \sqrt{2} + \sqrt{2}\,p \quad \text{Let}$$

$$G\left(p^2\right) = g\left(p\right)g\left(-p\right) = 2 - 2p^2$$

and

$$H\left(p^2\right) = h\left(p\right)h\left(-p\right) = 1 - p^2$$

In this case,

$$F\left(p^2\right) = f\left(p\right)f\left(-p\right) = G - H = 1 - p^2$$

Since $\widehat{g}\left(p\right) = f\left(-p\right)g\left(p\right)$ must be strictly Hurwitz, $f\left(p\right) = 1 - p$.

At this point, we have to be careful since $f\left(p\right)$ is neither even nor odd in the complex variable p Rather, it is a full polynomial of degree 1. Therefore, it cannot belong to a reciprocal lossless network.

However, by augmentation with the product $\frac{1+p}{1+p}$, $S_{21}(p)$ can be written as

$$S_{21}\left(p\right) = \frac{1-p}{\sqrt{2}+\sqrt{2}p}\frac{1+p}{1+p} = \frac{1-p^2}{\sqrt{2}+2\sqrt{2}p+\sqrt{2}p^2} = \frac{\widehat{f}(p)}{\widehat{g}(p)}$$

In this case, $\widehat{f}(p) = 1 - p^2$, which is an even polynomial, and $\widehat{g}(p) = \sqrt{2} + 2\sqrt{2}p + \sqrt{2}p^2$ become the strictly Hurwitz common denominator polynomials. Eventually, augmented $S_{11}(p)$ and $S_{22}(p)$ are given as follows

$$S_{11}(p) = \frac{\widehat{h}(p)}{\widehat{g}(p)} = \frac{1-p^2}{\sqrt{2} + 2\sqrt{2}p + \sqrt{2}p^2}$$

$$S_{11}(p) = -\mu\frac{\widehat{h}(-p)}{\widehat{g}(p)} = -\frac{1-p^2}{\sqrt{2} + 2\sqrt{2}p + \sqrt{2}p^2}$$

2.6 Blashke Products or All Pass Functions

As it is understood from the above discussions, the complex function $\eta(p) = \frac{f(p)}{f(-p)}$ must define a regular function in the closed right half plane (i.e. $p = \sigma \geq 0$) since the denominator polynomial $\widehat{g}(p) = f(-p)g(p)$ must be strictly Hurwitz to make $S_{22}(p)$ regular in the closed RHP[10]. Furthermore, on the $j\omega$ axis, the amplitude function $|\eta(j\omega)|^2 = 1$. Thus, a rational-regular (or analytic) function with unity amplitude is called a "**Blashke Product**" or an "**All Pass**" function.

Definition 5: Proper Polynomial $f(p)$[11]
A real polynomial in complex variable "p" is called proper if it yields a regular Blashke product $\eta(p) = \frac{f(p)}{f(-p)}$.

At this point it is crucial to note that for a reciprocal lossless two-port, the numerator polynomial $f(p)$ of $S_{21} = S_{12} = \frac{f(p)}{g(p)}$ must be proper to end up with the BR scattering parameters.

2.7 Possible Zeros of a Proper Polynomial $f(p)$

In general, zeros of a proper polynomial $f(p)$ can be classified as in the following cases.

Case A. If $f(p)$ includes a real positive zero, which is denoted by $\alpha > 0$ such that $f(p) = (\alpha - p)\widehat{f}(p)$, then the term $(\alpha - p)$ in $f(p)$ appears

---

[10]A complex variable function $\eta(p)$ is called analytic in a complex region R if it has no singularity in R In other words, it has to be differentiable in R. If $\eta(p)$ is rational and analytic in R, then it should be free of poles in R because poles lying in R introduce singularities for which $\eta(p)$ becomes infinity.

[11]To our knowledge, this is the first introduction of the "proper polynomial definition of $f(p)$.

as a factor $(\alpha + p)$ in $f(-p)$ yielding $\widehat{g}(p) = (\alpha + p)\widehat{f}(-p)g(p)$, which preserves the scattering Hurwitz property of $\widehat{g}(p)$. Therefore, real RHP zeros of $f(p)$ are called proper zeros. In this case, the Blashke product $\eta(p)$ must include the term $\frac{\alpha-p}{\alpha+p}$ or in general

$$\eta(p) = \frac{\alpha - p}{\alpha + p}\frac{\widehat{f}(p)}{\widehat{f}(-p)}$$

Obviously, in this representation, it is assumed that the term $\frac{\widehat{f}(p)}{\widehat{f}(-p)}$ is regular in the closed RHP to make $S_{22}(p)$ regular (or bounded real) in the closed RHP. This can only be achieved with possible cancellations in $\frac{\widehat{f}(p)}{\widehat{f}(-p)}$ as it is discussed in Case C.

Case B. If $f(p)$ includes a complex zero (such as $p = \alpha + j\beta$ with $\alpha > 0$ in the closed RHP), it must be accompanied with its conjugate in order to make $S_{21}(p)$ a real function. In this case, $f(p)$ takes the following form

$$f(p) = [(\alpha+j\beta)-p][(\alpha-j\beta)-p]\widehat{f}(p)$$

which is a real polynomial in p. Again, assuming a regular term $\frac{\widehat{f}(p)}{\widehat{f}(-p)}$ with possible cancellations, the Blashke product $\eta(p)$ is written as

$$\eta(p) = \frac{(\alpha-p)^2+\beta^2}{(\alpha+p)^2+\beta^2}\frac{\widehat{f}(p)}{\widehat{f}(-p)}$$

Therefore, a complex closed RHP zero of $f(p)$ must be matched with its conjugate pair in order to make $f(p)$ real and proper.

Case C. Assume that $f(p)$ includes a pure real closed LHP zero such that $p_{z1} = \delta < 0$ t. Obviously, this zero is not proper since it introduces a closed RHP zero in $f(-p)$ yielding un-regular $\eta(p)$[12].

However, if $p_{z1} = \delta$ is matched with its mirror image $p_{z2} = -\delta$ in the closed RHP, then $\eta(p)$ must include the term $\left[\frac{\delta-p}{\delta+p}\right]\left[\frac{-\delta-p}{-\delta+p}\right]$, which is equal to 1 by cancellation. Thus, we say that a proper $f(p)$ may include a purely real closed LHP zero if it is paired with its mirror image to make $\eta(p)$ regular with cancellations.

---

[12]In this context, the term "un-regular" is used to indicate a singularity in $\eta(p)$ *at a point* $p = -\delta > 0$ in the closed RHP.

Case D. Assume that $f(p)$ includes a complex closed LHP zero $p_{z1} = \delta + j\beta$ with $\delta < 0$.

As explained in the previous case, this zero cannot be proper unless it is paired with its mirror image, which results cancellations in (p).

Furthermore, in order to assure the reality of the polynomial $f(p)$, complex LHP zeros must be accompanied by their conjugate pairs yielding quadruple symmetry, as shown in Figure 2.10.

Thus, we say that complex LHP zeros of a proper $f(p)$ must have quadruple symmetry.

In this case, $f(p)$ should take the following form:

$$f(p) = \left[(\delta-p)^2+\beta^2\right]\left[(\delta+p)^2+\beta^2\right]\widehat{f}(p)$$

where $\delta < 0$ and $\widehat{f}(p)$ a describes a regular all pass function $\dfrac{\widehat{f}(p)}{\widehat{f}(-p)}$

Case E. Multiple DC zeros of $f(p)$, which is specified by p^k in $f(p) = p^k\widehat{f}(p)$, are always proper since they cancelled in $\eta(p)$.

Case F. Finite non-zero $j\omega$ zeros of $f(p)$ are always proper as long as they are paired with their complex conjugates. In this case, $f(p)$ is expressed as $f(p) = \left(p^2 + \omega_k^2\right)\widehat{f}(p)$, where ω_k denotes a non-zero but finite $j\omega$ zero of $f(p)$.

All the above zeros of $f(p)$ are summarized in Figure 2.10.

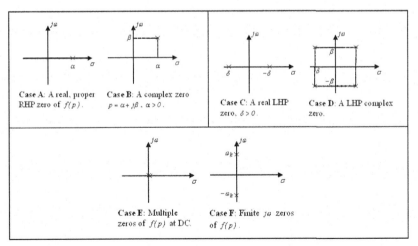

Figure 2.10 Zeros of $f(p)$.

2.8 Transmission Zeros[13]

Zeros of $f(p)$ coincide with the finite zeros of $S_{21}(p)$. Moreover, they also coincide with the zeros of the transducer power gain $TPG = |S_{21}(j\omega)|^2$ over the real frequency axis $j\omega$.

Obviously, signal transmission stops at zeros of $TPG = |S_{21}(j\omega)|^2$. Therefore, zeros of $S_{21}(p) S_{21}(-p)$ are called the **"transmission zeros"** of the lossless two-port. Transmission zeros of the lossless two-ports may introduce finite values of p or at infinity.

The circuit structure (or the circuit topology) of the lossless two-port imposes the transmission zeros. Therefore, in the following properties, some practical circuit topologies are introduced in connection with transmission zeros.

Property 8: In the complex p-plane, the general form of $f(p)$ may be written as follows.

Let $f_{DC}(p) = p^k$ be the term with multiple DC zeros.

Let $f_\omega(p) = \prod_{r=1}^{n_\omega} (p^2 + \omega_r^2)$ be the term with finite $j\omega$ zeros.

Let $f_\sigma(p) = \prod_{r=1}^{n_\omega} (\sigma_r - p)$ be the term with pure real RHP zeros[14].

Let $f_{MLHP}(p) = (-1)^{n_\delta} \prod_{r=1}^{n_\delta} (\delta_r^2 - p^2)$ be the term with mirror image zero pairs.

Let $f_{CRHP}(p) = \prod_{r=1}^{n_R} \left[(\sigma_r - p)^2 + \beta_r^2 \right]$ be the term with complex RHP zeros[15].

Let $f_{MCLHP}(p) = \prod_{r=}^{n_L} \left[(\delta_r - p)^2 + \beta_r^2 \right] \left[(\delta_r + p)^2 + \beta_r^2 \right]$ be the term with complex LHP zeros which are matched with their mirror image pairs.

Then, the general form of $f(p)$ may be specified as

$$f(p) = f_{DC}(p)\, f_\omega(p) f_\sigma(p)\, f_{MLHP}(p)\, f_{CRHP}(p)\, f_{MCLHP}(p)$$

This form of $f(p)$ is not proper due to terms with unmatched RHP zeros. However, by augmentation, one can add LHP mirror image zeros to be paired with RHP zeros; which makes the terms $f_\sigma(p)$ and $f_{CRHP}(p)$ even in

---

[13]Definition of the transmission zeros introduced in this book is somewhat slightly different from that introduced in the existing literature by Fano, Youla, Carlin, and Yarman.

[14]This form of $f_\sigma(p)$ is not proper. However, it may be augmented by its mirror image.

[15]This form of $f_{CRHP}(p)$ is not proper. However, it may be augmented by its mirror image.

complex variable p. In this case, proper generic form of $f(p)$ is given by

$$f(p) = p^k \prod_{r=1}^{n_\omega} (p^2 + \omega_r^2) \prod_{r=1}^{n_\delta} (\delta_r^2 - p^2) \prod_{r=}^{n_L} \left[(\delta_r - p)^2 + \beta_r^2 \right] \left[(\delta_r + p)^2 + \beta_r^2 \right]$$

(2.12a)

Full form of $f(p)$ corresponds to a highly complicated network structure, which may be impractical to realize. In practice, however, we generally deal with lossless ladder forms as summarized in Properties 9 and 10.

Property 9: Low-Pass Structures

The simplest proper form of $f(p)$ includes no zero. In this case, $f(p)$ is a constant and may be normalized at unity such that $f(p) = 1$. This form corresponds to a low-pass L-C ladder, as shown in Figure 2.11.

In this case, transmission zeros of the system will be at infinity of degree $2n$, where "n" is the degree of the denominator polynomial $g(p) = g_1 p^n + g_2 p^{n-1} + \cdots + g_n p + g_{n+1}$, which also specifies the total number of reactive elements[16] in the lossless two-port under consideration.

Property 10: Bandpass Structures

A slightly more complicated form of $f(p)$ may have only multiple zeros at DC such that

$$f(p) = p^k$$

This form corresponds to a band-pass structure, as shown in Figure 2.12.

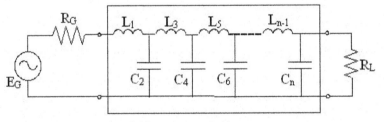

Figure 2.11 Low-pass ladder structure.

---

[16]In this case, reactive elements will be the elements of the low-pass ladder (LPL) structure, which is constructed as cascade connections of series inductors and shunt capacitors as depicted in Figure 2.10.

The lossless ladder of Figure 2.12 with $f(p) = p^k$ includes zero of transmissions at DC of order $2k$ and at infinity of order $2(n-k)$. The integer "k" is the count of total number of series capacitors and shunt inductors, which introduce transmission zeros at DC of degree "**2k**". Furthermore, the difference $(\mathbf{n-k})$ is the total number of series inductors and shunt capacitors, which introduce transmission zeros at infinity of degree $2(n-k)$.

Property 11: Finite real frequency or $j\omega$ zeros of $f(p) = \prod_{r=1}^{n_\omega}(p^2 + \omega_r^2)$ may be realized as either a parallel resonance circuit in series configuration or a series resonance circuit in shunt configuration or as a Darlington C section with coupled coils, as depicted in Figure 2.13.

Figure 2.2 shows the realization of a finite frequency transmission zero ω_r, which is specified by $f(p) = p^2 + \omega_r^2$.

If ω_r satisfies the Fujisawa conditions[17], it is realized, either as a parallel resonance circuit in the series arm, as shown in Figure 2.13a, or it is realized as a series resonance circuit in the shunt arm, as shown in Figure 2.13b.

If the finite frequency transmission zero ω_r does not satisfy the Fujisawa conditions, then it is realized as a Darlington C-Section, as shown in Figure 2.13c.

Details of the synthesis are not the focus of this book. Therefore, it is omitted here. However, interested readers are referred to relevant synthesis literature[18].

Now, let us run the following example.

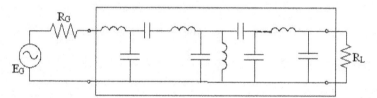

Figure 2.12 Bandpass ladder structure.

---

[17]Fujisawa's theorem indicates that for a given positive real impedance function $z_{in}(p)$ if $f(p) = \prod_{r=1}^{n_\omega}(p^2 + \omega_r^2)$ and if $z_{in}(0) = c > 0$ or equivalently if $y_{in}(0) = c > 0$ and if $z_{in}(p)$ has either a pole or a zero at infinity, then the zero of transmission ω_r can be realized as either a parallel resonance circuit in mid-series configuration or a series resonance circuit in mid-shunt configuration by extracting the term $\frac{2L_r p}{L_r C_r p^2 + 1}$ during the long division (or equivalently continuous fraction expansion) process of the driving point impedance.

[18]B.S. Yarman, "Computer Aided Darlington Synthesis of All Purpose Immittance Function", Istanbul University, Journal of Electrical-Electronics Engineering (IU-JEEE), Vol. 16(1), pp. 2027–2037, 2016.

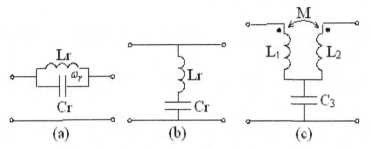

Figure 2.13 Possible realization of a finite real-frequency transmission zero.

Example 2.7: For a lossless two-port, let the unit normalized bounded real input reflection coefficient be given by

$$S_{11}(p) = \frac{h(p)}{g(p)} = \frac{-p^4 + p^3 + 2p}{p^4 + 3p^3 + 4p + 3p + 1}$$

o Derive $f(p)$ and check if it is proper.
o Find the transmission zeros of the lossless two-port.
o Comment on the possible network topology of the two-port.
o Construct the full scattering parameters form $S_{11}(p)$.
o Synthesize $S_{11}(p)$ as a Darlington lossless two-port in resistive termination which yields the transmission zeros of $T(p^2) = S_{21}(p) S_{21}(-p)$.

Solution

(a) Transmission zeros of the two-port are determined from

$$S_{21}(p) S_{21}(-p) = 1 - S_{11}(p) S_{11}(-p) = \frac{F(p^2)}{G(p^2)}$$
$$= \frac{g(p) g(-p) - h(p) h(-p)}{g(p) g(-p)}$$

where

$$h(p) = -p^4 + 3p^2 + 2p + 0$$

and

$$g(p) = p^4 + 3p^3 + 4p + 3p + 1$$

Then,

$$G\left(p^2\right) = p^8 - p^6 + 0p^4 - p^2 + 1$$
$$H\left(p^2\right) = p^8 - p^6 - p^4 + p^2 + 0$$

Or

$$F\left(p^2\right) = G\left(p^2\right) - H\left(p^2\right) = p^4 - 2p^2 + 1 = \left(p^2 + 1\right)^2$$

Thus,

$$f\left(p\right) = f\left(-p\right) = p^2 + 1$$

$f(p)$ is proper with one finite zero of transmissions located $\omega_{1,2} = \mp j1$. Furthermore, one can readily see that $\eta\left(p\right) = \frac{f(p)}{f(-p)} = 1$.

Since the degree of the denominator $g(p)$ is $n = 4$ and the degree of the proper polynomial $f\left(p\right)$ is $n_f = 2$, then the lossless two-port must have transmission zeros of order $2\left(n - n_f\right) = 2\left(4 - 2\right) = 4$ at infinity. The real frequency zero can be realized as either a parallel resonance circuit in series configuration or a series resonance circuit in shunt configuration if and only if $z_{in}\left(p\right) = \frac{1+S_{11}}{1-S_{11}}$ or $y_{in}\left(p\right) = \frac{1-S_{11}}{1+S_{11}}$ satisfies the Fujisawa theorem. If this is the case, the term $\left(p^2 + 1\right)$ appears as a multiplying factor in denominator polynomials of either in the impedance or in the admittance functions as we go alone with long division synthesis process of the driving point impedance function. This fact can be exercised in the course of synthesis. Otherwise, finite transmission zero can be extracted using zero shifting technique yielding a Darlington C-section, as cited in footnote 18.

Since $f\left(p\right) = p^2 + 1$ and $\eta\left(p\right) = 1$, then,

$$S_{21}\left(p\right) = \frac{p^2 + 1}{p^4 + 3p^3 4p + 3p + 1}$$

and

$$S_{22}\left(p\right) = -\frac{-p^4 + p^3 + 2p^2}{p^4 + 3p^3 4p + 3p + 1}$$

Let us first generate the driving point input impedance of the lossless two-port when the output port is terminated in 1 ohm.

$$z_{in}\left(p\right) = \frac{1 + S_{11}}{1 - S_{11}} = \frac{g\left(p\right) - h(p)}{g\left(p\right) + h(p)} = \frac{2p^3 + 2p^2 + 2p + 1}{p^4 + p^3 + p^2 + p + 1}$$

which satisfies the condition of the Fujisawa's theorem to be synthesized as a Low-pass ladder (LPL) since $z_{in}\left(0\right) = 1 > 0$.

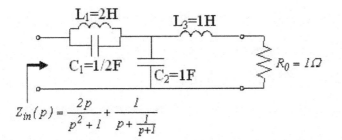

Figure 2.14 Synthesis of the input impedance $z_{in}(p)$ of Example 2.7.

Furthermore, it is found that

$$z_{in}(p) = \frac{2p^3 + 2p^2 + 2p + 1}{p^4 + p^3 + p^2 + p + 1} = \frac{2p^3 + 2p^2 + 2p + 1}{(p^2 + 1)(p^2 + p + 1)}$$

which includes the term (p^2+1) in the denominator. This term can be immediately extracted from the input impedance yielding

$$z_{in}(p) = \frac{2p}{(p^2 + 1)} + \frac{p + 1}{p^2 + p + 1}$$

Then, we can continue with the long division process. Hence,

$$z_{in}(p) = \frac{2p}{(p^2 + 1)} + \frac{1}{p + \frac{1}{p+1}}$$

Hence, the above form of the driving point impedance indicates that the first term $\frac{2p}{(p^2 + 1)}$ is a parallel resonance circuit with $L_1 = 2H$ and $C_1 = \frac{1}{2}F$ in series configuration. It is followed with a shunt capacitor of $C_2 = 1F$ and, then a series inductor $L_3 = 1H$ follows. Eventually, the lossless ladder is terminated in 1 ohm resistance, as shown in Figure 2.14.

2.9 Lossless Ladders

If a lumped element reciprocal lossless two-port is free of coupled coils, then it exhibits a special kind of circuit topology called lossless ladder. In this configuration, simple sections of Figure 2.10 are in tandem connection, as shown in Figure 2.15. In this figure, Z_i and Y_i designate the series-arm impedance and the shunt-arm admittance, respectively. For lossless ladders, it is straightforward to see that the proper $f(p)$ is either even or odd polynomial in the complex variable $f(p) = \sigma + j\omega$ having all its zeros on the $j\omega$ axis, yielding $\eta(p) = \frac{f(p)}{f(-p)} = \mu = \mp 1$. This fact can easily be seen by erasing the real and complex zeros of $f(p)$ from (2.11a). Thus, for a lossless ladder

$$f(p) = p^k \prod_{r=1}^{n_k} \left(p^2 + \omega_r^2\right) \tag{2.13a}$$

Obviously, in (2.13a), $f(p)$ is either an even or an odd polynomial in p depending on the value of the integer "k", which is associated with the multiple zeros at DC. Furthermore, all the zeros of $f(p)$ must appear as poles of Z_i and Y_i such that

$$\{Z_i \text{ or } Y_i\} = \left\{ \begin{array}{c} a_i p \\ \frac{1}{b_i} \\ \frac{c_i}{p^2 + \omega_i^2} \end{array} \right\} \tag{2.13b}$$

Let the input impedance of the lossless ladder be $Z_{in}(p)$. Then, it is expressed via continuous fraction expansion as

$$Z_{in}(p) = Z_1 + \cfrac{1}{Y_2 + \cfrac{1}{Z_2 + \frac{1}{Y_3} + \cfrac{1}{Z_4 + .. + \cfrac{1}{\cdots + \frac{1}{R}}}}} \tag{2.14}$$

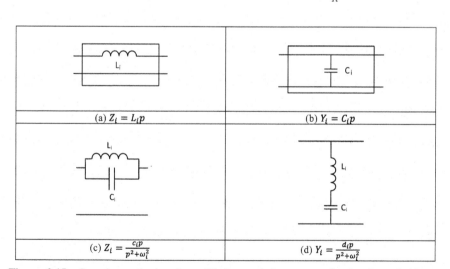

Figure 2.15 Generic synthesis of possible transmission zeros of a bandpass ladder as described by the driving point input impedance of (2.14).

In (2.14), R is the termination resistor of the LC-bandpass ladder. At the end of synthesis, R may be replaced with a transformer of $n^2 : 1$ turn ratio. Generic synthesis of each $Z_i(p)$ or $Y_i(p)$ is depicted in Figure 2.15.

As far as synthesis of ladders is concerned, interested readers are referred to [1–5].

2.10 Further Properties of the Scattering Parameters of the Lossless Two-Ports

For many practical manipulations, it is interesting to evaluate the determinant of a para-unitary scattering matrix ΔS. In the following, this derivation is introduced.

Derivation of ΔS

Let

$$S = \begin{bmatrix} S_{11} & S_{12} \\ S_{21} & S_{22} \end{bmatrix}$$

and be a para-unitary scattering matrix of a linear, reciprocal, lossless two-port.

Losslessness condition requires that

$$SS^\dagger = I = \begin{bmatrix} 1 & 0 \\ 0 & 1 \end{bmatrix}$$

And the reciprocity reveals that $S_{21} = S_{12}$

Determinant ΔS is given by

$$\Delta S = S_{11}S_{22} - S_{21}S_{12} \tag{2.15a}$$

Para-unitary condition yields that

$$S_{11}S_{11*} + S_{21}S_{21*} = 1 \tag{2.15b}$$

$$S_{22} = -\frac{S_{21}}{S_{21*}}S_{11*} \tag{2.15c}$$

Using (2.15b) and (2.15c) in (2.15a), we have

$$\Delta S = S_{11}S_{22} - S_{21}S_{12} = -S_{11}\frac{S_{21}}{S_{21*}}S_{11*} - S_{21}S_{12}$$

$$= -\frac{S_{21}}{S_{12*}}\left(S_{11}S_{11*} + S_{21}S_{21*}\right)$$

or

$$\Delta S = -\frac{S_{21}}{S_{12*}} = -\frac{S_{21}}{S_{21*}} \qquad (2.15d)$$

The above result can be summarized in the following property.

Property 12: For a lossless, reciprocal two-port, determinant $\Delta S = S_{11}$ $S_{22} - S_{21}S_{12}$ of the scattering matrix $S = \begin{bmatrix} S_{11} & S_{12} \\ S_{21} & S_{22} \end{bmatrix}$ is an all pass function η_Δ such that[19]

$$\Delta S = -\frac{S_{21}}{S_{12*}} = -\frac{S_{21}}{S_{21*}} = \frac{f(p)}{f(-p)} \frac{g(-p)}{g(p)} \qquad (2.15e)$$

Obviously, on the real frequency axis, ΔS has unitary amplitude. That is $|\Delta S(j\omega)| = 1$.

2.11 Transfer Scattering Parameters

Most of the communication systems are constructed by tandem connection of subsystems. Each of these may be described by means of scattering parameters. In order to assess the overall performance such as transducer power gain or phase shifting performance, we should be able to generate the scattering parameters of the complete structure. Thus, the transfer scattering parameters are introduced to facilitate the computation of the overall performance of the communication systems as follows.

Referring to Figure 2.16, transfer scattering parameters are defined in such a way that at the input, reflected b_1 and incident a_1 waves are related to the output port's incident a_2 and reflected b_2 such that

$$\begin{pmatrix} b_1 \\ a_1 \end{pmatrix} = \begin{pmatrix} \tau_{11} & \tau_{12} \\ \tau_{21} & \tau_{22} \end{pmatrix} \begin{pmatrix} a_2 \\ b_2 \end{pmatrix} = T \begin{pmatrix} a_2 \\ b_2 \end{pmatrix} \qquad (2.16a)$$

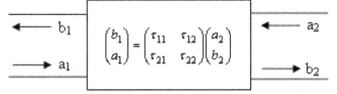

Figure 2.16   Definition of transfer scattering parameters.

---

[19]In this expression, it is assumed that $S_{11}(p) = \frac{h(p)}{g(p)}$ ans $S_{12}(p) = S_{21}(p) = \frac{f(p)}{g(p)}$

In (2.15a), transfer scattering parameter matrix T is defined as

$$T = \begin{pmatrix} \tau_{11} & \tau_{12} \\ \tau_{21} & \tau_{22} \end{pmatrix} \tag{2.16b}$$

From (2.15), a relation between scattering parameters and transfer scattering parameters can easily be derived.

Let us refresh our mind. Input reflected wave b_1 was defined as

$$b_1 = S_{11}a_1 + S_{12}a_2$$

However, by (2.15a)

$$b_1 = \tau_{11}a_2 + \tau_{12}b_2$$

Similarly,

$$b_2 = S_{21}a_1 + S_{22}a_2$$

or

$$a_1 = \frac{b_2 - S_{22}a_2}{S_{12}} = -\frac{S_{22}}{S_{21}}a_2 + \frac{1}{S_{21}}b_2$$

$$a_1 = -\frac{S_{22}}{S_{21}}a_2 + \frac{1}{S_{21}}b_2 \triangleq \tau_{21}a_2 + \tau_{22}b_2 \tag{2.16c}$$

On the other hand, employing $a_1 = -\frac{S_{22}}{S_{21}}a_2 + \frac{1}{S_{21}}b_2$ of (2.15c) in $b_1 = S_{11}a_1 + S_{12}a_2$ of (2.1), it is found that

$$b_1 = \left(S_{12} - \frac{S_{11}S_{22}}{S_{21}}\right)a_2 + \left(\frac{S_{11}}{S_{21}}\right) \triangleq \tau_{11}a_2 + \tau_{12}b_2 \tag{2.16d}$$

Hence, the relationship between the transfer scattering parameters and scattering parameters is given by

$$\tau_{11} = \left(S_{12} - \frac{S_{11}S_{22}}{S_{21}}\right) \quad \tau_{12} = \left(\frac{S_{11}}{S_{21}}\right)$$

$$\tau_{21} = -\frac{S_{22}}{S_{21}} \qquad\qquad \tau_{22} = \frac{1}{S_{21}} \tag{2.16e}$$

It is interesting to note that, in (2.15e), τ_{22} is directly related to S_{21}, which is a measure of phase shifting performance of a lossless two-port.

2.12 Cascaded (or Tandem) Connections of Two-ports

Referring to Figure 2.17, let us consider cascaded connection of two-port N_1 and N_2.

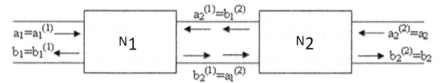

Figure 2.17   Tandem connection of two-ports.

Let the following be the transfer scattering parameters of N_1 and N_2, respectively.

$$\begin{pmatrix} b_1^{(1)} \\ a_1^{(1)} \end{pmatrix} = \begin{pmatrix} \tau_{11}^{(1)} & \tau_{12}^{(1)} \\ \tau_{21}^{(1)} & \tau_{22}^{(1)} \end{pmatrix} \begin{pmatrix} a_2^{(1)} \\ b_2^{(1)} \end{pmatrix} = T_1 \begin{pmatrix} a_2^{(1)} \\ b_2^{(1)} \end{pmatrix} \tag{2.17a}$$

and

$$\begin{pmatrix} b_1^{(2)} \\ a_1^{(2)} \end{pmatrix} = \begin{pmatrix} \tau_{11}^{(2)} & \tau_{12}^{(2)} \\ \tau_{21}^{(2)} & \tau_{22}^{(2)} \end{pmatrix} \begin{pmatrix} a_2^{(2)} \\ b_2^{(2)} \end{pmatrix} = T_2 \begin{pmatrix} a_2^{(2)} \\ b_2^{(2)} \end{pmatrix} \tag{2.17b}$$

From Figure 2.17a, it is observed that connection at port-2 of $[N_1]$ to port-1 of $[N_2]$ reveals that

$$\begin{pmatrix} a_2^{(1)} \\ b_2^{(1)} \end{pmatrix} = \begin{pmatrix} b_1^{(2)} \\ a_1^{(2)} \end{pmatrix} = T_2 \begin{pmatrix} a_2^{(2)} \\ b_2^{(2)} \end{pmatrix} \tag{2.17c}$$

Thus, inserting (2.17c) into (2.17a), we end up with

$$\begin{pmatrix} b_1^{(1)} \\ a_1^{(1)} \end{pmatrix} = [T_1 T_2] \begin{pmatrix} a_2^{(2)} \\ b_2^{(2)} \end{pmatrix} = [T] \begin{pmatrix} a_2^{(2)} \\ b_2^{(2)} \end{pmatrix} \tag{2.17d}$$

Therefore, we conclude that the overall transfer scattering matrix T of the network N, which is formed on the cascaded connection of N_1 and N_2, is given by

$$T = \begin{pmatrix} \tau_{11} & \tau_{12} \\ \tau_{21} & \tau_{22} \end{pmatrix} = \begin{pmatrix} \tau_{11}^{(1)} & \tau_{12}^{(1)} \\ \tau_{21}^{(1)} & \tau_{22}^{(1)} \end{pmatrix} \begin{pmatrix} \tau_{11}^{(2)} & \tau_{12}^{(2)} \\ \tau_{21}^{(2)} & \tau_{22}^{(2)} \end{pmatrix}$$

or

$$\tau_{11} = \tau_{11}^{(1)} \tau_{11}^{(2)} + \tau_{12}^{(1)} \tau_{21}^{(2)} \quad \tau_{12} = \tau_{11}^{(1)} \tau_{12}^{(2)} + \tau_{12}^{(1)} \tau_{22}^{(2)}$$

$$\tau_{21} = \tau_{21}^{(1)} \tau_{11}^{(2)} + \tau_{22}^{(1)} \tau_{21}^{(2)} \quad \tau_{22} = \tau_{21}^{(1)} \tau_{12}^{(2)} + \tau_{22}^{(1)} \tau_{22}^{(2)} \tag{2.17e}$$

At this point, it is interesting to evaluate $\tau_{22} = \frac{1}{S_{21}}$ of the composite structure

$$\tau_{22} = \tau_{21}^{(1)}\tau_{12}^{(2)} + \tau_{22}^{(1)}\tau_{22}^{(2)}$$

where

$$\tau_{21}^{(1)} = -\frac{S_{22}^{(1)}}{S_{21}^{(1)}} \quad \tau_{21}^{(1)} = -\frac{S_{22}^{(1)}}{S_{21}^{(1)}}$$

$$\tau_{12}^{(2)} = \frac{S_{11}^{(2)}}{S_{12}^{(2)}} \quad \tau_{22}^{(1)} = \frac{1}{S_{21}^{(1)}}$$

and

$$\tau_{22}^{(2)} = \frac{1}{S_{21}^{(2)}}$$

Then,

$$\tau_{22} = \frac{1 - S_{22}^{(1)}S_{11}^{(2)}}{S_{21}^{(1)}S_{21}^{(2)}} \tag{2.17f}$$

or the transfer scattering parameter of the cascaded structure is given by

$$S_{21} = \frac{S_{21}^{(1)}S_{21}^{(2)}}{1 - S_{22}^{(1)}S_{11}^{(2)}} \tag{2.17g}$$

We should stress that (2.17f) is quite an important equation since it directly reveals the phase performance of the composite structure.

2.13 Construction of an n-Bit Phase Shifter by Cascading Phase-Shifting Cells

At the given operation frequency ω_0, an n-bit phase shifter is constructed by cascading n units of phase-shifting cells with phase-shifts of $\varphi_k = 2^k \frac{360°}{2^n}$; $k = 0, 1, 2, \ldots, (n-1)$ such that $\varphi_n = \pi$, as shown in Figure 2.18.

At the given operating frequency ω_0, perfect match condition is satisfied such that

$$S_{11}^{(k)}(j\omega_0) = S_{22}^{(k)}(j\omega_0) = 0 \tag{2.18a}$$
$$; k = 1, 2, \ldots, n$$

$$S_{21}^{(k)}(j\omega_0) = 1e^{j\varphi_{21}^{(k)}}(j\omega_0) \tag{2.18b}$$

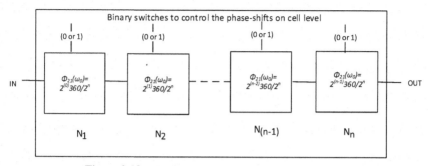

Figure 2.18   Tandem connection n-phase shifting cell.

At the operating frequency ω_0, employing (2.17g) in (2.18), generic form of the transfer scattering parameter of an n-bit phase shifter is derived as

$$S_{21}\left(j\omega_0\right) = S_{21}^{(1)} S_{21}^{(2)} \ldots S_{21}^{(n-1)} S_{21}^{(n)} = 1 e^{j\varphi_{21}(\omega_0)} = 1.e^{j\left[\sum_{k=1}^{n}\varphi_{21}^{(k)}(\omega_0)\right]}$$

(2.19a)

$$\varphi_{21}\left(\omega_0\right) = \sum_{i=1}^{n}\varphi_{21}\left(\omega_0\right)$$

(2.19b)

Phase of each unit $[N_k]$ in the cascade structure $[N]$ is controlled by means of binary switches. In Figure 2.18, phase of each phase-shifting cell is changed between two-states, per se State-A and State-B or equivalently logical state-one or state-zero, respectively.

For the kth stage, if the position of phase control switch is in logical State-1 (or equivalently State A), then phase of the transfer scattering parameter $S_{21A}^{(k)}$ becomes $\varphi_{21A}^{(k)}\left(\omega_0\right)$. When the switch is in logical State-0 (or equivalently State-B), the phase of $S_{21B}^{(k)}$ becomes $\varphi_{21B}^{(k)}\left(\omega_0\right)$.

Depending on the position of the digital switches, net phase shift of the cascade structure may be determined by means of a logical truth table, which is detailed in the following example.

Example 2.8: Referring to Figure 2.18, assume that for each phase-shifting cell $\varphi_{21B}^{(k)}\left(\omega_0\right) = 0$.

Design a three-bit digital phase shifter (i.e. $n = 3$ bits) in cascade connection.

Depending on the position of the switches, generate a bit-phase table to determine the total phase shift of the cascaded structure.

Solution

In the design, we will have total of $n = 3$ phase shifting cells in cascade connection.

In State-A, at the angular operating frequency ω_0, the phase shift $\varphi_{21A}^{(0)}$ of the first cell ($k = 0$) is given by

$$\varphi_{21A}^{(k)}(\omega_0) = 2^k \frac{360}{2^n} = \varphi_{21A}^{(0)}(\omega_0) = \frac{360°}{8} = 45°$$

For the second cell ($k = 1$), State A phase is given by

$$\varphi_{21A}^{(k)}(\omega_0) = 2^1 \frac{360°}{2^3} = \varphi_{21A}^{(1)}(\omega_0) = 90°$$

Finally, in State A, phase of the third phase shifting unit is defined for $k = 2$ as

$$\varphi_{21A}^{(2)}(\omega_0) = 2^2 \frac{360°}{2^3} = 180°$$

Based on the positions, we can generate the following Table 2.1

Table 2.1 Phase state table of a three-bit digital phase shifter

	Cascade Connection of Phase Shifting Cells (3-Phase Bit Cascade Connection)			
Switch Position Count	$k = 0$ Phase Cell-1 (Least Significant Bit)	$k = 1$ Phase Cell-2	$k = 2$ Phase Cell-3 (Most Significant Bit)	Total Phase Shift $\varphi_{21} = \varphi_{21}^{(0)} + \varphi_{21}^{(1)} + \varphi_{21}^{(2)}$
0	0 (State-B)	0 (State-B)	0 (State-B)	$0 + 0 + 0 = 0°$
1	0 (State-B)	0 (State-B)	1 (State-A)	$0 + 0 + 180 = 180°$
2	0 (State-B)	1 (State-A)	0 (State-B)	$0 + 90 + 0 = 90°$
3	0 (State-B)	1 (State-A)	1 (State-A)	$0 + 90 + 180 = 270°$
4	1 (State-A)	0 (State-B)	0 (State-B)	$45 + 0 + 0 = 45°$
5	1 (State-A)	0 (State-B)	1 (State-A)	$45 + 0 + 180 = 225°$
6	1	1	0	$45 + 90 + 0 = 135°$
7	1	1	1	$45 + 90 + 180 = 315°$

References

[1] S. Darlington, "Synthesis of Reactance 4-Poles Which Produce Prescribed Insertion Loss Characteristics", *J. Math. and Phys*, vol. 18, no. 9, pp. 257–353, 1939.

[2] B. S. Yarman, et al., "Computer Aided High Precison Darlington Synthesis for Real Frequency Matching," in *IEEE Benjamin Franklin Symposium on Microwave and Antenna Sub-Systems*, Philadelphia, 2014.

[3] B. S. Yarman, et al., "High Precison Synthesis of a Richards Immittance via Parametric Approach," *IEEE TCAS Part I: Regular Papers*, vol. 61, no. 4, pp. 1055–1067, 2014.

[4] A. Kilinc and B. Yarman, "High Precison LC Ladder Synthesis Part I: Lowpass Ladder Synthesis via Parametric Approach," *IEEE TCAS I: Regular Papers*, vol. 60, no. 8, pp. 2074–2083, 2013.

[5] B. S. Yarman and A. Kilinc, "High Precison LC Ladder Synthesis Part II: Immittance Synthesis with Transmission zeros at DC and Infinity," *IEEE TCAS Part I*, vol. 60, no. 10, pp. 2719–2729, 2013.

3

Transmission Lines as Phase Shifter

Summary

In this chapter, we cover the fundamentals of transmission lines as they are used circuit elements to construct perhaps inductors, capacitors, and loaded transmission phase shifters.

The content of this chapter is partially based on our book titled "Power Transfer Networks" [1].

3.1 Ideal Transmission Lines

Transmission lines are very simple conducting geometric structures that guide electromagnetic energy. For example, two long parallel conducting plates constitute the simplest version of such of structures yielding voltage versus current or electric field versus magnetic field relationships. Similarly, two conducting parallel wire or coaxial cables are also excellent examples of electromagnetic energy guiding systems.

Let us now consider two parallel long plates made of perfect conductors as shown in Figure 3.1. This structure guides the electromagnetic field energy as it propagates from the source-end down to the load-end in a loss free medium. Therefore, it is called the ideal transmission line.

Assume that a time-varying current is initiated on the top plate towards the positive z direction and it returns on the bottom plate. For this geometry, straightforward application of Gauss and revised Ampere's laws reveals that the electromagnetic field must be confined within the plates over the distance "d" uniformly. The conduction current $I(t, z)$ must be on the surface of the perfect conductors which in turn yields in a uniform magnetic field $\vec{H}_y$ in the y-direction; and it is aligned with the width "W" of conducting plates. In this case, say over a unit length $L = 1$ m, closed-path integration of the uniform magnetic field reveals that

63

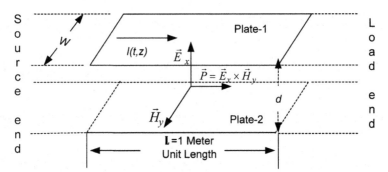

Figure 3.1   A parallel-plate transmission line made of perfect conductors.

$$\oint \vec{H} d\vec{l} = H_y W = I(t, z) \tag{3.1a}$$

or

$$H_y = \frac{I(t, z)}{W} \tag{3.1b}$$

Obviously, this magnetic field generates a uniform magnetic field density vector $\vec{B}_y = \mu \vec{H}_y$. Since the magnetic field density vector is outward the cross-sectional area $A = d\,(meter) \times L\,(meter) = d\,(meter) \times (1 meter) = d m^2$, the magnetic flux is given by

$$\emptyset = \iint_A \vec{B}.d\vec{S} = \mu H_y A = \mu H_y d \tag{3.1c}$$

In this case, we can define the inductance "L" over the unit length such that

$$L = \frac{\emptyset(t, z)}{I(t, z)} = \mu \frac{d}{W} \tag{3.1d}$$

Furthermore, we can assume that the displacement or leakage current due to uniform electric field $\vec{E}_x$ is generated by the uniform displacement vector $\vec{D}_x = \varepsilon \vec{E}_x$. Then, over the surface $A_W = W\,(meter) \times L\,(meter) = W \times 1 = W m^2$, the Gauss law reveals that

$$\iint_{A_W} \vec{D}(t, z).d\vec{S} = Q(t, z) \tag{3.2a}$$

or

$$\varepsilon E_x(t, z) W = Q(t, z) \tag{3.2b}$$

On the other hand, at any point "z", the potential difference $V(t, z)$ between the palates can be expressed as $V(t, z) = E_x(t, z).d$.

Hence, we can define the capacitance per unit length as

$$C = \frac{Q(t, z)}{V(t, z)} = \varepsilon \frac{W}{d} \qquad (3.2c)$$

which depends on the geometry and the electric property of the medium.

In summary, this ideal parallel-plate transmission line can be modeled by means of infinitely many cascaded T-sections of [L/2-C-L/2} or equivalently PI-Sections of [C-L-C] per unit length, as depicted in Figure 3.2.[1] In this configuration, letters L and C denote the inductor and capacitor per unit length, respectively.

It must be mentioned that plane waves with electric and magnetic field components on the same plane; perpendicular to the direction of propagation, just like the ones described above; are called **transverse-electric-magnetic field waves or in short TEM waves**.

For this line, we presume that the electric field vector is in positive x-direction (from plate-2 up to pate-1), the magnetic field vector is in positive y-direction and the electromagnetic wave energy propagates in positive z-direction.

Certainly, the solution of Maxwell equation yields the propagation pattern of the fields.

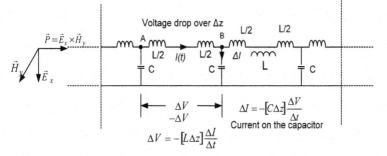

Figure 3.2 T or equivalently PI section model for an ideal transmission line.

---

[1] An ideal transmission line may consist of "two parallel perfect conductors", "two parallel wires" made of perfect conductors, or it may be a lossless coaxial cable made of perfect conductors. They all possess the same PI or T-section-type low pass L-C models with different values of inductors and capacitors.

On the other hand, one should be able to come up with wave equations when the descriptive voltage and current relations for inductors and capacitors are written. In fact, considering the PI-Section of C-L-C, the voltage drop $(-\triangle V)$ on the inductance $[L\triangle z]$ over an infinitesimally small length $\triangle z$ is given by

$$-\triangle V = V_L = [L\triangle z]\frac{\triangle I}{\triangle t}$$

or as $\triangle z \longrightarrow 0$

$$\lim_{\triangle z \to 0} \frac{\triangle V}{\triangle z} = \frac{\partial V}{\partial z} = -L\frac{\partial I}{\partial t} \tag{3.3a}$$

which is Faraday's law and it must be consistent with

$$\nabla \times \vec{E} = \frac{\partial \vec{E}_x}{\partial z} = -\mu \frac{\partial \vec{H}}{\partial t}$$

In a similar manner, over an infinitesimally small length $\triangle z$, the leakage or the displacement current $-\triangle I = I_C$ through the capacitor $[C\triangle z]$ is expressed as

$$-\triangle I = I_C = [C\triangle z]\frac{\triangle V}{\triangle t}$$

or as $\triangle z \longrightarrow 0$

$$\lim_{\triangle z \to 0} \frac{\triangle I}{\triangle z} = \frac{\partial I}{\partial z} = -C\frac{\partial V}{\partial t} \tag{3.3b}$$

Must be consistent with the revised version of Ampere's law.

$$\nabla \times H = \frac{\partial \vec{H}_y}{\partial t} = \vec{J}_C + \varepsilon\frac{\partial \vec{E}_x}{\partial t}$$

With conduction current density

$$\vec{J}_C = 0$$

By taking the derivative with respect to z, (3.3) can be written as

$$\frac{\partial^2 V}{\partial z^2} = -L\frac{\partial^2 I}{\partial z \partial t} \tag{3.4a}$$

Similarly, by taking the derivative with respect to time t, (3.2b) takes the following form

$$\frac{\partial^2 I}{\partial z \partial t} = -C\frac{\partial^2 V}{\partial t^2} \tag{3.4b}$$

Hence, (3.4a) and (3.4b) reveal that

$$\frac{\partial^2 V(z,t)}{\partial z^2} = LC\frac{\partial^2 V(z,t)}{\partial t^2}$$

or

$$\frac{\partial^2 V(z,t)}{\partial z^2} = \frac{1}{v^2}\frac{\partial^2 V(z,t)}{\partial t^2} \qquad (3.4c)$$

where

$$v = \frac{1}{\sqrt{LC}} \qquad (3.4d)$$

is the velocity of the wave propagation.

In fact, as shown above, for parallel conducting plates, which constitute an ideal TEM line, where the capacitor C and the inductor L per unit length are given by $C = \varepsilon\frac{W}{d}$ and $L = \mu\frac{d}{W}$, respectively. Thus, the velocity of wave propagation $v = \frac{1}{\sqrt{LC}} = \frac{1}{\sqrt{\mu\varepsilon}}$ is verified.

In as similar manner to Equation (3.5), the current wave equation $I(z,t)$ and the magnetic field wave equation are written as

$$\frac{\partial^2 I(z,t)}{\partial z^2} = \frac{1}{v^2}\frac{\partial^2 I(z,t)}{\partial t^2} \qquad (3.4e)$$

$$\frac{\partial^2 \vec{H}_y(z,t)}{\partial z^2} = \frac{1}{v^2}\frac{\partial^2 \vec{H}_y(z,t)}{\partial t^2} \qquad (3.4f)$$

with $LC = \mu\varepsilon$ and $v = \frac{1}{\sqrt{LC}}$

Here, it should be clear that uniform electric field $E_x(z,t)$ can easily be obtained from the voltage wave $V(z,t)$ by

$$V(z,t) = -\int_{Line-2}^{Line-1} E(z,t)\,dx$$

which simply yields

$$V(z,t) = E_x(z,t).d$$

or

$$E_x(t,z) = \frac{V(t,z)}{d} \qquad (3.4g)$$

Furthermore, Ampere's law gives relation between magnetic field and current by

$$H_y\left(z,t\right) = \frac{1}{W}I(z,t).$$

For historical reasons, Equations (3.1) and (3.2) are called Telegrapher's equations and they are summarized as below.

$$\frac{\partial V}{\partial z} = -L\frac{\partial I}{\partial t} \tag{3.5a}$$

$$\frac{\partial I}{\partial z} = -C\frac{\partial V}{\partial t} \tag{3.5b}$$

3.2 Time Domain Solutions of Voltage and Current Wave Equations

It can be shown that solution of voltage wave equation is given by

$$V\left(z,t\right) = V_i\left(t-\frac{z}{v}\right) + V_r\left(t+\frac{z}{v}\right) \tag{3.6a}$$

where

$$V_i\left(t-\frac{z}{v}\right)$$

refers to incident wave and the term

$$V_r\left(t+\frac{z}{v}\right)$$

is the reflected wave.

The current wave solution can be obtained by taking the derivative of (3.5) with respect to z (i.e. $\frac{\partial V}{\partial z} = -L\frac{\partial I}{\partial t}$)

$$-L\frac{\partial I(z,t)}{\partial t} = -\frac{1}{v}V_i'\left(t-\frac{z}{v}\right) + \frac{1}{v}V_r'\left(t+\frac{z}{v}\right)$$

and integrating the above equation over time, one obtains the following solution

$$I\left(z,t\right) = \frac{1}{Lv}\left[V_i\left(t-\frac{z}{v}\right) - V_r\left(t+\frac{z}{v}\right)\right] \tag{3.6b}$$

$$I\left(z,t\right) = \frac{1}{Lv}\left[V_i\left(t-\frac{z}{v}\right) - V_r\left(t+\frac{z}{v}\right)\right] \tag{3.6c}$$

or by setting $Z_0 = Lv = \sqrt{\frac{L}{C}}$

$$I\left(z,t\right) = \frac{1}{Z_0}\left[V_i\left(t-\frac{z}{v}\right) - V_r\left(t+\frac{z}{v}\right)\right] \qquad (3.6d)$$

3.3 Model for a Two-Pair Wire Transmission Line as an Ideal TEM Line

For a two-pair parallel wire transmission line, the unit length capacitor is given by

$$C = \frac{2\pi\varepsilon}{ln\left[\frac{d}{a}\right]} \qquad (3.7a)$$

On the other hand, unit length inductor is given by

$$L = \frac{\mu}{2\pi}ln\left[\frac{d}{a}\right] \qquad (3.7b)$$

In these expressions, "a" is the radius of parallel wires and "d" is the distance between them. Thus, we see that for this uniform TEM line $LC = \mu\varepsilon$, which verifies the result, obtained for ideal parallel plates. For this case, characteristic impedance is given by

$$Z_0 = \sqrt{\frac{L}{C}} = \frac{ln\left[\frac{d}{a}\right]}{2\pi}\sqrt{\frac{\mu}{\varepsilon}} \qquad (3.7c)$$

3.4 Model for a Coaxial Cable as an Ideal TEM Line

Coaxial cables also guide uniform electric and magnetic fields, which in turn yield in the voltage and current wave equations as above. In this regard, the unit length capacitance and inductance, respectively, are

$$C_{Coax} = \frac{2\pi\varepsilon}{ln\left[\frac{b}{a}\right]} \qquad (3.8a)$$

$$L_{Coax} = \frac{\mu}{2\pi}ln\left[\frac{b}{a}\right] \qquad (3.8b)$$

Hence, velocity of propagation verifies the above results

$$v = \frac{1}{\sqrt{L_{Coax}.C_{Coax}}} = \frac{1}{\sqrt{\mu\varepsilon}}$$

In a similar manner to Equation (3.10), characteristic impedance of the coaxial line is given by

$$Z_0 = \sqrt{\frac{L_{Coax}}{C_{Coax}}} = \frac{ln\left[\frac{b}{a}\right]}{2\pi}\sqrt{\frac{\mu}{\varepsilon}} \tag{3.8c}$$

3.5 Field Solutions for TEM Lines

For any TEM line, regardless of whether it is a parallel plate, two-pair wire, or coax cable, the time domain solutions of electric and magnetic fields are given by

$$E_x\left(t,z\right) = E_{xi}\left(t-\frac{z}{v}\right) + E_{xr}\left(t+\frac{z}{v}\right) \tag{3.9a}$$

$$H_y\left(t,z\right) = \frac{1}{Z_0}\left[E_{xi}\left(t-\frac{z}{v}\right) - E_{xr}\left(t+\frac{z}{v}\right)\right] \tag{3.9b}$$

where $v = \frac{1}{\sqrt{LC}}$ and $Z_0 = \sqrt{\frac{L}{C}}$

Furthermore, uniform electric field vector intensity is specified in terms of the potential difference between lines $V(t,z)$ at any point "z" as

$$E_x\left(t,z\right) = \frac{V\left(t,z\right)}{d}; E_{xi} = \frac{V_i\left(t-\frac{z}{v}\right)}{d}; E_{xr} = \frac{V_r\left(t+\frac{z}{v}\right)}{d} \tag{3.10a}$$

where "d" is the distance between the lines.

Similarly,

$$H_y\left(t,z\right) = \frac{1}{W}I\left(t,z\right) = \frac{1}{W Z_0}\left[V_i\left(t-\frac{z}{v_p}\right) - V_r\left(t+\frac{z}{v_p}\right)\right] \tag{3.10b}$$

3.6 Phasor solutions for ideal TEM Lines

Phasor forms of Telegrapher's equations can be obtained by replacing

$$\frac{\partial}{\partial t} \text{ by } j\omega \text{ and } \frac{\partial^2}{\partial t^2} \text{ by } -\omega^2 [2pt]$$

Hence, (3.6) reveals that

$$\frac{\partial V(z, j\omega)}{\partial z} = -j\omega I(z, j\omega) \tag{3.11a}$$

$$\frac{\partial I(z, j\omega)}{\partial z} = -j\omega V(z, j\omega) \tag{3.11b}$$

or by taking the derivative of the first equation with respect to z and using it in the second equation, we end up with

$$\frac{\partial^2 V(z, j\omega)}{\partial z^2} = -\omega^2 LCV(z, j\omega) = (j\omega L)(j\omega C) V(z, j\omega)$$

$$\frac{\partial^2 I(z, j\omega)}{\partial z^2} = -\omega^2 LCI(z, j\omega) = (j\omega L)(j\omega C) I(z, j\omega)$$

Let

$$Z(j\omega) = j\omega L$$
$$Y(j\omega) = j\omega C$$
$$\gamma^2 = -\omega^2 LC = Z(j\omega) Y(j\omega)$$

or

$$\gamma = j[\omega LC] = j\left(\frac{\omega}{v}\right) = j\beta(\omega) = \sqrt{Z(j\omega) Y(j\omega)} \tag{3.12a}$$

$$\beta(\omega) = \frac{\omega}{v_p} \tag{3.12b}$$

$$v_p = \frac{1}{\sqrt{LC}} \tag{3.12c}$$

Then, phasor solution for voltage and current is given by

$$V(z, j\omega) = Ae^{-\gamma z} + Be^{+\gamma z} \tag{3.13a}$$

$$I(z, j\omega) = \frac{1}{Z_0}\left[Ae^{-\gamma z} - Be^{+\gamma z}\right] \tag{3.13b}$$

where

$$\gamma = j\beta$$

is the propagation constant; and

$$Z_0 = \sqrt{\frac{Z(j\omega)}{Y(j\omega)}} = \sqrt{\frac{L}{C}}$$

is the characteristic impedance of the TEM line.

3.7 Steady-State Time Domain Solutions for Voltage and Current at any Point z on the TEM Line

Assume that a TEM line is excited by a sinusoidal voltage source

$$E_G(t) = E_m Cos(\omega t) = real\left\{E_m e^{j\omega t}\right\}.$$

Then, time domain solution for voltage and current at any point z on the line is given by

$$V(t, z) = real\left\{V(z, j\omega) e^{j\omega t}\right\}$$
$$I(t, z) = real\left\{I(z, j\omega) e^{j\omega t}\right\}$$

In the above equations, voltage and current phasors are given by

$$V(z, j\omega) = V_m(z, \omega) e^{-j\varphi_v(\omega)}$$

$$I(z, j\omega) = I_m(z, \omega) e^{-j\varphi_i(\omega)}$$

In this case, at any point z, sinusoidal steady-state solutions for voltage and current are given by

$$V(t, z) = V_m(z, \omega) \cos?(\omega t - \varphi_v)$$

$$I(t, z) = I_m(z, \omega) \cos?(\omega t - \varphi_i)$$

In the following, some practical definitions and notations for transmission lines are given.

Transmission lines as circuit elements.

3.8 Definition of the Major Parameters of a Transmission Line

Propagation Constant γ

In the above equation, $\gamma(j\omega)$ is called the propagation constant. For ideal lossless TEM lines, the propagation constant $\gamma = \sqrt{Z(j\omega) Y(j\omega)}$ is purely imaginary and it is specified as $\gamma(j\omega) = j\beta(\omega)$. However, referring to Figure 3.2, for a lossy line, a series loss resistance r and a shunt conductance G are connected with inductor L and capacitor C, respectively. In this case, the propagation constant takes the following form

$$\gamma(j\omega) = \alpha(\omega) + j\beta(\omega) = \sqrt{(r+j\omega L)(G+j\omega C)} \qquad (3.14)$$

Phase Constant β

The quantity *imaginary part of* $\{\gamma\} = \beta(\omega) = \frac{\omega}{v}$ is called the phase constant. For lossless TEM lines, it is interesting to observe that phase constant is a linear function of the frequency of operation yielding

$$\frac{d\beta}{d\omega} = \frac{1}{v_p} \qquad (3.15a)$$

For a fixed operation frequency f, the wavelength λ is given by

$$\lambda = \frac{v_p}{f} = \frac{2\pi v_p}{2\pi f} = \frac{2\pi v_p}{\omega}$$

In this case, the phase constant β is expressed as

$$\beta = \frac{2\pi}{\lambda} \ \{radian/meter\} \qquad (3.15b)$$

Attenuation Constant α

The real part of γ, which is $\alpha(\omega) = real\ \{\gamma(j\omega)\}$, is called the attenuation constant.

Phase Velocity v_p

The quantity $\left[\frac{d\beta}{d\omega}\right]^{-1}$ is called the phase velocity and it is designate by v_p. For ideal TEM lines $v_p = \frac{1}{\sqrt{\mu\varepsilon}}$.

Delay Length τ Let us feed a TEM line of length "L" with a sinusoidal source of frequency $f = \frac{1}{T}$ period $T = \frac{1}{f}$ and wave length $\lambda = \frac{v}{f} = Tv_p$. In this case, it takes $\tau = \frac{L}{v}$ *seconds* for a sinusoidal waveform of one period to travel from one end of the line to the other end. Hence, τ is called the delay length of the line.

Phase delay

At any point $z = l$ on the line, the phase delay is measured by the multiplication of $\varphi = \beta l$ which has the unit either degree or radian. Usually,

βl is measured in radian. It is straight forward to see that the phase delay

$$\varphi = \beta l = \frac{\omega l}{v_p} = \omega \tau.$$

Incident Wave $A\ (z, j\ \omega)$

In (3.13), the term

$$A(z, j\omega) = Ae^{-\gamma z}$$

is called the incident wave at point "z".

Reflected Wave $B\ (z, j\ \omega)$

In (3.13), the term

$$B(z, j\omega) = Be^{+\gamma z}$$

is called the reflected wave.

3.9 Voltage and Current Expression in Terms of Incident and Reflected Waves

Based on the definitions of incident and reflected waves, voltage and current expressions of (3.13) are written as

$$V(z, j\omega) = A(z, j\omega) + B(z, j\omega) \qquad (3.16a)$$

$$I(z, j\omega) = \frac{1}{Z_0}[A(z, j\omega) - B(z, j\omega)] \qquad (3.16b)$$

3.10 Reflection Coefficient $S(z, j\ \omega)$

At any point "z" on the line, the ratio of reflected wave to incident wave is called the "reflection coefficient" and it is given by

$$S(z, j\omega) = \frac{B}{A}e^{2\gamma z} \qquad (3.17)$$

3.11 TEM Lines as Circuit or "Distributed" Elements

In the classical microwave circuit literature, a typical TEM line is called a "distributed circuit element" with physical length "L" as depicted in Figure 3.3.

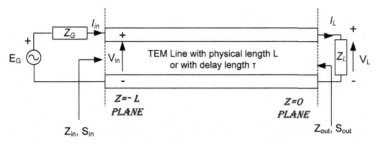

Figure 3.3 A typical TEM transmission line as a "distributed circuit element".

In this physical layout, the conducting plates or wires are designated by two long rectangular boxes. At microwave frequencies, these conducting boxes or planes may be realized as micro strip lines or coplanar lines on a selected substrate such as alumina, silicon using metal deposition techniques.

In Figure 3.3, the line is driven by a voltage source E_G with internal impedance Z_G. At the end, the line is terminated by complex load impedance called $Z_L(j\omega)$. The line has a physical length "L". For convenience, the origin of the $z - axis$ is selected at the far end of the line and geometrically; it is designated by $z = 0$ $plane$. The line is physically placed on the negative $z - axis$. With this convention, the voltage source at the input is connected to the line at $z = -L$ $plane$.

As a distributed circuit element, a TEM line is considered as a two-port, which introduces a physical length for the actual designs.

Now, let us take a look at the voltage and current relations at the input (or source –end; z $= -$ L plane) and the output (or at the load end; z $= 0$ Plane)

3.12 Voltage and Current Expressions at the Load End; Load Reflection Coefficient on the z $=$ 0 Plane

First of all, at the far end of the line $(on \ the \ z = 0 \ plane)$, the reflection coefficient is given by

$$S(0, j\omega) = S_L = \frac{B}{A} \tag{3.18}$$

S_L is called the load reflection coefficient

Equation (3.18) reveals a boundary condition such that voltage across the load must be equal to line voltage at $z = 0$ plane. Therefore,

$$V_L = Z_L I_L = A + B = A\left(1 + \frac{B}{A}\right) = A(1 + S_L)$$

On the other hand, the current on the load-end or at z = 0 plane is given by

$$I_L = \frac{1}{Z_0}[A - B] = \frac{A}{Z_0}(1 - S_L)$$

Then, by setting, $Z_L = \frac{V_L}{I_L}$, it is found that

$$Z_L = Z_0 \frac{1 + S_L}{1 - S_L} \tag{3.19a}$$

or

$$z_L = \frac{Z_L}{Z_0} = \frac{1 + S_L}{1 - S_L} \tag{3.19b}$$

In (3.19), z_L is called the normalized load impedance with respect to given standard impedance Z_0. The characteristic impedance Z_0 could be either a real or complex impedance. If the line is lossless, then $z_0 = \sqrt{\frac{L}{C}}$ is real, otherwise it is a complex impedance.

Utilizing (3.19), the reflection coefficient S_L may be expressed in terms of the normalized load impedance z_L such that

$$S_L = \frac{Reflected\ Wave}{Incident\ Wave} = \frac{B}{A} = \frac{z_L - 1}{z_L + 1} = \frac{Z_L - Z_0}{Z_L + Z_0} \tag{3.20}$$

3.13 Voltage and Current Expressions at the Source-End; Input Reflection Coefficient on the z = −L Plane

The source-end or z = − L plane of the transmission line may be named as the input. In this case, at z = − L, the reflection coefficient of (3.19) takes the following form

$$S(-L, j\omega) = S_{in} = \frac{Reflected\ Wave}{Incidient\ Wave} = \frac{Be^{-\gamma l}}{Ae^{\gamma L}} = S_L e^{-2\gamma L} \tag{3.21}$$

On the other hand, voltage and current relations of (3.16) become

$$V(-L, j\omega) = V_{in} = Ae^{\gamma L} + Be^{-\gamma L} = Ae^{\gamma L}\left(1 + \frac{B}{A}e^{-2\gamma L}\right)$$

or

$$V_{in} = Ae^{\gamma L}\left(1 + S_L e^{-2\gamma L}\right)$$

or

$$V_{in} = Ae^{\gamma L} \left[1 + S_{in} \right]$$

Similarly,

$$I(-L, j\omega) = I_{in} = \frac{1}{Z_0} Ae^{\gamma L} \left[1 - S_L e^{-2\gamma L} \right]$$

or

$$I_{in} = \frac{1}{Z_0} Ae^{\gamma L} \left[1 - S_{in} \right]$$

Then, the input impedance $Z_{in} = \frac{V_{in}}{I_{in}}$ is written as
Then,

$$Z_{in} = Z_0 \frac{1 + S_{in}}{1 - S_{in}} \tag{3.22a}$$

Let the normalized input impedance be

$$z_{in} = \frac{Z_{in}}{Z_0}$$

Then,

$$S_{in} = \frac{z_{in} - 1}{z_{in} + 1} = \frac{Z_{in} - Z_0}{Z_{in} + Z_0} \tag{3.22b}$$

or

$$z_{in} = \frac{1 + S_{in}}{1 - S_{in}} \tag{3.22c}$$

Furthermore, in a similar manner to (3.20), we can define the reflection coefficient for the internal impedance Z_G as

$$S_G = \frac{Z_G - Z_0}{Z_G + Z_0} = \frac{z_G - 1}{z_G + 1} \tag{3.22d}$$

where

$$z_G = \frac{Z_G}{Z_0}$$

is the normalized generator impedance.

3.14 Output Reflection Coefficient at z = 0 Plane

It should be noted that if the transmission lines is driven at the $z = 0$ plane by a voltage source with an internal impedance Z_L and if it is terminated by an impedance Z_G at the $z = -L$ plane, then one can drive the expressions for output reflection coefficient S_{out} and output impedance Z_{out} and normalized output impedance z_{out} as above.

The normalized impedance of actual the impedance and the output reflection coefficient is given by

$$S_{out} = \frac{Reflected\ Wave}{Incident\ Wave} = \frac{Z_{out} - Z_o}{Z_{out} + Z_o} = \frac{z_{out} - 1}{z_{out} + 1} = S_G e^{-2\gamma L} \quad (3.23a)$$

where

$$Z_{out} = Z_0 \frac{1 + S_{out}}{1 - S_{out}}$$

or the normalized output impedance is given by

$$z_{out} = \frac{Z_{out}}{Z_0} = \frac{1 + S_{out}}{1 - S_{out}} \quad (3.23b)$$

3.15 Voltage Standing Wave Ratio: VSWR

At any point on the ideal TEM line, maximum of voltage phasor can be expressed as

$$V_{max} = \left| Ae^{-j\beta l} \right| + \left| Be^{-j\beta l} \right|$$

or

$$V_{max} = A + B = A\left(1 + \frac{B}{A}\right) = A\left[1 + |S_L|\right]$$

Similarly, the minimum of the voltage is given by

$$V_{min} = \left| Ae^{-j\beta l} \right| - \left| Be^{-j\beta l} \right|$$

or

$$V_{min} = A - B = A\left[1 - |S_L|\right]$$

Then, the voltage standing wave ratio (VSWR) of the line is defined as the ratio of the maximum voltage amplitude to the minimum voltage amplitude. Thus,[2]

$$VSWR = \frac{1 + |S_L|}{1 - |S_L|} \quad (3.23c)$$

---

[2]It should be noted that the VSWR is a positive real number and it varies between zero and infinity since the load reflection coefficient is bounded by 1 (i.e. $|S_L| \leq 1$).

3.16 Open Expressions for the Input and Output Reflection Coefficients and Impedances

Equations (3.21) and (3.23) specify the input and output reflection coefficients by means of the term $e^{-2\gamma l}$ as $S_{in} = S_L e^{-2\gamma l}$ and $S_{out} = S_G e^{-2\gamma l}$.

On the other hand, the term $e^{-2\gamma l}$ can be expressed in terms of tangent hyperbolic function as follows

$$tanh\,(\gamma l) = \frac{e^{\gamma l} - e^{-\gamma l}}{e^{\gamma l} + e^{-\gamma l}} = \frac{e^{\gamma l}\left[1 - e^{-2\gamma l}\right]}{e^{\gamma l}\left[1 + e^{-2\gamma l}\right]} = \frac{1 - e^{-2\gamma l}}{1 + e^{-2\gamma l}} \qquad (3.24a)$$

or

$$e^{-2\gamma l} = \frac{1 - \tanh\,(\gamma l)}{1 + \tanh\,(\gamma l)} \qquad (3.24b)$$

Let us define a new complex variable "λ", which is called the "Richards variable", as

$$\lambda = \tanh\,(\gamma l) = \sum +j\Omega \qquad (3.24c)$$

Then, in terms of the Richards variable,

$$e^{-2\gamma l} = \frac{1 - \lambda}{1 + \lambda} \qquad (3.24d)$$

Using the above identity, input and output reflection coefficients are given by

$$S_{in}\,(\lambda) = \frac{1 - \lambda}{1 + \lambda} S_L \qquad (3.24e)$$

$$S_{out}\,(\lambda) = \frac{1 - \lambda}{1 + \lambda} S_G \qquad (3.24f)$$

Using (3.24e) in (3.22a), we have

$$Z_{in}(\lambda) = Z_0 \frac{(1+\lambda) + (1-\lambda)\,S_L}{(1+\lambda) - (1-\lambda)\,S_L} \qquad (3.24g)$$

or employing $S_L = \frac{Z_L - Z_0}{Z_L + Z_0}$ in the above equation of $Z_{in}(\lambda)$, it is found that

$$Z_{in}(\lambda) = Z_0\frac{Z_L+Z_0\lambda}{Z_0+Z_L\lambda} \tag{3.24h}$$

or in open form

$$Z_{in} = Z_0\frac{Z_L+Z_0tanh(\gamma l)}{Z_0+Z_Ltanh(\gamma l)} \tag{3.24i}$$

In a similar manner to (3.24h), the output impedance Z_{out} is given by

$$Z_{out}(\lambda) = Z_0\frac{Z_G+Z_0\lambda}{Z_0+Z_G\lambda} \tag{3.24j}$$

or in open form

$$Z_{out} = Z_0\frac{Z_G+Z_0tanh(\gamma l)}{Z_0+Z_Gtanh(\gamma l)} \tag{3.24k}$$

For an ideal TEM transmission line, the attenuation constant $\alpha(\omega)$ must be zero. Therefore, the propagation constant has only imaginary part and it is specified as $= j\beta$. On the other hand, the characteristic impedance is given as a positive real number such that $Z_0 = \sqrt{\frac{L}{C}}$.

In this case, for an ideal TEM line, Equation (3.24) is simplified as

$$\lambda = j\Omega \tag{3.25a}$$

$$\Omega = tan(\beta l) = tan(\omega\tau) \tag{3.25b}$$

$$Z_{in} = Z_0\frac{Z_L+jZ_0tan(\beta l)}{Z_0+jZ_Ltan(\beta l)} \tag{3.25c}$$

$$Z_{out} = Z_0\frac{Z_G+jZ_0tan(\beta l)}{Z_0+jZ_Gtan(\beta l)} \tag{3.25d}$$

Equation (3.25a–f) may as well be expressed in terms of admittances. In this case, all impedances, which are designated by letter "Z_{xx}", are replaced by corresponding admittances, which are denoted by letters "Y_{xx}". In this notation, subscripts "xx" refer to input, output, load, and generator and the characteristic impedances and corresponding admittances respectively.

$$Y_{in} = Y_0\frac{Y_L+jY_0tan(\beta l)}{Z_0+jY_Ltan(\beta l)} \tag{3.25e}$$

$$Y_{out} = Y_0\frac{Y_G+jY_0tan(\beta l)}{Z_0+jZ_Gtan(\beta l)} \tag{3.25f}$$

Ideal TEM lines may be utilized as a distributed capacitors and inductors by properly choosing the proper real characteristic impedance $Z_0 = R_0 = \sqrt{\frac{L}{C}}$, the line phase length $\varphi_L = \beta L = \omega\tau$, and termination impedances Z_G and Z_L. In the following, we present some useful distributed one- or two-port components that are useful to design power transfer networks.

3.17 An Open-end TEM line as a Capacitor

When the far end of a transmission line is left open, it is considered to be terminated with an infinite impedance load, *i.e.*, $Z_L = \infty$. In this case, the input impedance acts like a capacitor C in $\lambda - domain$ such that

$$Z_{in}(\lambda) = \lim_{z_L \to \infty} Z_0 \frac{Z_L + Z_0\lambda}{Z_0 + Z_L\lambda} = \frac{1}{\lambda C}$$

Or equivalently

$$Y_{in}(\lambda) = \lim_{Y_L \to 0} Y_0 \frac{Y_L + Y_0\lambda}{Y_0 + Y_L\lambda} = \lambda C$$

where the capacitor is given by

$$C = \frac{1}{Z_0} = Y_0$$

$Y_0 = \frac{1}{Z_0}$ is the characteristic admittance.
   The open form of the admittance equation is given by

$$Y_{in}(\lambda) = \lambda C = j\Omega C = j\left[Y_0 tan(\omega\tau)\right]$$

This is a typical capacitive admittance defined in the transform frequency domain $\Omega = \tan(\omega\tau)$. However, one should never forget that, here, we are dealing with transmission lines. The frequency Ω is expressed as a tangent function in actual frequency ω. In this case, if $\omega\tau$ is small enough, then one can approximate $\tan(\omega\tau)$ *as* $\tan(\omega\tau) \approx \omega\tau$. Hence, a lumped element capacitor C_{Lump} may be approximated with a short-length open-end transmission line of characteristic admittance Y_0 such that

$$j\left[Y_0 tan(\omega\tau)\right] \approx j\omega\left[Y_0\tau\right] = j\omega C_{Lump}$$

Hence, for a small length, open-ended TEM line, equivalent lumped capacitance can be approximated as

$$C_{Lump} \cong \tau Y_0 \qquad (3.26)$$

With small-length TEM line, what we mean is that the physical length "L" of the line must be much smaller than that of the wave length "λ" specified at the operation frequency. Therefore, we say that $\omega\tau = \beta L \ll \beta\lambda$. For many practical situations, it may be sufficient to choose the physical length of the line as

$$L < 0.1\lambda; \quad \lambda = \frac{v_p}{f}$$

Certainly, selection of $L \cong 0.01\lambda$ is a good choice.

Now, let us run an example to compute the capacitor of a short line.

Example 3.1: At the operating frequency of $f = 1\ GHz$, compute the capacitance of an open-ended short transmission line of length $L \cong 0.01\lambda$ with characteristic impedance $Z_0 = 20\ \Omega$.

Solution

If $f = 1GHz$ then, $L = 0.01 \times \frac{3\times10^8}{10^9} = 0.003\ m = 3mm$ is a good choice to realize the capacitor as an open-end TEM line. In this case, the delay length is $\cong \frac{3\times10^{-3}}{3\times10^8} = 1 \times 10^{-11}\ second = 10\ pico\ seconds$.

Hence, the lumped equivalent of the capacitance is given by[3]

$$C_{Lump} = 10Y_0\ pico\ Farad$$

By setting $Y_0 = \frac{1}{20} = 0.05$, we have[4]

$$C_{Lump} = 10 \times 0.05\ PF = 0.5pF\ at1\ GHz.$$

In fact, considering a wide microstrip patch, a short-length open-end transmission line can be considered as a parallel-plate capacitor, as shown in Figure 3.4.

For the above structure, we can roughly estimate the characteristic admittance Y_0 as follows.

The value of the capacitor is given by $C_{Lump} = \varepsilon\frac{W\times L}{h} \cong \tau Y_0$. Hence,

---

[3]It is noted that in order to get a big value of the capacitance, characteristic admittance must be a big value or equivalently, the characteristic impedance must be as small as possible. For a selected dielectric material, employing microstrip line technology, we can safely achieve a small characteristic impedance of $Z_0 = 20\ ohm$. Low-value characteristic impedances exhibit wide width patch on microstrips.

[4]Here, as the velocity of propagation, we used free space velocity of light, but it should be modified depending on the selected dielectric material as $= \frac{c}{\sqrt{\mu_r \varepsilon_r}}$

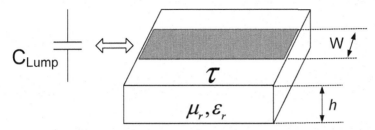

Figure 3.4 An open-end and short-length microstrip line as a capacitor.

$$Y_0 \cong \varepsilon \left(\frac{W}{h}\right) \left(\frac{L}{\tau}\right)$$

where h is the thickness of the substrate. It should be noted that, usually, the permeability μ of a good dielectric material is equal to $\mu = \mu_r \mu_0 \cong \mu_0$ (i.e., $\mu_r \cong 1$). On the other hand, the phase velocity is $v_p = \frac{L}{\tau} = \frac{1}{\sqrt{\mu\varepsilon}}$ and the characteristic impedance of free space is $\eta_0 = \sqrt{\frac{\mu_0}{\varepsilon_0}} \cong 120\pi$.

With these figures, the characteristic admittance can be approximated as

$$Y_0 \cong \sqrt{\frac{\varepsilon}{\mu}} \left(\frac{W}{h}\right) = \sqrt{\varepsilon_r} \eta_0^{-1} \left(\frac{W}{h}\right)$$

or the rough estimate on the characteristic impedance is given by

$$Z_0 \cong \frac{120\pi}{\sqrt{\varepsilon_r}} \left(\frac{h}{W}\right) \tag{3.27}$$

Example 3.2: For an aluminum substrate of thickness $h = 0.635mm$, compute the top surface gold-conducting patch width W for a characteristic impedance of $Z_0 = 20\ \Omega$.

Solution

For aluminum $\varepsilon_r \cong 9.8$. Then, $\left(\frac{h}{W}\right) = \frac{\sqrt{\varepsilon_r} Z_0}{120\pi} = \frac{3.1305 \times 20}{120 \times \pi} = 0.1661$ or $W = \frac{0.635}{0.1661} = 3.8235mm$.

3.18 A Shorted TEM Line as an Inductor

A shorted TEM line (i.e., $Z_L = 0$) acts like an inductor L in $\lambda - domain$ such that

$$\lim_{z_L \to 0} Z_{in}(\lambda) = Z_0 \frac{Z_L + Z_0 \lambda}{Z_0 + Z_L \lambda} = \lambda L$$

where the inductance L is given by

$$L = Z_0$$

As shown in the previous section, we should remember that we are dealing with a transmission line. Therefore, the input impedance of a shorted transmission line is expressed as a function of $\Omega = tan(\omega\tau)$. In this case, a lumped inductance can be realized as the input impedance of a small length transmission line when it is shorted at the other end.

Hence,

$$Z_{in} = jZ_0\tan(\omega\tau) \cong j\omega[\tau Z_0] = jL_{Lump}$$

or

$$L_{Lump} \cong \tau Z_0 \tag{3.28}$$

where τ is the delay length of the line, which is selected as the time delay of the physical length of $L \cong 0.01\lambda = 0.01\frac{v}{f}$

As it is seen from the above equation, a high-value inductance can be realized with a high-value characteristic impedance Z_0 of a shorted small length line.

In practice, a shorted line can be realized on a dielectric substrate as a very thin-plate conductor with width " W " as a microstrip conducting patch, as shown in Figure 3.5.

The series inductance between conducting plates of "$h = meter\ a\ part$" with length "L" can roughly be estimated as in Section 3.1 such that

$$L_{Plate} = \mu\left(\frac{h}{W}\right)(L)$$

On the other hand, this inductance must be equal to the input impedance of the shorted line. Therefore,

$$L_{Plate} = \mu\left(\frac{h}{W}\right)L$$

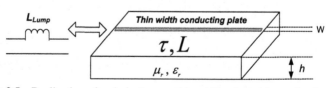

Figure 3.5   Realization of an inductor as a thin width microstrip conducting patch.

$$L_{Lump} = Z_0 \tau = L_{Plate} = \mu \left(\frac{h}{W}\right)(L)$$

Using $v_p = \frac{L}{\tau}$ and $\eta_0 = \sqrt{\frac{\mu_0}{\varepsilon_0}}$, the characteristic impedance of the line is roughly estimated as

$$Z_0 \cong \frac{\eta_0}{\sqrt{\varepsilon_r}} \left(\frac{h}{W}\right) = \frac{120\pi}{\sqrt{\varepsilon_r}} \left(\frac{h}{W}\right) \tag{3.29}$$

Let us run a practical example to compute a possible high value of an inductance manufactured on alumina substrate.

Example 3.3: On alumina, we can safely build a high characteristic impedance of $Z_0 = 100\ ohm$. For the sake of simplicity, let us choose the relative dielectric constant as $\varepsilon_r \cong 9.8$ and relative permeability $\mu_r \cong 1$. In this case, the propagation or phase velocity is given by $v_p \approx \frac{3 \times 10^8}{\sqrt{9}} = 10^8\ m/s$. At $f = 1\ GHz$, the wavelength is given by $\lambda = \frac{v}{f} = 0.1\ meter = 10\ cm$. Then, let us choose the physical length as $L = 0.01\lambda = 10^{-3}meter$. The time delay of the line is $\tau = \frac{L}{v} = 10^{-11}second.$[5] Thus, the characteristic impedance of $Z_0 = 100\ \Omega$ yields an equivalent lumped inductance of $L_{Lump} = 100 \times 10^{-11} = 10^{-9}\ Henry = 1nH$.

For this example, estimated $\left(\frac{h}{W}\right)$ is compute as

$$\left(\frac{h}{W}\right) = \frac{\sqrt{\varepsilon_r}}{120\pi} Z_0 = \frac{3 \times 100}{120\pi} = 0.7958 \cong 0.8$$

It should be noted that all the above derivations are carried out for the ideal parallel-plate TEM lines, which may approximate the behavior of microstrip lines. However, microstrip lines are not ideal. Regarding the practical implementation of lumped element inductors and capacitors, the reader is encouraged to refer the books of Inder Bahl and Prakash Bhartia.[6]

---

[5]It should be noted that, when the delay length of the line L is selected as the fraction δ of the wavelength λ, then the time delay is given by $\tau = \frac{\delta}{f}$, which is independent of the propagation velocity. For the example under consideration, $\delta = 0.01$ and $f = 1\ GHz$ are selected. Thus, $\tau = 10^{-11}$ sec is found.

[6]Iner Bahl, "Lumped Elements for RF and Microwave Circuits", Artech House, ISBN 1-58053-309-4, 2003Iner Bahl and Prakash Bhartia, "Microwave Solid State Circuit Design", Wiley-InterScience, ISBN 0-471-20755-1, 2003.

3.19 A Quarter Wavelength TEM Line at Resonance Frequency

On an ideal TEM line, when the phase-shift is $\varphi = \beta L = \frac{\pi}{2} = \frac{2\pi}{\lambda}L$ or $L = \frac{\lambda}{4}$, the Richard variable becomes infinity (i.e., $\lambda = jtan\,(\beta L) = \infty$). In this case, the input impedance Z_{in} becomes

$$Z_{in}\,(\lambda) = \lim_{\lambda\to\infty} Z_0 \frac{Z_L+Z_0\lambda}{Z_0+Z_L\lambda} = \frac{Z_0^2}{Z_L}$$

or

$$Z_{in}Z_L = Z_0^2 \tag{3.30}$$

From the design point of view, (3.30) may be utilized to adjust the input impedance of the line for matching purposes like a transformer by playing the characteristic impedance Z_0.

So far, we investigated the behavior of open- and short-ended transmission lines with very small physical lengths as compared to the wavelength of operations. However, we can directly investigate the input impedance or equivalently input admittance of (3.30) by varying the termination immittances (i.e., impedance Z_L or admittance Y_L) starting from small values down to zero. This way, we can have the chance to investigate the limit cases and perhaps, develop simple lump circuit models for TEM lines, which are basic resonance circuits. These models may as well be utilized to understand basic operation of dipole antennas as open- or short-ended TEM lines.

Let us first work for the limit case where the termination impedance approaches to infinity (i.e., $Z_L \to \infty$) or equivalently the termination admittance approaches to zero, which is much easier to work (i.e., $Y_L \to 0$).

3.20 Open-ended TEM Line with Arbitrary Length

Using normalized admittance values with respect to characteristic admittance Y_0 (*i.e.,* $Y_0 = 1$ *selected*) and setting $Y_L = 0.1$, let us plot real and imaginary parts of input admittance $Y_{in}\,(j\omega\tau) = G_{in}\,(\omega\tau) + jB_{in}\,(\omega\tau)$ as shown in Figures 3.6 and 3.7, respectively. A close examination of these figures reveals that operation of low admittance loaded or shorted transmission line (i.e., *normalized* $Y_L = 0.1$) resembles the operation of a series resonance circuit, as shown in Figure 3.8. As Y_L approaches to zero, we reach to ideal case, that is to say, we reach to lossless series resonance circuit configuration. In this case, as determined above, the equivalent lumped element capacitor, inductor, and the resistor are given by $C_{Lump} = \tau Y_0$, $L_{Lump} = \tau Z_0$ and a

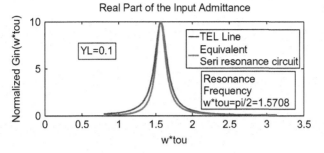

Figure 3.6 Real part of the input admittance.

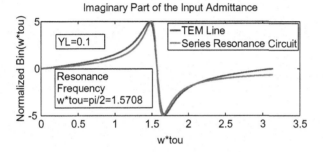

Figure 3.7 Imaginary part of the input admittance.

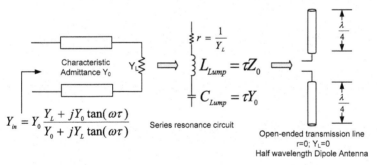

Figure 3.8 Low admittance loaded transmission line, its lumped element, and open-ended transmission line as half-wavelength dipole antenna.

series resistance $r = \dfrac{1}{G_{in}\left(\frac{\pi}{2}\right)} = \dfrac{1}{Y_L}$, respectively. The resonance frequency of the line is located at $\omega\tau = \frac{\pi}{2}$ or the resonance circuit resonates at $= \frac{\pi}{2\tau}$.

Now, let us consider a half-wavelength dipole antenna. This structure can be obtained by wide opening the arms of the TEM line as shown in Figure 3.8.

So, this antenna should act like a series resonance circuit and it should resonate at the resonance frequency such that angular resonance frequency is given by

$$\omega = 2\pi f = \frac{\pi}{2\tau}$$

or actual resonance frequency is

$$f = \frac{1}{4\tau}$$

At the resonance frequency, obviously, we must see a resistive input admittance, which is associated with the series loss resistance of the wires specified by $r = \frac{Y_0^2}{Y_L} = \frac{1}{Y_L}$ (*normalized with respect to* Y_0). However, for the antenna case, at the resonance, input resistance must also include the radiation resistance of the dipole antenna.

3.21 Shorted TEM Line with Arbitrary Length

In a similar manner to that of open-ended lines, it is straightforward to show that a shorted TEM line acts like a parallel resonance circuit at the operating frequency of $f = \frac{1}{4\tau}$. In this case, it is appropriate to work with the input impedance.

Now, let us normalize all the impedances with respect to characteristic impedance Z_0. Then, as above, we set $Z_0 = 1$ and the termination resistance to a small value such as $Z_L = 0.1$ to investigate the impedance variation versus frequency. Under these assumptions, we developed a MatLab program to analyze the input impedance of $Z_{in} = Z_0 \frac{Z_L + j Z_0 tan(\beta l)}{Z_0 + j Z_L tan(\beta l)}$ called "TRLINE" as listed at the end of this section. This program was tested for different values of Z_L. As Z_L approaches to zero, fit between the line impedance and the model impedance improves as expected.

Hence, we end up with the variation of the real part R_{in} and the imaginary part X_{in} of the input impedance as a function of $\omega \tau$ as depicted in Figures 3.9 and 3.10, respectively.

On the same figures, we also plot the parallel resonance circuit model for the line under consideration.

As in the previous subsection, circuit elements of the parallel resonance circuit are given as follows

$$Resonance\ frequency\colon f = \frac{1}{4\tau}$$
$$Parallel\ Capacitor\ C_p = \tau Y_0$$

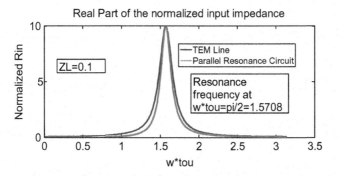

Figure 3.9 Real part of the input impedance when it is terminated in normalized load impedance $Z_L = 0.1$.

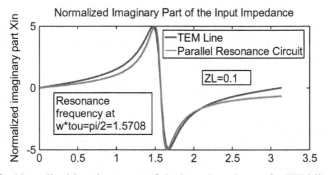

Figure 3.10 Normalized imaginary part of the input impedance of a TEM line, which is terminated in $Z_L = 0.1$.

$$Parallel\ Inductor\ L_p = \tau Z_0$$

$$Parallel\ Resistor\ R_p = \frac{Z_0^2}{Z_L} = \frac{1}{Z_L} \tag{3.31}$$

We should also mentioned that the parallel resonance circuit model can also be used to model loop antennas with arbitrary length as shown in Figure 3.11.

The antenna resonates at the frequency of $f = \frac{1}{4\tau}$ and it sees resistive input impedance R_{in}, which is partially related to the loss of the wires specified by $R_P \frac{Z_0^2}{Z_L} = \frac{1}{Z_L}$ *(normalized with respect to Z_0)* and the radiation resistance.

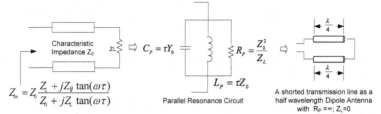

$$Z_{in} = Z_0 \frac{Z_L + jZ_0 \tan(\omega\tau)}{Z_0 + jZ_L \tan(\omega\tau)}$$

Figure 3.11   Shorted transmission line model as a parallel resonance circuit, which resembles a shorted dipole antenna.

3.22 Remarks: $tan(\omega\tau)$ as a Foster Function

(1) We should note that for an open- or short-ended transmission line, the input immittance $F_{in} = jX_{in}(\omega) = jtan(\omega\tau)F_0 = j\Omega F_0$ is a Foster function for all ω since $\frac{dX_{in}(\omega)}{d\omega} = \frac{[\tau F_0]}{cos^2(\omega\tau)} \geq 0; \; \forall\omega.$ Therefore, it is lossless and a "Foster function". In this nomenclature, the letter F_0 denotes either the characteristic impedance Z_0 or characteristic admittance Y_0.

(2) Over the interval $0 \leq \omega\tau < \frac{3\pi}{4}$, the function $X(\omega) = \tan(\omega\tau)$ exhibits resonance circuit type of an immittance variation with the resonance frequency placed at $\omega\tau = \frac{\pi}{2}$ as shown in Figure 3.12. Therefore, it may be meaning full to build approximate models for open or shorted transmission lines employing resonance circuits as we have shown in above sections. However, one should bear in mind that, at the operation frequency f the line delay length τ should not violate the inequality of $\omega\tau < \frac{3\pi}{4}$. Therefore, it may always be appropriate to choose the fixed length of the transmission to satisfy the inequality of $0 \leq \omega\tau < \frac{3\pi}{4}$.

We would like to take the attention of the reader to Figures 3.7 and 3.10. These are the plots of imaginary parts of the input immittance. A close examination of these figures reveals that for TEM lines, the zero cross point is placed at $\omega\tau = \frac{3\pi}{4} \cong 2.356$; on the other hand, for resonance structures, imaginary parts of the immittance continue toward infinity below zero, as shown in Figure 3.12.

The "resonance circuit type models" of transmission lines provides very rough fit between the model and the actual data. In practice, we prefer to work with precise models, which probably include three or more reactive elements. Chapter 11 presents an elegant procedure to model the measured data as a positive real function, which in turn yields the desired circuit.

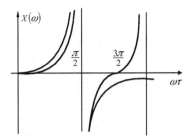

Figure 3.12 Frequency behavior of $X(\omega) = \tan(\omega\tau)$ function. Dashed lines denote the immittance variation of a resonance circuit.

3.23 Ideal TEM Lines with No Reflection: Perfectly Matched and Miss-Matched Lines

Reflection free TEM lines have special importance in designing power transfer networks.

When the line is terminated in complex load impedance $Z_L(j\omega)$, the average power delivered to this load is expressed is

$$P_L = real\ part\ \{V_L(j\omega)\,I_L(-j\omega)\}$$

As shown before, in terms of the voltage incident and reflected wave, the voltage and current are given by

$$V_L(j\omega) = A(0,\,j\omega) + B(0,\,j\omega) = A(0,\,j\omega)\,[1+S_L(j\omega)]$$
$$I_L(j\omega) = \frac{1}{Z_0}[A(0,\,j\omega) - B(0,\,j\omega)] = \frac{A(0,\,j\omega)}{Z_0}[1-S_L(j\omega)]$$

With

$$S_L(j\omega) = \frac{B(0,\,j\omega)}{A(0,\,j\omega)} = \frac{Z_L - Z_0}{Z_L + Z_0}; Z_L = \frac{V_L}{I_L} = Z_0\frac{1+S_L}{1-S_L}$$

For lossless TEM lines, the characteristic impedance is real. In this case, $Z_0(j\omega) = Z_0(-j\omega)$ *and it is a positive real number.* Therefore,

$$P_L = Real\ \{V_L(j\omega)\,I_L(-j\omega)\} = \frac{|A(0,j\omega)|^2}{Z_0}\left\{\left[1 - \left|\frac{B(0,j\omega)}{A(0,j\omega)}\right|^2\right]\right\}$$

or

$$P_L = \left[\frac{|A(0,j\omega)|^2}{Z_0} - \frac{|B(0,j\omega)|^2}{Z_0}\right]$$

or

$$P_L = \frac{|A(0,j\omega)|^2}{Z_0}\left[1-|S_L|^2\right] \tag{3.32a}$$

It should be noted that in (3.32), $0 \le |S_L|^2 = \frac{(R_L-Z_0)^2+(X_L)^2}{(R_L+R_0)^2+(X_L)^2} \le 1$. Therefore, maximum value of the average load power is obtained when $S_L = 0$, which requires $Z_L = Z_0$. Then, maximum of the average power delivered to load is obtained as

$$(P_L)_{max} = \frac{|A(0,j\omega)|^2}{Z_0} \tag{3.32b}$$

Furthermore, at the generator-end, if $Z_G = Z_0$, then (3.25) yields that

$$Z_{in} = Z_0 \tag{3.32c}$$

$$Z_{out} = Z_0 \tag{3.32d}$$

Definition 1: An ideal TEM line, which satisfies Equation (3.32), is called perfectly matched at the input and the output, respectively.

For a loss-free line, at the generator-end, amplitude of the incident wave $|A(l,j\omega)|$ must be equal to $|A(0,j\omega)|$. Therefore, at the generator-end, maximum deliverable input power is expressed as

$$(P_{in})_{max} = \frac{|A(l,j\omega)|^2}{Z_0} = \frac{|A(0,j\omega)|^2}{Z_0} \tag{3.32e}$$

Definition 2: We should note that $(P_{in})_{max}$ is also known as the available power of the generator and it is denoted by P_A.

For a "loss-free and reflection-free" TEM line, the above electrical conditions are pictured by means in the Thevenin equivalent circuits at the input and the output ends of the TEM line as shown in Figure 3.13.

Hence, by (3.32), it is seen that

$$P_L = P_A\left[1-|S_L|^2\right] \tag{3.33a}$$

Definition 3: A TEM transmission line can be regarded as a two-port, which has a physical length "l". In this case, the transducer power gain (TPG) of this two-port can be defined as the ratio of the power delivered to the load (P_L) to the available power of the generator (P_A). Thus, by definition,

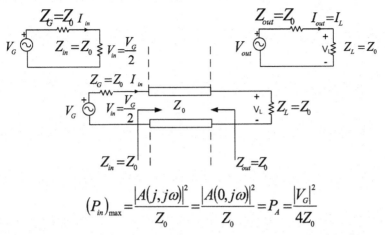

$$\left(P_{in}\right)_{max}=\frac{|A(j,j\omega)|^2}{Z_0}=\frac{|A(0,j\omega)|^2}{Z_0}=P_A=\frac{|V_G|^2}{4Z_0}$$

Figure 3.13 Thevenin equivalent circuits of input and out ends of the line.

$$TPG \triangleq \frac{P_L}{P_A}=1-|S_L|^2 \qquad (3.33b)$$

Consequently, we can make the following statements and definitions.

Statement 1: When a TEM line is terminated in its characteristic impedance, (i.e. $Z_L = Z_0$), the refection coefficient S_L vanishes (i.e., $S_L = 0$), which makes the line free of reflection. Thence, the reflected voltage wave at the load-end $B(0, j\omega)$ becomes zero, which requires no reflection on the line. Therefore, $B(z, j\omega)$ must be zero for all z.

Statement 2: If the line is reflection free and lossless; in other words, if the ideal TEM line is terminated in its characteristic impedance, then the average power delivered to load is maximum; and by (3.32), it is given as

$$(P_L)_{max} = \frac{|A(0, j\omega)|^2}{Z_0} = P_A; \; Z_0 = Z_L$$

which yields

$$|A(0, j\omega)|^2 = |V_L(j\omega)|^2_{max}$$

Definition 4: For a perfectly matched line, the input incident wave power $|a_{in}|^2$ is defined as

$$P_A \triangleq |a_{in}|^2 = P_{incident} = \frac{|A(-l, j\omega)|^2}{Z_0} \qquad (3.33c)$$

If the TEM line is mismatched at the input, i.e., if $Z_{in} \neq Z_0$, then the total power delivered to input (P_{in}) is given in a similar manner to Equation (3.32)

$$P_{in} = real \left\{ V_{in} (j\omega) I_{in} (-j\omega) \right\}$$

or

$$P_{in} = \frac{|V_{in}|^2}{Z_0}$$

or

$$P_{in} = \frac{1}{Z_0} [A (-l, j\omega) + B (-l, j\omega)] [A (-l, -j\omega) + B (-l, -j\omega)]$$

$$P_{in} = \frac{|A (-l, j\omega)|^2}{Z_0} - \frac{|B (-l, j\omega)|^2}{Z_0} \tag{3.33d}$$

Definition 5: (3.33d) suggests the reflected wave power definition as,

$$P_{reflected} \triangleq |b_{in}|^2 \triangleq \frac{|B (-l, j\omega)|^2}{Z_0} \tag{3.33e}$$

Statement 3: Based on Definitions 4 and 5, we say that power delivered to the input port of the transmission line is equal to the difference between incident and reflected powers. That is,

$$P_{in} = |a_{in}|^2 - |b_{in}|^2 \tag{3.33f}$$

Based on the definitions of incident and reflected powers, Equation (3.48) may be re-written as

$$P_{in} = \frac{1}{Z_0} [A (-l, j\omega) + B (-l, j\omega)] [A (l, -j\omega) + B (-l, -j\omega)]$$

Or emphasizing the positive realness of the characteristic impedance Z_0 by setting $R_0 = Z_0$,

$$P_{in} = real \left\{ \left[\frac{1}{\sqrt{R_0}} A (-l, j\omega) + \frac{1}{\sqrt{R_0}} B (-l, j\omega) \right] \right.$$
$$\left. \left[\frac{1}{\sqrt{R_0}} A (-l, -j\omega) - \frac{1}{\sqrt{R_0}} B (-l, -j\omega) \right] \right\}$$

or

$$P_{in} = \left[\frac{V_{in}}{\sqrt{R_0}}\right]\left[\frac{V_{in}}{\sqrt{R_0}}\right]^* \tag{3.33g}$$

where * denotes the complex conjugate of a complex quantity.

Then, Equation (3.33f) suggests the definition of the following voltage-based normalized quantities.

Definition 6: Normalized input voltage v_{in}
Referring to Equation (3.33g), the quantity

$$v_{in} = \frac{V_{in}}{\sqrt{R_0}} = \frac{1}{\sqrt{R_0}}A\left(-l, j\omega\right) + \frac{1}{\sqrt{R_0}}B\left(-l, j\omega\right) \tag{3.33h}$$

is called the normalized input voltage with respect to normalization resistance or number R_0.

The above normalized voltage definition leads us to define voltage-based normalized incident a_{in} and reflected b_{in} waves as follows.

Definition 7: Voltage-based incident wave a_{in}
The quantity

$$a_{in} = \frac{A(l, j\omega)}{\sqrt{R_0}} \tag{3.34a}$$

is called the voltage-based normalized incident wave or in short the **"voltage-based incident wave"**.

Definition 8: Voltage-based reflected wave b_{in}
The quantity

$$b_{in} = \frac{B(-l, j\omega)}{\sqrt{R_0}} \tag{3.34b}$$

is called the voltage-based normalized reflected wave or in short the **"voltage-based reflected wave"**.

Definitions of normalized incident and reflected waves lead us to the following statement.

Statement 4: Normalized input voltage v_{in} is expressed as the sum of incident and reflected waves such that

$$v_{in} = a_{in} + b_{in} \tag{3.35a}$$

Input current I_{in} of the transmission line can as well be expressed in terms of the actual input and reflected waves as

$$I_{in} = \frac{1}{R_0}[A(l, j\omega) - B(-l, j\omega)]$$

Let us multiply both side of the above equation by $\sqrt{R_0}$, then we end up with

$$\sqrt{R_0}I_{in} = \frac{\sqrt{R_0}}{R_0}[A(l, j\omega) - B(-l, j\omega)] = \frac{A(l, j\omega)}{\sqrt{R_0}} - \frac{B(-l, j\omega)}{\sqrt{R_0}}$$

or

$$\sqrt{R_0}I_{in} = a_{in} - b_{in} \qquad (3.35b)$$

Definition 9: Normalized input current i_{in}
Equation (3.35) leads to the definition of the normalized input current.
   The quantity

$$i_{in} = \sqrt{R_0}I_{in} \qquad (3.35c)$$

is called the normalized input current with respect to normalization number R_0.

Statement 5: The above definition of input current reveals that normalized input current of a TEM line can be expressed as the difference between incident and reflected waves such that

$$i_{in} = a_{in} - b_{in} \qquad (3.35d)$$

Statement 6: It should be noted that, all the above definitions can be made for the output port of a TEM line when it is terminated in Z_G at the input-end and driven by a voltage source V_L of internal impedance Z_L at the back-end, as shown in Figure 3.14.
   In this case, normalized output voltage v_{out} and out current i_{out} are described in terms of the voltage-based output incident a_{out} and reflected b_{out} waves as follows.

$$v_{out} = a_{out} + b_{out} \qquad (3.35e)$$

and

$$i_{out} = a_{out} - b_{out} \qquad (3.35f)$$

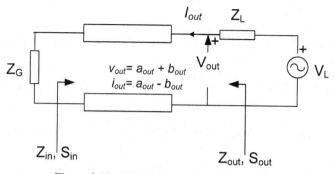

Figure 3.14 TEM line driven on the back-end.

Or equivalently output incident and reflected waves are given in terms of the measured output voltage and current as

$$a_{out} = \frac{1}{2}\left[\frac{V_{out}}{\sqrt{R_0}}+\sqrt{R_0}I_{out}\right] = \frac{1}{2}\left[v_{out}+i_{out}\right]$$

$$b_{out} = \frac{1}{2}\left[\frac{V_{out}}{\sqrt{R_0}}-\sqrt{R_0}I_{out}\right] = \frac{1}{2}\left[v_{out}-i_{out}\right] \qquad (3.35g)$$

Statement 7: Based on the above definitions and derivations, in terms of voltage-based normalized incident and reflected waves, actual powers delivered to the input and the output ports of a TEM line are expressed as

$$P_{in} = |a_{in}|^2-|b_{in}|^2 = |a_{in}|^2\left[1-|S_{in}|^2\right] \qquad (3.36a)$$

$$P_{out} = |a_{out}|^2-|b_{out}|^2 = |a_{out}|^2\left[1-|S_{out}|^2\right] \qquad (3.36b)$$

where

$$S_{in} = \frac{Z_{in} - R_0}{Z_{in} + R_0}$$

and

$$S_{out} = \frac{Z_{out} - R_0}{Z_{out} + R_0}$$

such that by (3.25)

$$Z_{in}\left(\lambda\right) = R_0\frac{Z_L + R_0\lambda}{R_0 + Z_L\lambda}$$

$$Z_{out}\left(\lambda\right) = R_0\frac{Z_G + R_0\lambda}{R_0 + Z_G\lambda}$$

3.24 Conclusion

In this chapter, on the extension of electromagnetic wave theory, basic properties of transverse electric and magnetic field lines or in short TEM lines are summarized. The next chapter is devoted to the circuits built using the equal-length ideal TEM lines, the so-called "Commensurate Lines".

Appendix 3: MatLab Programs for Chapter 3

Program List (3.1): Computation of the input immittance of a transmission line with arbitrary length

```
% TEM Line Input impedance computation
%
%                    ZL+jZ0*tan(beta*L)
%          Zin=Z0 ------------------
%                    Z0+jZL*tan(beta*L)
%
% Note that phase=beta*L; varies between 0 and pi
% or beta*L=(w/vp)*L where vp is the phase velocity\newline
%
% Input
%       ZL: Load termination\newline
%       Z0: Characteristic Impedance
%       phase=beta*L=w*tou; we sweep over w*tou axis\newline
% The phase is being scanned
%
% Output
%       Zin=Rin(w)+jXin(w)
% Program TEM Line: Computation of input impedance
ZL=0.1 % INPUT: Normalized with respect to 50 ohm
Z0=1;  % INPUT: Normalized with respect to 50 ohms
N=501; % Sampling number
delta=pi/N; % Step size for sweeping over w*tou axis
phase=0.0;% phase=w*tou axis

% Quarter wavelength equivalent resonance circuit components
L=2/pi; % Normalized inuctance with respect to tou
C=L;    % Normalized Capacitance
R=1/ZL; % Resistance at resonance frequency: w*tou=pi/2
j=sqrt(-1);
for i=1:N
    Nin=ZL+j*tan(phase); % Numerator of Zin(w*tou)
    Din=Z0+j*ZL*tan(phase);% Denominator of Zin(w*tou)
    Zin=Nin/Din; % TEM Line input imedance
    Rin(i)=real(Zin);
    Xin(i)=imag(Zin);
    fi(i)=phase;% array for plot purpose
 % Equivalent parallel resonance circuit impedance computations
    Yres=(1/L/phase/j)+j*phase*C+ZL;% Admittance
```

```
    Zres=1/Yres;% Impedance of the resonance circuit
    Rres(i)=real(Zres);
    Xres(i)=imag(Zres);
    Yin=1/Zin;
    Gin(i)=real(Yin);
    Bin(i)=imag(Yin);
    phase=phase+delta;
end
figure
plot(fi,Rin,fi,Rres); % plots of real parts
figure
plot(fi,Xin,fi,Xres); % plots of imaginary parts
```

Reference

[1] B. Yarman, Design of Ultra Wide-Band Matching Networks via Real Frequency Techniques, London: Wiley, 2010.

4

Loaded Line Digital Phase Shifters

Summary

In this chapter, various practical loaded line digital phase shifter topologies are introduced.

Explicit design equations and related design algorithm for each phase shifter topology is presented.

Electrical performance analysis of the loaded line digital phase shifters are studied by means of examples and S/W tools specifically developed for this book.

It is shown that larger the phase shift, smaller the bandwidth.

Loaded line phase shifters are good for small phase shifts. Moreover, it is not possible to construct a 180° phase shifting cell employing the loaded line phase shifters.

In the phase shifter designs, we employed digital loads either in series or in parallel configurations. Solid-state switches used in the designs, could be PIN, varactor diodes or any kind of CMOS etc.

Ideally, when the switches are forward biased, they are ideally short circuits. When they are reverse biased, they act like capacitors. Immittance switching from one state to other results in the desired phase change at the design frequency.

They are two kinds of loaded line digital phase shifters (LL-DPS), namely perfectly matched and intrinsically miss-matched LL-DPS. Perfectly matched LL-DPS has no loss at the design frequency. On the other hand, in the intrinsically miss-matched LL-DPS, there is an un-equalized loss at both digital states. Therefore, they provide smaller phase range and narrower frequency band over the perfectly matched digital phase shifters. Perfectly matched LL-DPS can cover much wider digital phase range up to 180° phase range. However, frequency band becomes narrower as the digital phase shift increases.

4.1 Loaded Line Phase Shifters with Single Reactive Elements

An ideal transmission line, of length $\theta_{T0} = \omega_0\tau$ can be used as a digital phase shifter (DPS) when it is symmetrically loaded with series or parallel digital lossless immittances at both ends as shown in Figure 4.1 [1, 2]. In this representation, the line length θ_{T0} is specified in degree or in radians for a given angular center frequency ω_0 and the delay length τ is specified in seconds.

Let "A" and "B" represent different switching states. State-A and State-B may be called as BIT-IN and BIT-OUT states respectively.

Referring to Figure 4.1a, in State-A, the transmission line is loaded with series reactive loads X_A (or shunt susceptive loads Y_A). Similarly, in State-B, it is loaded with X_B (or Y_B).

In Figure 4.1, generically, the series reactance X may be considered as a lossless two-port $[X]$.

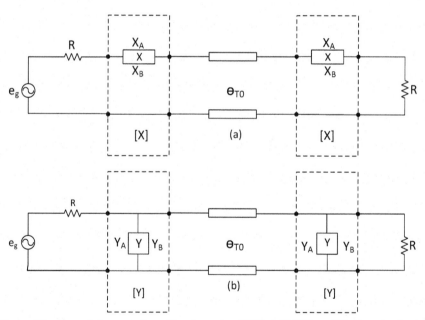

Figure 4.1   Cascaded connection of lossless trio of $[X] - [TRL] - [X]$ (a) series loaded line, (b) parallel loaded line.

On the real frequency axis $j\omega$, let it be described by means of its scattering parameters S_L as

$$S_L(j\omega) = \begin{bmatrix} S_{11L}(j\omega) & S_{12L}(j\omega) \\ S_{21L}(j\omega) & S_{22L}(j\omega) \end{bmatrix} = \begin{bmatrix} \rho_{11L}e^{j\varphi_{11L}} & \rho_{12L}e^{j\varphi_{12L}} \\ \rho_{21L}e^{j\varphi_{21L}} & \rho_{22L}e^{j\varphi_{22L}} \end{bmatrix} \quad (4.1)$$

Similarly, let S_T designates the scattering parameters of the unloaded ideal transmission line $[TRL]$. Then,

$$S_T(j\omega) = \begin{bmatrix} S_{11T}(j\omega) & S_{12T}(j\omega) \\ S_{21T}(j\omega) & S_{22T}(j\omega) \end{bmatrix} = \begin{bmatrix} \rho_{11T}e^{j\varphi_{11T}} & \rho_{12T}e^{j\varphi_{12T}} \\ \rho_{21T}e^{j\varphi_{21T}} & \rho_{22T}e^{j\varphi_{22T}} \end{bmatrix} \quad (4.2a)$$

Since both $[X]$ and $[TRL]$ are reciprocal-lossless two-ports, then their transfer scattering parameters S_{21} and S_{12} must be equal. In this case, S_L may be detailed as

$$S_{11L}(j\omega) = S_{22L}(j\omega) = \frac{(R+jX_a)-R}{(R+jX_a)+R} = \frac{jX_a}{2R+jX_a}$$

$$= \frac{X_a}{\sqrt{4R^2+X_a^2}} e^{+jtan^{-1}\left(\frac{2R}{X_a}\right)} \quad (4.2b)$$

$$\rho_{11L}(\omega) = \frac{X_a}{\sqrt{4R^2+X_a^2}}; \quad \varphi_{11L}(\omega) = tan^{-1}\left(\frac{2R}{X_a}\right) \quad (4.2c)$$

$$S_{12L}(j\omega) = S_{21L}(j\omega) = \frac{2R}{2R+jX_a} = \rho_{21}e^{j\varphi_{21L}}$$

$$= \frac{2R}{\sqrt{4R^2+X_a^2}} e^{-jtan^{-1}\left(\frac{X_a}{2R}\right)} \quad (4.2d)$$

$$\rho_{21L}(\omega) = \frac{2R}{\sqrt{4R^2+X^2}}; \quad \varphi_{21L}(\omega) = -tan^{-1}\left(\frac{X_a}{2R}\right)$$

$$= 180° - tan^{-1}\left(\frac{X_a}{2R}\right) \quad (4.2e)$$

In the above formulas, $X_a(\omega)$ represents the actual value of the series loading reactance. However, (4.3a) can be simplified using the normalized impedances.

Let $X = X_a/R$ be the normalized loading reactance. Then, (4.3a) is expressed in terms of the normalized reactance X as

$$S_{11L}(j\omega) = S_{22L}(j\omega) = \frac{jX}{2 + jX} = \frac{X}{\sqrt{4 + X^2}}e^{+jtan^{-1}\left(\frac{2}{X}\right)} \tag{4.3a}$$

$$\rho_{11L}(\omega) = \frac{X}{\sqrt{4 + X^2}}; \quad \varphi_{11L}(\omega) = tan^{-1}\left(\frac{2}{X}\right) \tag{4.3b}$$

$$S_{12L}(j\omega) = S_{21L}(j\omega) = \frac{2}{2 + jX} = \rho_{21}e^{j\varphi_{21L}} = \frac{2}{\sqrt{4 + X^2}}e^{-jtan^{-1}\left(\frac{X}{2}\right)} \tag{4.3c}$$

$$\rho_{21L}(\omega) = \frac{2}{\sqrt{4 + X^2}}; \quad \varphi_{21L}(\omega)$$
$$= -tan^{-1}\left(\frac{X}{2}\right) = 180° - tan^{-1}\left(\frac{X}{2}\right) \tag{4.3d}$$

It should be noted that if the reactance X is capacitive, then its actual value must be a negative real number, which in turn results in negative phase $\varphi_{11L}(\omega) = tan^{-1}\left(\frac{2}{X}\right)$.

The unloaded ideal transmission line θ_{T0} possesses a physical length "l", which is a positive quantity and its scattering parameters $\{S_{Tij}; i, j = 1, 2\}$ are given by

$$S_{11T} = S_{22T} = 0 \tag{4.4a}$$

$$S_{12T}(j\omega) = S_{21T}(j\omega) = \rho_{21T}e^{j\varphi_{21T}(\omega)} = 1e^{-j\theta_T(\omega)} \tag{4.4b}$$

$$\rho_{21T} = 1; \quad \varphi_{21T} = -\theta_T(\omega) = \omega\tau \tag{4.4c}$$

$$\theta_{T0} = \omega_0\tau = 2\pi f_0\tau; \quad \tau = \frac{l}{v} = \frac{\theta_{T0}}{\omega_0} \tag{4.4d}$$

where f_0 is the actual center frequency and "v" is the actual speed of propagation. In (4.4), to end up with realizable positive delay length τ, θ_{T0} must be a positive quantity in radians.

Based on the above nomenclature, scattering parameters $S = \begin{bmatrix} S_{11} & S_{12} \\ S_{21} & S_{22} \end{bmatrix}$ of the cascaded connection of $[X] - [TRL] - [X]$ is computed at $\omega = \omega_0$ as

$$S_{11}(j\omega_0) = \frac{(z_{in} + jX) - 1}{(z_{in} + jX) + 1} \tag{4.5a}$$

$$z_{in}(j\omega_0) = z_0\frac{z_L + z_0\left[jtan\left(\theta_{T0}\right)\right]}{z_0 + z_L\left[jtan\left(\theta_{T0}\right)\right]} = z_0\frac{(1 + jX) + z_0\left[jtan\left(\theta_{T0}\right)\right]}{z_0 + (1 + jX)\left[jtan\left(\theta_{T0}\right)\right]} \tag{4.5b}$$

$$S_{11}(j\omega) = j\frac{2X - \mu X^2}{2(1 - \mu X) + j(2\mu + 2X - \mu X^2)} \tag{4.5c}$$

where

- X is the normalized loading reactance, which is given by

$$X = \frac{X_a}{R} \tag{4.6a}$$

- $z_L = 1 + jX$ is the normalized termination of the transmission line θ_{T0} at the back-end;
- z_{in} is the normalized driving point input impedance of the transmission line θ_{T0} when the back-end is terminated in the load impedance z_L
- z_0 is the normalized characteristic impedance of the transmission line defined with respect to R; it is usually selected as unity (i.e. actual characteristic impedance of the line is selected as $Z_0 = R$ and $z_0 = \frac{Z_0}{R} = 1$).

and

$$\mu = tan(\theta_{T0}) \tag{4.6b}$$

is called the major design parameter. Depending on the physical length of the transmission line, μ could be either positive or negative.

It can be shown that the transfer scattering parameter, S_{21} can be written as in [3, 4].

$$S_{21} = \frac{[S_{21L}^2][S_{21T}]}{(1 - S_{22L}S_{11T})\left(1 - \hat{S}_{22T}S_{11L}\right)} \tag{4.7a}$$

where

$$S_{11L} = S_{22L}$$
$$S_{11T} = S_{22T} = 0$$
$$\hat{S}_{22T} = S_{22T} + \frac{S_{21T}^2}{1 - S_{22T}S_{11L}}S_{11L} = [S_{21T}^2][S_{11L}] \tag{4.7b}$$

Hence, S_{21} is simplified as

$$S_{21} = \frac{[S_{21L}^2][S_{21T}]}{1 - S_{21T}^2 S_{11L}^2} = \rho_{21}e^{j\varphi_{21}} \tag{4.7c}$$

In (4.7c), S_{11L} and S_{21L} may be written in terms of the normalized load X as

$$S_{11L} = j\frac{X}{2 + jX} \tag{4.7d}$$

and

$$S_{21L} = \frac{2}{2 + jX} \tag{4.7e}$$

For an ideal phase shifter, transmission loss must be zero. Therefore, we demand that

$$|S_{11}|^2 = |S_{22}|^2 = 1 - |S_{21}|^2 = 0 \tag{4.8}$$

and employing (4.5), one ends up with

$$2X - \mu X^2 = 0 \tag{4.9}$$

or

$$\mu = \tan(\theta_{To}) = \frac{2}{X(\omega_0)} \tag{4.10a}$$

$$\theta_{T0} = \tan^{-1}\left[\frac{2}{X(\omega_0)}\right] + k\pi \ (in \ radians) \tag{4.10b}$$

or

$$\theta_{T0} = \tan^{-1}\left[\frac{2}{X(\omega_0)}\right] + k \times 180° \ (in \ degree) \tag{4.10c}$$

Remarks: In (4.9), $X = 0$ is the trivial solution, which is useless.

In (4.10), if X is capacitive, in other words, if X is negative, then μ must be negative and the phase term $\tan^{-1}\left[\frac{2}{X(\omega_0)}\right]$ becomes negative. If this is the case, we can always add $180°$ or equivalently π radians to the term $\tan^{-1}\left[\frac{2}{X(\omega_0)}\right]$ to make θ_{T0} positive, which results in longer line preserving the value of μ. Therefore, in (4.10b), we select $k = 0$ if $\mu > 0$ or $k = 1$ for $\mu < 0$.

Desired phase shift $\Delta\theta$ is defined as the phase difference between $\varphi_{21A}(\omega)$ and $\varphi_{21B}(\omega)$. In other words,

$$\Delta\theta(\omega) = \varphi_{21B}(\omega) - \varphi_{21A}(\omega) \tag{4.11}$$

where φ_{21A} and φ_{21B} are the corresponding phases of S_{21} at State-A and State-B, respectively.

For a given reactance $X(\omega_0)$, the phase φ_{21} of S_{21} is given by

$$\varphi_{21}(\omega_0) = 2\varphi_{21L} - \theta_{To} - phase \ of \ \{1 - S_{21T}^2 S_{11L}^2\} \tag{4.12a}$$

Let us investigate the phase of $\left\{S_{21T}^2 S_{11L}^2\right\}$ at $\omega = \omega_0$

$$phase\ of\ \left\{S_{21T}^2 S_{11L}^2\right\} = 2\left[\varphi_{21T} + \varphi_{11L}\left(\omega_0\right)\right] = 2\left[-\theta_{T0} + \varphi_{11L}\left(\omega_0\right)\right]$$

$$(4.12b)$$

Employing perfect match condition of (4.9) and (4.5), by means of the normalized reactance $X\left(\omega_0\right)$, we have,

$$\varphi_{11L}\left(\omega_0\right) = +tan^{-1}\left[\frac{2}{X\left(\omega_0\right)}\right]$$

$$phase\ of\ \left\{S_{21T}^2 S_{11L}^2\right\} = 2\left\{-tan^{-1}\left[\frac{2}{X\left(\omega_0\right)}\right] + tan^{-1}\left[\frac{2}{X\left(\omega_0\right)}\right]\right\} = 0$$

$$(4.12c)$$

Therefore, the term $1 - S_{21T}^2 S_{11L}^2$ is a real number, and using (4.3b), we have

$$1 - S_{21T}^2 S_{11L}^2 = 1 - \rho_{21T}^2 \rho_{11L}^2 = 1 - 1 \times \frac{X^2}{4 + X^2}$$

$$= \frac{4 + X^2 - X^2}{4 + X^2} = \frac{4}{4 + X^2} \qquad (4.12d)$$

Hence, at $\omega = \omega_0$, $|S_{21}|^2$ can be determined in terms of the normalized reactance values of (4.7), such that

$$|S_{21}|^2 = \left|\frac{[S_{21L}^2][S_{21T}]}{1 - S_{21T}^2 S_{11L}^2}\right|^2 = \left|\frac{\frac{4}{4+X^2}}{\frac{4}{4+X^2}}\right|^2 = 1 \qquad (4.12e)$$

as expected. Furthermore, at $\omega = \omega_0$, the phase of S_{21} is given by

$$\varphi_{21}\left(\omega_0\right) = 2\varphi_{21L} + \varphi_{21T} = 2tan^{-1}\left(\frac{X}{2}\right) - \theta_{T0} \qquad (4.12f)$$

Considering (4.11) together with (4.12), under perfect matched conditions, the phase difference between State-A and State-B is given by

$$\Delta\theta = \left[2tan^{-1}\left(\frac{X_B}{2}\right) - \theta_{T0}\right] - \left[2tan^{-1}\left(\frac{X_A}{2}\right) - \theta_{T0}\right]$$

$$= 2\left[tan^{-1}\left(\frac{X_B}{2}\right) - tan^{-1}\left(\frac{X_A}{2}\right)\right] \qquad (4.13)$$

Remarks:

- It is important to note that under perfect transmission conditions, if $X_A \neq 0$, (4.10) defines two different electrical lengths $\theta_{oA} = tan^{-1}\left(\frac{X_A}{2}\right)$ and $\theta_{oB} = tan^{-1}\left(\frac{X_B}{2}\right)$ for the same transmission line θ_{To}, which is a contradiction. For small phase changes however, the transmission loss at both states A and B can be neglected and θ_{oA} and θ_{oB} will be close to each other. In this case, (4.13) is safe to use and the length θ_{To} can be chosen between θ_{oA} and θ_{oB} (or θ_{oB} and θ_{oA}) to minimize and balance the insertion loss at both states. Our practical experience indicates that, when $X_A \neq 0$, for small phase shifts such as $|\Delta\theta| \leq 45°$ it may be convenient to choose θ_{T0} as

$$\theta_{T0} \equiv \frac{\theta_{0A} + \theta_{0B}}{2} \tag{4.14}$$

or

$$\theta_{T0} \equiv \sqrt{\theta_{0A} \cdot \theta_{0B}} \tag{4.15}$$

- However, if $X_A = 0$, then $\theta_{T0} = tan^{-1}\left(\frac{2}{X_B}\right) + k\pi$ of (4.10b) directly specifies the required length of the loaded transmission line $[TRL]$.

Based on the above analysis, it is concluded that, there is an inevitable mismatch loss built in the loaded line phase shifters since the center line θ_{To} can never satisfy the perfect transmission condition of (4.10) simultaneously if both X_A and X_B are different from zero (or Y_A, $Y_B \neq \infty$). Therefore, it is appropriate to name these phase shifters as "intrinsically mismatched" loaded line digital phase shifters.

It should be noted that generic topology of the series load $[X]$ (or equivalently shunt load $[Y]$) should be somewhat digital to perform State-A and State-B operations. In other words, the load must include a switch in it. When the switch is closed (i.e. BIT-IN case), the load must act as reactance X_A (or Y_A). Similarly, when the switch is opened (i.e. BIT-OUT case), it is X_B (or Y_B).

For example, a simple series load may be realized using a single PIN-Diode $[D]$ with reverse biased capacitance C_D. Ideally, when $[D]$ is forward biased (i.e. D is on), it is shorted (State-A). When $[D]$ is reversed biased (i.e. $[D]$ is off), it exhibits a reverse biased capacitor C_D. In this case, ideally, in switching State-A, the series reactance is $X_A = 0$ and in switching State-B, series reactance $X_B = -\frac{1}{\omega C_D}$ as shown in Figure 4.2.

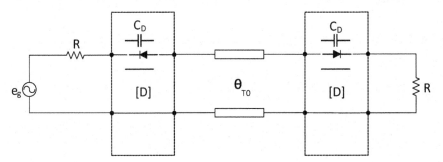

Figure 4.2 A simple loaded line with a series diode as a switching element.

Let us propose the following algorithm to design a simple loaded line – digital phase shifter (LL-DPS) utilizing PIN diodes as switching elements.

Algorithm 1. Design of a Simple Loaded Line Digital Phase Shifter (LL-DPS) using PIN diodes as a switching element

This algorithm designs a simple digital phase shifter using an ideal transmission line of physical length *"l"* loaded with series switching PIN diodes with reverse biased capacitance C_D at both ends as shown in Figure 4.2. The ideal transmission line can be realized on a selected substrate perhaps, employing microstripline technology. In practice, actual port terminations R and characteristic impedance of the line may be selected as $R = Z_0 = 50$ ohms. In this case, series diodes are accessed over 50 *ohm* lines at both ends and they are soldered on the top surface conductors. If microstrip technology is utilized, then propagation velocity is determined as

$$v_p = \frac{v_0}{\sqrt{\mu_r \varepsilon_r}} \cong \frac{v_0}{\sqrt{\varepsilon_r}} \tag{4.16}$$

where μ_r is the relative susceptibility which is usually almost equal to unity, ε_r is the relative permittivity of the microstrip substrate and v_0 is the speed of light in free space which is $v_0 = 3 \times 10^8$ m/sec. Eventually, for specified inputs, algorithm determines the physical length of the line as well as the reverse biased capacitance of the switching PIN diodes for specified phase shift $\Delta\theta$.

Inputs:

- Desired phase shift $\Delta\theta$.

In this algorithm, we are dealing with simple capacitive loads using PIN-diodes. Therefore, $\Delta\theta$ must be selected as a negative quantity.

- Actual Design Frequency f_{0a}
- Normalized angular Frequency ω_0 (it is usually selected as unity)
- Port normalization number R (It is usually selected as $R = 50$ ohms)
- Z_0: Characteristic impedance of the transmission line (usually, Z_0 is selected as the port normalization number $R = 50\ ohms$. Then, normalized characteristic impedance of (4.5) is $z_0 = \frac{Z_0}{R} = 1$.
- v_p: Propagation velocity of the transmission line medium or relative permittivity ε_r of the circuit substrate on which LL-DPS is placed.

Computational Steps:

Step 1. Compute the normalized value of the loading reactance X_B in State-B by setting $X_A = 0$ as in (4.13).

$$X_B\left(\omega_0\right) = 2tan\left(\frac{\Delta\theta}{2}\right)$$

Note that X_B is the State-B reactance when the diode is reverse biased. Therefore, it must be given as

$$X_B = -\frac{1}{\omega_0 C_D} < 0$$

Therefore, in the input, $\Delta\theta$ must be specified as a negative phase shift.

Step 2. Employing (4.10), compute the physical length of the transmission line.

$$\theta_{T0} = tan^{-1}\left[\frac{2}{X_B\left(\omega_0\right)}\right] + k\pi > 0$$

If X_B is positive, k is set to 0. For negative values of X_B, k is set to unity. Therefore, for this algorithm, $k = 1$ is selected. Then, we compute delay length τ and physical length "l" as

$$\tau = \frac{\theta_{T0}(in\ radian)}{2\pi f_{0a}}$$

$$l = \tau v_p = \tau\frac{v_0}{\sqrt{\varepsilon_r}}$$

Note: If the transmission line is realized as a microstrip line, then v_p is computed employing (4.16).

Step 3. Compute the normalized value of the required reverse biased capacitance C_D of the PIN diode.

$$C_D = \left| \frac{1}{\omega_0 X_B} \right|$$

Step 4. Compute the actual value of the reverse biased capacitance C_{Da}

$$C_{Da} = \frac{C_D}{2\pi f_0 R}$$

Step 5. Analyze the electrical performance of the designed phase shifter LL-DPS

Electrical performance analysis can be accomplished by plotting the following functions:

- Actual phase shift versus actual frequency: $\Delta\theta\left(f_a\right) = \varphi_{21B}\left(f_a\right) - \varphi_{21A}\left(f_a\right)$
- Gain of State-A $TPG_A(dB) = 10log_{10}\rho_{21A}^2\left(f_a\right)$ versus actual frequency f_a
- Gain of State-B $TPG_B(dB) = 10log_{10}\rho_{21B}^2\left(f_a\right)$ versus actual frequency f_a

This step concludes the design algorithm. If desired, electrical performance of the completed design may be improved using a commercially available optimization tool such as AWR and ADS.

Let us run an example to exhibit the utilization of the above algorithm.

Example 4.1A: In this example, we will design a simple loaded line phase shifter using PIN diodes as switching elements to achieve $\Delta\theta_0 = -45°$ at the center frequency $f_0 = 3GHz$. In the course of design, transmission line is realized using microstrip technology. Inputs to the design algorithm are given as follows:

Inputs:

Phase shift: $\Delta\theta = -45°$,
Actual center frequency: $f_0 = 3GHz$,
Normalized angular center frequency: $\omega_0 = 1$,
Port normalization numbers: $R = 50 \, \Omega$,
Characteristic impedance of [TRL]: $Z_0 = 50 \, \Omega$ or normalized characteristic impedance: $z_0 = 1$

Relative permittivity of the selected substrate for the microstrip line: $\varepsilon_r = 3.4$.

(a) Compute X_B

(b) Compute θ_{T0}, delay length τ, velocity of propagation v_p physical length of the microstrip line "l"

(c) Compute the actual reverse biased capacitance C_{Da} of the switching PIN diodes.

(d) Analyze the phase performance of the LL-DPS.

Solution

Algorithm 1 is programmed under a MatLab program called

$$\text{"Main\_DPS\_LoadedTRL\_Example\_3\_1.m"}$$

Results are given as follows.

(a)

$$X_B(\omega_0) = 2tan\left(\frac{\Delta\theta}{2}\right) = -0.8284$$

(b)

$$\theta_{T0} = tan^{-1}\left[\frac{2}{X_B(\omega_0)}\right] + 180° = -67.5 + 180 = 112.5 \; degree > 0$$

$$\theta_{T0-Rad} = \frac{\theta_{T0} \times 180}{180} = 1.9635 \; radians > 0$$

$$\tau = \frac{\theta_{T0}(in \; radian)}{2\pi f_0} = 1.0417 \times 10^{-10} \; sec$$

Normalize delay length:

$$\tau_n = \frac{\theta_{T0-Rad}}{\omega_0} = 1.9635$$

Velocity of propagation:

$$v_p = \frac{v_0}{\sqrt{\varepsilon_r}} = 1.6270 \times 10^8 \; m/sec$$

Physical length of the transmission line:

$$l = \tau v_p = 1.0417 \times 10^{-10} \times 1.6270 \times 10^8 = 0.0169 \; m = 1.69 \; cm$$

(c) Normalized value of the reverse biased diode capacitance:

$$C_D = \left| \frac{1}{\omega_0 X_B} \right| = 1.2071$$

Actual value of the reverse biased diode capacitance:

$$C_{Da} = \frac{C_D}{2\pi f_0 R} = 1.2808 \, pF$$

(d) Phase shifting performance of the capacitive loaded transmission line is depicted in Figure 4.3.

A close examination of Figure 4.3(a) indicates that from $0.876 \times 3 = 2.628 \, GHz$ to $1.124 \times 3 = 3.72 \, GHz$ (i.e. over 1.092 GHz Bandwidth), the phase shift variation Δ_θ is about $\mp 10\%$. In the neighborhood of 3 GHz, bandwidth may be considered as 1 GHz.

Gain performance of the LL-DPS is depicted in Figure 4.3b. Since the TRL is lossless, in State-A, we observe no loss, as expected. However, in

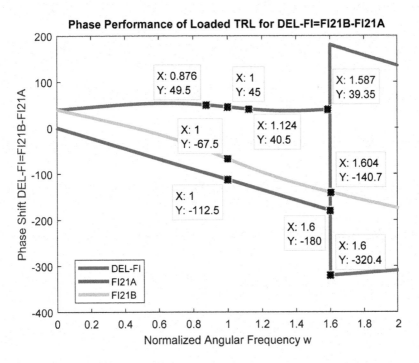

Figure 4.3a Phase shifting performance of the ideal loaded line digital phase shifter.

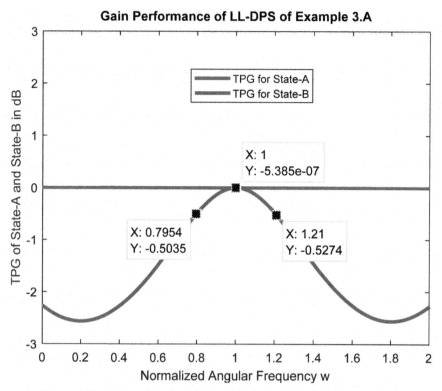

Figure 4.3b   Gain performance of the ideal loaded line digital phase shifter.

State-B, over the 1 GHz bandwidth, loss is better than 0.6 dB, which may be considered as acceptable loss. Certainly, loss characteristics will be worsened with the inclusion of diodes and microstrip losses.

Remark: We solved the same problem for positive phase shift $\Delta\theta_0 = +45°$. In this case, we ended up with the same electrical performance but shorter center line length (i.e. $l = 1.02\ cm\ versus\ 1.69\ cm$)

Now, let us investigate the phase shift capability of an LL-DPS for $\Delta\theta = -90°$

Example 4.1B: Repeat Example 3.1.A for $\Delta\theta = -90$.

Solution

Execution "*Main_DPS_LoadedTRL_Example_3_1B.m*" reveals the following results.

(e) $X_B(\omega_0) = 2\tan\left(\frac{\Delta\theta}{2}\right) = -2$

(f)

$$\theta_{T0} = \tan^{-1}\left[\frac{2}{X_B(\omega_0)}\right] + 180° = 135\ degree > 0$$

$$\theta_{T0-Rad} = \frac{\theta_{T0} \times 180}{180} = 2.3562\ radians > 0$$

$$\tau = \frac{\theta_{T0}(in\ radian)}{2\pi f_0} = 1.25 \times 10^{-10}\ sec$$

Normalize delay length:

$$\tau_n = \frac{\theta_{T0-Rad}}{\omega_0} = 2.3562$$

Velocity of propagation:

$$v_p = \frac{v_0}{\sqrt{\varepsilon_r}} = 1.6270 \times 10^8\ m/sec$$

Physical length of the transmission line:

$$l = \tau v_p = 1.0417 \times 10^{-10} \times 1.6270 \times 10^8 = 0.0203\ m = 2.03\ cm$$

(g) Normalized value of the reverse biased diode capacitance:

$$C_D = \left|\frac{1}{\omega_0 X_B}\right| = 0.5$$

Actual value of the reverse biased diode capacitance:

$$C_{Da} = \frac{C_D}{2\pi f_0\ R} = 0.53052\ pF$$

(h) Phase shifting performance of the capacitive loaded transmission live is depicted in Figure 4.4.

Examination of the above phase shift curves reveals that in State-A, the phase shift simply the delay occurs on the transmission line which is $-135°$. In State-B, we have the total phase shift of the load and the transmission line which is $-135° + 90° = -45°$. Eventually, the net phase shift between the states is given by $\Delta\theta = \theta_B - \theta_A = 90°$ as it should be 10% phase fluctuation is achieved over 2.89–3.1 GHz. Obviously, the bandwidth has narrowed as compare to the DPS with $\Delta\theta = 45°$ (1 GHz versus 200 MHz).

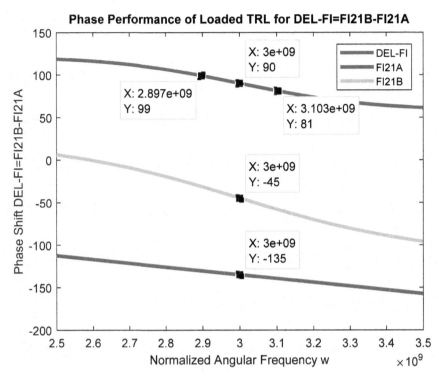

Figure 4.4a   Phase shifting performance of the ideal loaded line digital phase shifter.

Loss characteristic of the designed $90°$ phase shifter is depicted in Figure 4.4b. Ideal gain characteristics is above -6 dB which is reasonable.

Remark 1: We solve the same problem for the positive phase shift $\Delta\theta_0 = +90°$. In this case, we ended up with shorter line ($l = 6.8\ mm$ *versus* $2.08\ cm$) as expected.

Remark 2: Comparison of Example 3.1A with that of Example 3.1B reveals that $45-°$ phase shift results in wider bandwidth than the phase shifter designed for $90°$. This is the natural consequence. As we know, larger the phase-shift results in narrow the frequency bandwidth.

Remark 3: Unfortunately, phase shifters constructed with loaded transmission lines can never accomplish $180°$ phase shift. This conclusion can be derived from (4.13) by taking the tangent of

$$tan\left(\frac{\Delta\theta}{2}\right) = tan\left[tan^{-1}\left(\frac{X_B}{2}\right) - tan^{-1}\left(\frac{X_A}{2}\right)\right] = \frac{X_B - X_A}{4 + X_A X_B}$$

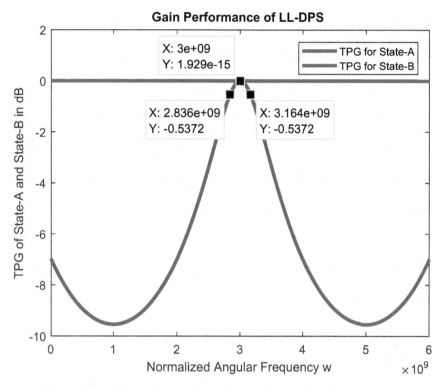

Figure 4.4b Gain performance of the ideal loaded line digital phase shifter.

The above equation forces capacitive loads to approach to zero or inductive loads to go infinity as $\Delta\theta$ reaches 180°. For the example under consideration, reverse biased diode capacitance must be zero (i.e. $X_B = -\frac{1}{\omega_0 C_D} \to \infty$ *when* $C_D = 0$) for $\Delta\theta = 180°$, which purges the loaded line phase shifter topology.

4.2 Inductively Series Loaded Line Digital Phase Shifter

In Figure 4.5, a simple "Inductively Series Load Line - Digital Phase Shifter" (in short ISLL-DPS) is shown. In this topology, when the diodes are on, ideally, they act as short circuits. This mode of operation is called State-B. In this mode, series normalized loads X_B is given by

$$X_B = \omega L \tag{4.17}$$

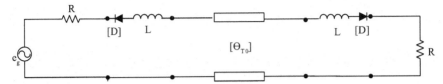

Figure 4.5a   Phase shifting performance of the ideal loaded line digital phase shifter.

where L designates the normalized value of the series inductor.

In State-A, the diodes are reversed biased and they act as series capacitors C_D. In this mode, inductors L resonate with the capacitors C_D at the normalized angular frequency ω_0, which perform short circuits. Thus, C_D is given by

$$C_D = \frac{1}{\omega_0^2 L} \tag{4.18}$$

In State-A, series load $X_A(\omega)$ is given by

$$X_A(\omega) = \omega L - \frac{1}{\omega C_D} \tag{4.19}$$

The value of the inductor L is given by (4.13) as

$$L = \frac{X_B(\omega_0)}{\omega_0} = \frac{2}{\omega_0} tan\left(\frac{\Delta\theta}{2}\right) \tag{4.20}$$

Eventually, centerline length θ_{T0} is given by (4.10b) as follows.

$$\theta_{T0} = tan^{-1}\left[\frac{2}{\omega_0 L}\right] + k\pi > 0 \tag{4.21}$$

Hence, one can construct the ISLL-DPS employing the following design algorithm.

Algorithm 2. Construction of Inductively Series Loaded Line – Digital Phase Shifter (ISLL-DPS)

Inputs:
f_{0a}: Actual Operating frequency at which ISLL-DPS is designed
ω_0: Normalized angular frequency at which ISLL-DPS is designed.
$\Delta\theta_0$: Desired phase shift, which is specified at the normalized angular frequency ω_0.

R: termination resistor, which is equal to the characteristic impedance Z_0 of the loaded line.

v_p: Propagation velocity or equivalently relative permittivity ε_r of the transmission medium.

v_0: Speed of light (Electromagnetic waves) in free space which is taken as $3 \times 10^8 \ m/sec$.

Computational Steps:

Step 1: Compute normalized value of the series load inductor L

$$L = \frac{2}{\omega_0} tan\left(\frac{\Delta\theta}{2}\right)$$

Step 2: Compute the normalized values of the reverse biased diode capacitor C_D.

$$C_D = \frac{1}{\omega_0^2 L}$$

Step 3: Compute the electrical length θ_{T0} of the loaded line.

$$\theta_{T0} = tan^{-1}\left[\frac{2}{\omega_0 L}\right] + k\pi > 0$$

Step 4: Compute the normalized delay length τ_n, actual delay length τ and physical length l of the transmission line.

$$\tau_n = \theta_{T0}/\omega_0$$
$$\tau = \frac{\theta_{T0}(in \ radian)}{2\pi f_{0a}}$$
$$l = \tau v_p = \tau \frac{v_0}{\sqrt{\varepsilon_r}}$$

Step 5: Compute the actual element values for the loading inductor L_a and the actual Capacitor C_{Da}.

Step 6: Analyze the ideal electrical performance of the ISLL-DPS
Now, let us run an example to design an ISLL-DPS.

Example 4.1C: Repeat Example 3.1A to design an ISLL-DPS for $\Delta\theta_0 45°$ (i.e. $R = Z_0 = 50\Omega$, $f_{0a} = 3\,GHz$, $\omega_0 = 1$, $\varepsilon_r = 3.4$)

Solution

Step 1: Compute normalized value of the series load inductor L

$$L = \left| \frac{2}{\omega_0} tan \left(\frac{\Delta\theta}{2} \right) \right| = 0.8284$$

Step 2: Compute the normalized values of the reverse biased diode capacitor C_D.

$$C_D = \frac{1}{\omega_0^2 L} = 1.2071$$

Step 3: Compute the electrical length θ_{T0} of the loaded line.

$$\theta_{T0} = tan^{-1} \left[\frac{2}{\omega_0 L} \right] + k\pi = 1.1781\,rad$$

It is note that, in this step, $k = 0$ taken since $tan\left(\frac{\Delta\theta}{2} \right) > 0$.

Step 4: Compute the normalized delay length τ_n, actual delay length τ and physical length l of the transmission line.

$$\tau_n = \frac{\theta_{T0}}{\omega_0} = 1.1781$$

$$\tau = \frac{\theta_{T0}(in\,radian)}{2\pi f_{0a}} = 6.2500e - 11\,sec$$

$$l = \tau v_p = \tau \frac{v_0}{\sqrt{\varepsilon_r}} = 0.0102\,m = 1.02\,cm$$

Step 6: Analyze the ideal electrical performance of the ISLL-DPS
  For this purpose, we developed a MatLab program called

"Main_ISLL_DPS_with_Inductors_Example_3_1C.m"

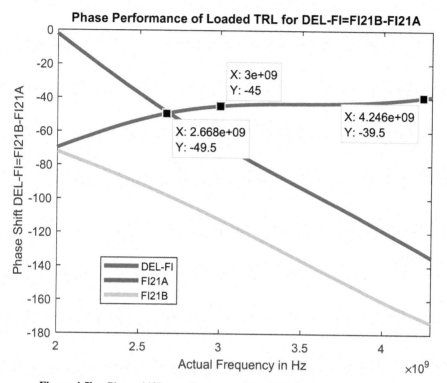

Figure 4.5b Phase shifting performance of the ISLL-DPS of Example 3.1C.

This program firstly designs the ISLL-DPS, then analyzes the electrical performance of the topology with the computed element values. Phase shifting performance of ISLL-DPS for 45° is depicted in Figure 4.5b.

A close examination of the above figure reveals that the phase bandwidth of ISLL-DPS is approximately 1.5 GHz, which is about 75% bandwidth. The loss performance is shown in Figure 4.5c.

The loss curve of Figure 4.5c indicates that more than one octave, TPG is better than −1.6 dB, which is very good.

In the following section, we introduce a Loaded Line-DPS topology with impedance transformer, which provides with degree of freedom to select the PIN diode reverse biased capacitors C_D in advance. This is a significant improvement in design to manufacture PIN diodes with optimum geometric dimensions.

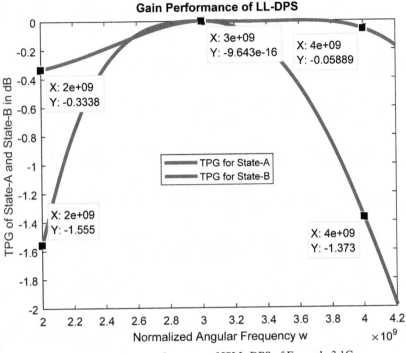

Figure 4.5c   Loss performance of ISLL-DPS of Example 3.1C.

4.3 Series Loaded Line Digital Phase Shifter

A typical "Series Loaded Line-Digital Phase Shifter (SLL-DPS)" config-uration is shown in Figure 4.6a. In this figure, the reactive impedances of identical diodes D_1 and D_2 are transformed on the centerline over the transformer lines $[\theta_T]$.

The desired phase shift $\Delta\theta$ is achieved when the diodes are switched on and off. Depending on the switching states of the diodes, series reactance X_A and X_B load the centerline θ_{To}.

For a given diode capacitance C_D and the phase shift $\Delta\theta_0$, the line lengths θ_{T0} and θ_T can be derived at the given normalized angular frequency ω_0, which is associated with the actual center frequency f_{oa}.

Let us introduce the new design parameters Z_L, β and μ_T as follows.

Z_L is the generic form of the diode impedance. In this regard, it is considered as the load of the transforming line θ_T. Ideally, switching diodes are either short circuits when they are forward biased (State-A) or they exhibit a capacitor C_D when they are reversed biased (State-B). In both switching

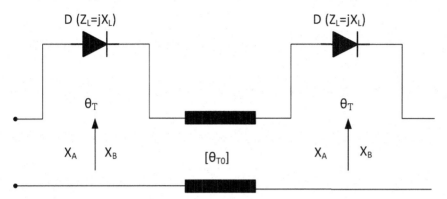

Figure 4.6a   Series loaded line digital phase shifter with impedance transformer.

states, ideally, the load impedance Z_L is pure reactive. Therefore, it can be represented as $Z_L = jX_L(\omega)$. In this case,

$$X_L = \begin{cases} 0 & when\ D\ is\ on\ (State-A) \\ -\dfrac{1}{\omega_o C_D} = -\dfrac{1}{C_D}\ (if\ \omega_o = 1) & when\ D\ is\ off\ (State-B) \end{cases}$$

(4.22)

The phase shift $\Delta\theta_0$ related parameter β is defined as the major design parameter such that

$$\beta = \frac{1}{2}tan\left(\frac{\Delta\theta_0}{2}\right) \qquad (4.23)$$

θ_T is the electrical length in degree or radians of the transformer with actual characteristic impedance Z_T and μ_T is defined as

$$\mu_T = tan\ \theta_T \qquad (4.24)$$

For small phase changes, (4.13) can be used to approximate β such that

$$\Delta\theta_0 = 2\left[tan^{-1}\left(\frac{X_B}{2}\right) - tan^{-1}\left(\frac{X_A}{2}\right)\right]$$

or

$$\frac{\Delta\theta_0}{2} = tan^{-1}\left(\frac{X_B}{2}\right) - tan^{-1}\left(\frac{X_A}{2}\right)$$

or

$$tan\left(\frac{\Delta\theta_0}{2}\right) = \frac{\left(\frac{X_B}{2}\right) - \left(\frac{X_A}{2}\right)}{1 + \left(\frac{X_A}{2}\right)\left(\frac{X_B}{2}\right)}$$

or

$$\beta = \frac{1}{2}tan\left(\frac{\Delta\theta_0}{2}\right) = \frac{X_B - X_A}{4 + X_A X_B} \tag{4.25}$$

Employing the transmission line input impedance formula, loading impedance Z_{inT} of the centerline can be determined by (4.5b) as follows:

$$Z_{inT} = Z_T \frac{Z_L + jZ_T tan(\theta_T)}{Z_T + jZ_L tan(\theta_T)} = jX_{in}$$

$$= \left\{ \begin{matrix} jX_A & when\ D\ is\ on\ such\ that\ Z_L = 0 \\ jX_B & when\ D\ is\ off\ such\ that\ Z_L = -j\frac{1}{\omega_o C_D} \end{matrix} \right\} \tag{4.26}$$

where Z_T is the actual characteristic impedance of the line θ_T which may be selected as the terminations resistors $R = Z_T = 50\ ohms$. In this case, normalized value of Z_T is unity (i.e. $z_T = 1$). Then, the normalized load reactance X_A of the center line θ_{T0} in State-A is found as

$$Z_{inT} = Z_0 \frac{Z_L + jZ_0 tan(\theta_T)}{Z_0 + jZ_L tan(\theta_T)} = jX_{inT} = jX_A;\ Z_L = jX_L = 0;\ if\ Z_T$$

$$= R,\ then\ z_T = \frac{Z_T}{R} = 1 = 1 \tag{4.27a}$$

or

$$X_A = tan(\theta_T) = \mu_T \tag{4.27b}$$

On the other hand, if X_B is extracted from (4.20), we have

$$X_B = \frac{X_A - 4\beta}{1 + \beta X_A} \tag{4.27c}$$

In State-B, diode impedance $Z_L = jX_L$ is transformed as a series load reactance X_B on the centerline θ_{T0} as in (4.21). In this case, by selecting normalized characteristic impedance $z_0 = 1$, X_B is found as

$$X_B = z_T \frac{X_L + z_T tan(\theta_T)}{z_T - X_L tan\ d(\theta_T)} = z_T \frac{X_L + z_T \mu_T}{z_T - X_L \mu_T} = z_T \frac{X_L + z_T X_A}{z_T - X_L X_A}$$

$$\tag{4.27d}$$

Using (4.27b) in (4.27d) and equating (4.27c) and (4.27d), we have

$$X_B = \frac{X_A - 4\beta}{1 + \beta X_A} = z_T \frac{X_L + z_T X_A}{z_T - X_A X_L} \tag{4.28}$$

We can solve (4.28) for X_A such that

$$(X_A - 4\beta)(z_T - z_T X_A X_L) = z_T(X_L + z_T X_A)(1 + \beta X_A)$$

or further simplification of the above equation yields

$$aX_A^2 + bX_A + c = 0 \qquad (4.29a)$$

for $z_T = 1$, the coefficients $\{a, b, c\}$ of the above equation is given by

$$a = (\beta + X_L) \qquad (4.29b)$$
$$b = -3\beta X_L \qquad (4.29c)$$
$$c = (4\beta + X_L) \qquad (4.29d)$$

Hence, solution of (4.29a) becomes

$$X_{A(1,2)} = \frac{-b \mp \sqrt{\Delta_a}}{2a} \qquad (4.30a)$$

where Δ_a is the discriminant of the (4.29a) and it is given by

$$\Delta_a = b^2 - 4ac > 0 \qquad (4.30b)$$
$$\Delta_a = 9\beta^2 X_L^2 - 4(\beta + X_L)(4\beta + X_L) \ if \ z_T = 1 \qquad (4.30c)$$

On the other hand, if the normalized characteristic impedance z_T is different from 1, then, (4.29) is revised such that

$$a = \left(z_T^2 \beta + X_L\right) \qquad (4.31a)$$
$$b = -\left(4\beta X_L - z_T \beta X_L + z_T - z_T^2\right) \qquad (4.31b)$$
$$c = (4z_T \beta + z_T X_L) \qquad (4.31c)$$

For a selected normalized angular center frequency ω_0 and reverse-biased diode capacitance C_D, X_L is fixed as

$$X_L(\omega_0) = -\frac{1}{\omega_0 C_D} \qquad (4.32)$$

Then, $X_A(\omega_0)$ is determined as in (4.30a). Thus, the transformer length θ_T is found as

$$\theta_T = \tan^{-1}[X_A(\omega_0)] + k_1 \pi \qquad (4.33)$$

In (4.33), if X_A is negative, k_1 set to 1 just to make $\theta_T > 0$. Otherwise, $k_3 = 0$.

Under the perfect transmission conditions of (4.10), the center line length is given by

$$\theta_{o_A} = \tan^{-1}\left(\frac{2}{X_A}\right) + k_2\pi > 0 \tag{4.34a}$$

As above, if $X_A < 0$, then $k_2 = 1$ selected. Otherwise we set $k_2 = 0$. Once X_A is determined, by (4.28), X_B is found either using

$$X_B = \frac{X_A - 4\beta}{1 + \beta X_A}$$

or using

$$X_B = z_T \frac{X_L + z_T X_A}{z_T - X_A X_L}$$

Similarly, the same centerline length must also equal

$$\theta_{o_B} = \tan^{-1}\left(\frac{2}{X_B}\right) + k_3\pi > 0 \tag{4.34b}$$

The integer k_3 is set to 1 if $X_B < 0$. Otherwise, $k_3 = 0$.

Physically, (4.34a) and (4.34b) cannot be satisfied simultaneously. Therefore, one needs to make and approximation such as

$$\theta_{T0,1} = \sqrt{\theta_{o_A}\theta_{o_B}} \tag{4.35a}$$

or

$$\theta_{T0,2} = \frac{\theta_{o_A} + \theta_{o_B}}{2} \tag{4.35b}$$

During the design process, one of (4.35a) or (4.35b) is picked to minimize and equalize the transmission loss of the loaded line digital phase shifter at each switching state.

It is important to note that, due to (4.34), a loaded line-digital phase shifter can never provide balanced loss characteristics, even on the design frequency f_0. Furthermore, it cannot deliver maximum power at the selected design frequency ω_0. Therefore, the loaded line phase shifters are called the "intrinsically mismatched digital phase shifter".

Remarks:

- In (4.30b), Δ_a must be non-negative ($\Delta_a \geq 0$). If this is not the case, the diode capacitance C_D or the characteristic impedance z_T may be adjusted to end up with a positive discriminant Δ_a. Usually, z_T is selected as the termination resistor R, which may be selected as the standard termination $R_0 = 50\,\Omega$. Nevertheless, designer is always free to adjust the value of z_T to make Δ_a positive.

- Scattering parameters of the center line is defined based on the termination resistors R. Therefore, characteristic impedance Z_0 of the center line must be equal to R which may be selected as $R_0 = R = 50\,\Omega$. Under this assumption, ideally, $S_{11T} = S_{22T} = 0$ and $S_{21T} = S_{12T} = 1e^{j\varphi_{21T}}$.

- Under many practical circumstances, loaded line phase shifters are more attractive to designers due to their cheap manufacturing cost with simple technology over the other phase shifter topologies, even though they provide narrow frequency band and phase range.

- Once the ideal element values of the loaded phase shifters are calculated, detailed analysis, including the diode and other parasitic, can be carried out using any commercially available software tools such as Matlab, AWR, ADS etc.

Let us combine the above derivations under a design algorithm.

Algorithm 3. Design of Series Loaded Line Digital Phase Shifter with Transformer Line θ_T

In this algorithm, the above derivations are organized to design a simple series loaded line with transmission line transformer θ_T step by step.

Inputs:

R: Termination resistors and normalization numbers for the scattering parameters

Z_T: Characteristic impedance of the transforming line θ_T

f_{0a}: Actual design frequency

$\Delta\theta_0$: Desired phase shift between the switching states, which is specified at the design frequency f_{0a}

ω_0: Normalized angular frequency at f_{0a}

C_{Da}: Actual reverse biased capacitance C_{Da} of the switching diodes

Computational Steps:

Step 1a: Normalize Z_T

$$z_T = \frac{Z_T}{R}$$

Step 1b: Normalize C_{Da}

$$C_D = R\omega_{0a}C_{Da} = R \times 2 \times \pi \times f_{0a} \times C_{Da}$$

Step 1c: Compute the major design parameter β

$$\beta = \frac{1}{2} tan \left(\frac{\Delta\theta_0}{2} \right)$$

Step 2: Compute the normalized load reactance X_L

$$X_L = -\frac{1}{\omega_0 C_D}$$

Step 3: Compute the coefficients $\{a, b, c\}$ of the quadratic Equation (4.29a) as in (4.31)

$$a = \left(z_T^2 \beta + X_L \right)$$
$$b = -\left(4\beta X_L - z_T \beta X_L + z_T - z_T^2 \right)$$
$$c = \left(4z_T \beta + z_T X_L \right)$$

Step 4: Compute the discriminant Δ_a of the quadratic Equation as in (4.30b)

$$\Delta_a = b^2 - 4ac \geq 0$$

Step 5: Check if $\Delta_a > 0$. If yes, continue with the following steps. If not, vary the design parameters ω_0 or C_D until you end up with positive discriminant Δ_a.

Step 6a: Compute the loading reactance X_A of State-A

$$X_{A(1,2)} = \frac{-b \mp \sqrt{\Delta_a}}{2a}$$

In this step, we prefer the positive value for X_A to make the centerline as short as possible. If both values of $X_{A(1,2)}$ are negative, then prefer the smallest value of $\left| X_{A(1,2)} \right|$ to minimize the length of the centerline.

Step 6b: Compute X_B from X_A

$$X_B = \frac{X_A - 4\beta}{1 + \beta X_A}$$

Step 7: Compute electrical length of the line transformer θ_T

$$\theta_T = \tan^{-1} \left[X_A \left(\omega_0 \right) \right] + k_1 \pi$$

In this step, if X_A is positive set $k_1 = 0$. If X_A is negative, then set $k_1 = 1$.

Step 8: Compute the electrical lengths θ_{oA} and θ_{oB} as in (4.34)

$$\theta_{oA} = \tan^{-1}\left(\frac{2}{X_A}\right) + k_2\pi > 0$$

$$\theta_{oB} = \tan^{-1}\left(\frac{2}{X_B}\right) + k_3\pi > 0$$

Step 9: Compute the mismatched lengths $\theta_{T0,1}$ and $\theta_{T0,2}$ of the centerline as in (4.35)

$$\theta_{T0,1} = \sqrt{\theta_{oA}\theta_{oB}}$$

$$\theta_{T0,2} = \frac{\theta_{oA} + \theta_{oB}}{2}$$

Step 10: Analyze the phase shift performance of the designed Series Loaded Line-Digital Phase Shifter (SLL-DPS) using MatLab both for $\theta_{T0,1}$ and $\theta_{T0,2}$, plot the results and identify the best value for θ_{T0} out of $\{\theta_{T0,1}, \theta_{T0,2}\}$.

Let us now run an example to show the utilization of the above algorithm to design.

Example 4.2: Referring to Figure 4.6a, it is desired to design a series loaded line-digital phase shifter with impedance transforming lines employing identical PIN diodes as switching elements.

Let the termination resistors, characteristic impedances of the transforming lines and the centerline be equal such that $R = Z_0 = Z_T = 50\ \Omega$.

Let the reverse biased capacitance of the switching PIN diodes be $C_{Da} = 2.8\ pF$.

Let the actual center frequency be $f_{0a} = 3\ GHz$.

At the center frequency $f_{0a} = 3\ GHz$, let the phase shift $\Delta\theta_0$ between the switching states be

$$\Delta\theta_0 = 22.5°$$

Determine the electrical lengths θ_T and θ_{T0} of the impedance transforming lines and the centerline, respectively.

Solution

Let us follow the computational logic of Algorithm 2 step by step to design the SLL-DPS.

Step 1a: Normalized characteristic impedance $z_T = \frac{Z_T}{R} = \frac{50}{50} = 1$.

Step 1b: Normalize C_{Da}

$$C_D = R\omega_{0a}C_{Da} = R \times 2 \times \pi \times f_{0a} \times C_{Da} = 2.6389$$

Step 1c: Compute the major design parameter β:

$$\beta = \frac{1}{2}tan\left(\frac{\Delta\theta_0}{2}\right) = 0.0995$$

Step 2: Compute the normalized load reactance X_L

$$X_L = -\frac{1}{\omega_0 C_D} = -0.3789$$

Step 3: Compute the coefficients $\{a, b, c\}$ of the quadratic equation (4.24a) as in (4.26)

$$a = \left(z_T^2\beta + X_L\right) = -0.2795$$
$$b = -\left(4\beta X_L - z_T\beta X_L + z_T - z_T^2\right) = 0.1131$$
$$c = (4z_T\beta + z_T X_L) = 0.0189$$

Step 4: Compute the discriminant Δ_a of the quadratic equation as in (4.30b)

$$\Delta_a = b^2 - 4ac = 0.0339 \geq 0$$

Step 5: Check if $\Delta_a > 0$. The answer is YES. There is no problem with CD.

Step 6a: Compute the loading reactance X_A of State-A

$$X_{A(1,2)} = \frac{-b \mp \sqrt{\Delta_a}}{2a}$$
$$X_{A1} = 0.5316$$
$$X_{A2} = -0.1271$$
$$X_A = -0.1271 \; is \; selected$$

Step 6b: Compute X_B from X_A

$$X_B = \frac{X_A - 4\beta}{1 + \beta X_A} = -0.5316$$

Step 7: Compute electrical length of the line transformer θ_T

$$\theta_T = \tan^{-1}\left[X_A\left(\omega_0\right)\right] + k_1\pi = 162.0481$$

It is noted that, in this step, we choose $k_1 = 1$ since $X_A = -0.1271$ is negative.

Step 8: Compute the electrical lengths θ_{0A} and θ_{0B} as in (4.29)

$$\theta_{o_A} = \tan^{-1}\left(\frac{2}{X_A}\right) + k_2\pi = 172.7568 > 0; \ with \ k_2 = 1$$

$$\theta_{o_B} = \tan^{-1}\left(\frac{2}{X_B}\right) + k_3\pi = 152.0031 > 0; \ with \ k_3 = 1$$

Step 9: Compute the mismatched lengths $\theta_{T0,1}$ and $\theta_{T0,2}$ of the centerline as in (4.30)

$$\theta_{T0,1} = \sqrt{\theta_{o_A}\theta_{o_B}} = 162.0481$$

$$\theta_{T0,2} = \frac{\theta_{o_A} + \theta_{o_B}}{2} = 162.3800$$

$$\theta_{T0} = 162.0481 \ is \ selected$$

Step 10: Analyze the phase shift performance of the designed Series Loaded Line Digital Phase Shifter (SLL-DPS) using MatLab program.

For this purpose, we developed a MatLab program called

"*Main_SLL_DPS_Example_3_2.m*".

Execution of this program results in the electrical performances depicted in Figure 4.6b and Figure 4.6c.

Phase shifting performance of the designed SLL-DPS is shown in Figure 4.5b. At $f_{0a} = 3 \ GHz$, the phase shift is equal to $22.5°$ as expected. At $f_{1a} = 2.98 \ GHz$, the phase shift is about $25.18°$. At $f_{2a} = 3.08 \ GHz$, the phase shift is $20.0°$. Therefore, 10% phase bandwidth is about $\Delta F = 100 \ MHz$. In this regard, we can say that SLL-DPS with transformer results in narrow-frequency band and smaller phase range than that of the previously introduced phase shifter.

The loss curves of the phase shifter are shown in Figure 4.6c.

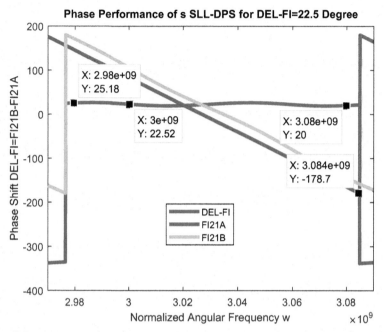

Figure 4.6b   Phase performance of SLL-DPS of Example 3.2A for $\Delta\theta_0 = 22.5°$ phase shift.

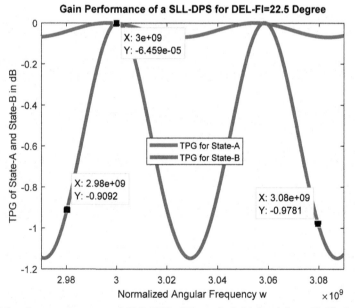

Figure 4.6c   Loss performance of SLL-DPS of Example 3.2A for $\Delta\theta_0 = 22.5°$ phase shift.

4.4 Parallel Load Line Digital Phase Shifters with Transformer

Analysis of the parallel loaded line phase shifter with transformers θ_T is very similar to that of series one. However, it is a little bit tricky as explained in the following.

For the case under consideration, we assume that the parallel loaded line digital phase shifter (PLL-DPS) is driven by a voltage source with internal resistor R and it is also terminated in the same resistor R. We further presume that characteristic impedance Z_T of the transforming line (θ_T) and characteristic impedance Z_0 of the centerline (θ_{T0}) are also equal to R.

Let us work with normalize admittances and normalized impedances to start with. In this case, the normalization resistor is selected as R. For many practical cases, $R = 50$ Ω is selected.

The generic form of the normalized loading shunt admittance of the centerline is designated by $Y_n\,(j\omega)$. Ideally, $Y_n(j\omega)$ is a digital load, which includes loss free switches and reactive elements. Therefore, on the real frequency axis, it is purely an imaginary quantity and it is represented by $jY\,(\omega)$ and expressed by

$$Y_n\,(j\omega) = jY\,(\omega) = \frac{1}{Z_n} \tag{4.36a}$$

where $Z_n = 1/Y_n$ is the normalized impedance equivalence of the shunt load admittance $Y_n(j\omega)$ and $Y(\omega)$ is the normalized susceptance.

Considering the shunt admittance Y_n as a separate lossless two-port, its scattering parameters are given by

$$S_Y = \begin{pmatrix} S_{11Y} & S_{12Y} \\ S_{21Y} & S_{22Y} \end{pmatrix} \tag{4.36b}$$

and in a similar manner to that of (4.5), we have,

$$S_{11Y} = S_{22Y} = \frac{1 - (Y_n + 1)}{1 + (Y_n + 1)} = -j\frac{Y\,(\omega)}{2 + jY\,(\omega)}$$

$$= \frac{Y}{-Y + j2} = \rho_{11Y}e^{j\varphi_{11Y}(\omega)} \tag{4.36c}$$

$$S_{21Y} = \frac{2}{2 + jY(\omega)} \tag{4.36d}$$

where

$$\varphi_{11Y}(\omega) = tan^{-1}\left(\frac{2}{Y}\right) \tag{4.36e}$$

$$\rho_{11Y} = \frac{Y(\omega)}{\sqrt{4+Y^2}}$$

$$\varphi_{21Y} = -tan^{-1}\left(\frac{Y}{2}\right) \tag{4.36f}$$

$$\rho_{21Y}^2 = 1 - \rho_{11Y}^2 = \frac{4}{4+Y^2} \tag{4.36g}$$

As far as generic forms of S_{11Y} and S_{21Y} are concerned, (4.36) and (4.3) reveal that

- S_{11Y} can be obtained from S_{11L} replacing X by Y within a negative sign such that

$$S_{11Y} = -S_{11L}|_{X=Y} \tag{4.36h}$$

and

- S_{21Y} can be obtained from S_{21X} simply replacing X by Y such that

$$S_{21Y}(Y) = S_{21X}(X)|_{X=Y} \tag{4.36i}$$

Referring to Figure 4.7a, when the diode D is forward biased (State-A), ideally, it is short circuited ($Z_L = 0$). Therefore, the diode susceptance Y_L approaches to infinity and θ_T is calculated as follows.

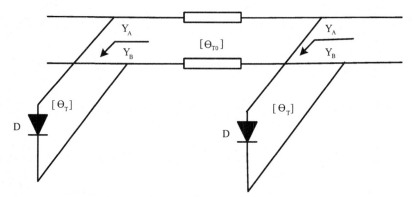

Figure 4.7a   Parallel loaded line digital phase shifter with admittance transformers θ_T.

Setting the normalized characteristic impedance $z_T = 1$, by (4.5), input impedance of the transmission line θ_T is given by

$$Z_{inA} = Z_A = \frac{1}{Y_{nA}} = \frac{1}{jY_A} = z_T \frac{z_L + jz_T tan\,(\theta_T)}{z_T + jz_L tan\,(\theta_T)} = jtan\,(\theta_T) \qquad (4.37a)$$

$$tan\,\theta_T = \mu_T = -\frac{1}{Y_A} \qquad (4.38b)$$

In State-B, the diode is reverse biased and it exhibits a capacitive admittance Y_{CD}

$$Y_{CD} = j\omega C_D = jY_L \qquad (4.38c)$$
$$Y_L = \omega C_D \qquad (4.38d)$$

In this case, the normalized input admittance Y_{nB} of the transmission line θ_T is $Y_{nB} = jY_B$ and it is given by

$$Y_{nB} = jY_B = \frac{jY_L + j\,[tan\,(\theta_T)]}{1 + (jY_L)\,[jtan\,(\theta_T)]} \qquad (4.38e)$$

or using (4.37b) in (4.38e). we have

$$Y_B = \frac{Y_L + \mu_T}{1 - Y_L \mu_T} = \frac{Y_L - \frac{1}{Y_A}}{1 + Y_L \frac{1}{Y_A}} = \frac{Y_L Y_A - 1}{Y_L + Y_A} \qquad (4.38f)$$

At this point, we should generate the scattering parameters $S_{11} = S_{22}$ and $S_{21} = S_{12}$ of the cascaded trio of $[Y] - [Centerline - \theta_T] - [Y]$ in a similar manner to that of (4.7) and (4.10) such that

$$S_{21} = \frac{S_{21Y}^2 S_{21T}}{1 - S_{21T}^2 S_{11Y}^2} = \rho_{21} e^{j\varphi_{21}} \qquad (4.38a)$$

$$\varphi_{21} = 2\varphi_{21Y} - \theta_{T0} = -2tan^{-1}\left(\frac{Y}{2}\right) - \theta_{T0} \qquad (4.39b)$$

$$S_{11}\,(j\omega) = j\frac{2Y - \mu Y^2}{2\,(1 - \mu Y) + j\,(2\mu + 2Y - \mu Y^2)} \qquad (4.39c)$$

$$\mu = tan\,(\theta_{T0}) \qquad (4.39d)$$

For perfect transmission $S_{11} = S_{22} = 0$ yields that

$$2Y - \mu Y^2 = 0 \qquad (4.39e)$$

Then, at $\omega = \omega_0$

$$\mu = tan\,(\theta_{T0}) = \frac{2}{Y};\qquad\qquad(4.39f)$$

$$\theta_{T0} = tan^{-1}\left(\frac{2}{Y}\right) + k_{T0}\pi\qquad\qquad(4.39g)$$

It should be note that, in a similar manner to that of (4.12b), in (4.39a), the term in the denominator $\left(1 - S_{21T}^2 S_{11Y}^2\right)$ is real since the phase of the complex term $\left(S_{21T}^2 S_{11Y}^2\right)$ is zero such that at $\omega = \omega_0$

Phase of $S_{21T} = -\theta_{T0} = -tan^{-1}\left(\frac{2}{Y}\right)$ and *phase of* $S_{11Y} = +tan^{-1}\left(\frac{2}{Y}\right)$. Thus, phase of the term is $2\left[-\theta_{T0} + tan^{-1}\left(\frac{2}{Y}\right)\right] = 2\left[-tan^{-1}\left(\frac{2}{Y}\right) + tan^{-1}\left(\frac{2}{Y}\right)\right] = 0$.

Moreover, at ω_0, the digital phase shift is given by

$$\Delta\theta_0 = |\varphi_{21B} - \varphi_{21A}| = 2\left[tan^{-1}\left(\frac{Y_B}{2}\right) - tan^{-1}\left(\frac{Y_A}{2}\right)\right]\qquad(4.40a)$$

On the other hand, major design parameter is given by

$$\beta = \frac{1}{2}tan\left(\frac{\Delta\theta_0}{2}\right)\qquad\qquad(4.40b)$$

By transmission line impedance or equivalently admittance equation, in State-B, the loading admittance $Y_{nB} = jY_B$ or Y_B of (4.38d) is

$$Y_B = \frac{Y_A Y_L - 1}{Y_A + Y_L}\qquad\qquad(4.40c)$$

On the other hand, by (4.40a), β is found as

$$\beta = \frac{1}{2}tan\left(\frac{\Delta\theta_0}{2}\right) = \frac{Y_B - Y_A}{4 + Y_A Y_B}\qquad(4.40d)$$

Extracting Y_B from (4.40d), we have,

$$Y_B = \frac{Y_A + 4\beta}{1 - \beta Y_A}\qquad\qquad(4.40e)$$

Then, from equation (4.40c) to (4.40e), we end up with a second order equation in Y_A such that

$$a_Y Y_A^2 + b_Y Y_A + c_Y = 0\qquad\qquad(4.40f)$$

where

$$a_Y = \beta Y_L + 1; \quad b_Y = 3\beta; \quad c_Y = 4\beta Y_L + 1 \tag{4.40g}$$

Thus, solution of (4.40f) yields

$$Y_A = \frac{-3\beta \mp \sqrt{\Delta_Y}}{2\left(1 + \beta Y_L\right)} \tag{4.40h}$$

where

$$\Delta_Y = b_Y^2 - 4a_y c_Y > 0 \tag{4.40i}$$

It is very interesting to note that in the above equation, Δ_Y must be positive to end up with real value for Y_A either it is positive or negative. To make Δ_Y positive, we must have

$$b_Y^2 = 9\beta^2 \geq 4a_y c_Y = 4\left(1 + \beta Y_L\right)\left(1 + 4\beta Y_L\right) \tag{4.41a}$$

We know by heart that $Y_L = \omega_0 C_D$ must be positive. If $\beta > 0$, then the maximum value of β is obtained for $Y_L = 0$. This is not practical at all. However, if β is negative, we can end up with reasonable values for the switch capacitor C_D. In this case, it is appropriate to select $\Delta\theta_0$ as a negative quantity to end up with positive Δ_Y.

Hence, (4.38b) yields the desired length of the transformer θ_T as

$$\theta_T = -tan^{-1}\left(\frac{1}{Y_A}\right) + k_T\pi \tag{4.41b}$$

Centerline length θ_{T0} can be computed using (4.39f) for both in State-A and State-B modes as

$$\theta_{OB} = \tan^{-1}\left(\frac{2}{Y_B}\right) + k_B\pi \tag{4.42}$$

and

$$\theta_{OA} = \tan^{-1}\left(\frac{2}{Y_A}\right) + k_A\pi \tag{4.43a}$$

where $k_{(A,B)}$ is set to 1 for $Y_{(A\ or\ B)} < 0$, it is zero otherwise. Then, the centerline length θ_{T0} is chosen as

$$\theta_0\left(A\ or\ B\right) \leq \theta_0 \leq \theta_0\left(B\ or\ A\right) \tag{4.43b}$$

The final length can be determined as in (4.35).

$$\theta_{T0,1} = \sqrt{\theta_{oA}\theta_{oB}} \tag{4.43c}$$

or

$$\theta_{T0,2} = \frac{\theta_{oA} + \theta_{oB}}{2}$$ (4.43d)

There are three major advantages of building parallel loaded line digital phase shifters.

1. They are easy to build using microstrip or co-planar line technology
2. Designer has the freedom to select the revere biased capacitor C_D.
3. They are suitable for small phase changes.

On the other hand, they suffer from poor tracking capability with high transmission loss, since there is an inevitable mismatch loss introduced at each switching state even if the ideal switches are used in the design. In addition, the effect of parasitic elements penalizes the transmission loss substantially as the operating frequency deviates from the center frequency f_{oa}.

Now, let us combine the above computational steps under the following design algorithm.

Algorithm 4: Design of Parallel Loaded Line Digital Phase Shifter with Transformer Line θ_T

In this algorithm, the above derivations are organized to design a simple parallel loaded line with transmission line transformer θ_T step by step.

Inputs:

R: Termination resistors and normalization numbers for the scattering parameters

f_{0a}:   Actual design frequency

$\Delta\theta_0$: Desired phase shift between the switching states, which is specified at the design frequency f_{0a}. It is appropriate to select $\Delta\theta_0$ as a negative quantity.

ω_0:   Normalized angular frequency at f_{0a}

C_{Da}: Actual reverse biased capacitance C_{Da} of the switching diodes

Computational Steps:

Step 1a: Normalize C_{Da}

$$C_D = R\omega_{0a}C_{Da} = R \times 2 \times \pi \times f_{0a} \times C_{Da}$$

Step 1b: Compute the major design parameter β

$$\beta = \frac{1}{2}tan\left(\frac{\Delta\theta_0}{2}\right)$$

Step 2: Compute the normalized load susceptance Y_L

$$Y_L = \omega_0 C_D$$

Step 3: Compute the discriminant a_Y, b_Y, c_Y and Δ_Y as in (4.40g) and (4.41)

$$a_Y = \beta Y_L + 1$$
$$b_Y = 3\beta$$
$$c_Y = 4\beta Y_L + 1$$

and

$$\Delta_Y = b_Y^2 - 4a_Y c_y$$

Step 4: Check if $\Delta_Y > 0$. If yes, continue with the following steps. If not, vary the design parameters ω_0 or C_D until you end up with positive discriminant Δ_Y.

Step 5: Compute the loading susceptance Y_A of State-A

$$Y_{A(1,2)} = \frac{-b_Y \mp \sqrt{\Delta_Y}}{2a_Y}$$

In this step, prefer the positive value for Y_A to make the centerline as short as possible. If both values of $Y_{A(1,2)}$ are negative, then prefer the smallest value of $|Y_{A(1,2)}|$ to minimize the length of the centerline.

Step 6: Compute Y_B from Y_A

$$Y_B = \frac{Y_A Y_L - 1}{Y_A + Y_L}$$

or

$$Y_B = \frac{Y_A + 4\beta}{1 - \beta Y_A}$$

Step 7: Compute electrical length of the line transformer θ_T

$$\theta_T = -\tan^{-1}\left(\frac{1}{Y_A}\right) + k_T \pi$$

In this step, if Y_A is positive, set $k_T = 1$. If Y_A is negative, then set $k_T = 0$.

Step 8: Compute the electrical lengths θ_{0A} and θ_{0B} as in (4.43)

$$\theta_{0A} = \tan^{-1}\left(\frac{2}{Y_A}\right) + k_A\pi > 0$$

$$\theta_{0B} = \tan^{-1}\left(\frac{2}{Y_B}\right) + k_B\pi > 0$$

In this step, k_A and k_B are selected as in k_T of Step 7.

Step 9: Compute the mismatched lengths $\theta_{T0,1}$ and $\theta_{T0,2}$ of the centerline as in (4.44)

$$\theta_{T0,1} = \sqrt{\theta_{0A}\theta_{0B}}$$

$$\theta_{T0,2} = \frac{\theta_{0A} + \theta_{0B}}{2}$$

Step 10: Analyze the phase shift performance of the PLL-DPS Loaded with line transformers terminated in identical switch diodes using MatLab and plot the results.

Let us now run an example to show the utilization of the above algorithm to design.

Example 4.3: Referring to Figure 4.7a, it is desired to design a parallel loaded line-digital phase shifter with impedance transforming lines employing identical PIN diodes as switching elements.

Let the termination resistors, characteristic impedances of the transforming lines and the centerline be equal such that $R = Z_0 = Z_T = 50\ \Omega$.

Let the reverse biased capacitance of the switching PIN diodes be $C_{Da} = 2.8\ pF$.

Let the actual center frequency be $f_{0a} = 3\ GHz$.

At the center frequency $f_{0a} = 3\ GHz$, let the phase shift $\Delta\theta_0$ between the switching states be

$$\Delta\theta_0 = -22.5°$$

Determine the electrical lengths θ_T and θ_{T0} of the impedance transforming lines and the centerline, respectively.

Solution

Let us follow the computational logic of Algorithm 2 step by step to design the SLL-DPS.

Step 1a: Normalized characteristic impedance $z_T = \frac{Z_T}{R} = \frac{50}{50} = 1$.

Step 1b: Normalize C_{Da}

$$C_D = R\omega_{0a}C_{Da} = R \times 2 \times \pi \times f_{0a} \times C_{Da} = 2.6389$$

Step 1c: Compute the major design parameter β:

$\beta = \frac{1}{2}tan\left(\frac{\Delta\theta_0}{2}\right) = -0.0995$

Step 2: Compute the normalized load susceptance Y_L

$$Y_L = \omega_0 C_D = 2.6389$$

Step 3: Compute the coefficients $\{a, b, c\}$ of the quadratic equation (4.24a) as in (4.26)

$$a_Y = (1 + \beta Y_L) = 0.7375$$
$$b_Y = 9\beta^2 = -0.2984$$
$$c_Y = (1 + 4\beta Y_L) = -0.0498$$

Step 4: Compute the discriminant Δ_a of the quadratic equation as in (4.30b)

$$\Delta_Y = b_Y^2 - 4a_Y c_Y = 0.2360 \geq 0$$

Step 5: Check if $\Delta_a > 0$. The answer is YES. There is no problem with C_D. Then, continue with the following steps. If not, go to the input revise the value of your switching capacitor

Step 6: Compute the loading reactance Y_A of State-A

$$Y_{A(1,2)} = \frac{-b_Y \mp \sqrt{\Delta_Y}}{2a_Y}$$
$$Y_{A1} = -0.1271$$
$$Y_{A2} = 0.5316$$
$$Y_A = -0.1271 \; is \; selected$$

Step 7: Compute Y_B from Y_A

$$Y_B = \frac{Y_A + 4\beta}{1 - \beta Y_A} = -0.5316$$

Step 8: Compute electrical length of the line transformer θ_T

$$\theta_T = -\tan^{-1}\left[1/Y_A\left(\omega_0\right)\right] + k_T\pi = 82.7568 \ \ degree$$

It is noted that, in this step, we choose $k_T = 0$ since $Y_A = -0.1271$ is negative.

Step 8: Compute the electrical lengths θ_{0A} and θ_{0B} as in (4.29)

$$\theta_{o_A} = \tan^{-1}\left(\frac{2}{Y_A}\right) + k_A \times 180 = 93.6361 > 0; \ \ with \ k_A = 1$$

$$\theta_{o_B} = \tan^{-1}\left(\frac{2}{Y_B}\right) + k_B \times 180 = 104.8861 > 0; \ \ with \ k_B = 1$$

Step 9: Compute the mismatched lengths $\theta_{T0,1}$ and $\theta_{T0,2}$ of the centerline as in (4.30)

$$\theta_{T0,1} = \sqrt{\theta_{o_A}\theta_{o_B}} = 99.1016$$

$$\theta_{T0,2} = \frac{\theta_{o_A} + \theta_{o_B}}{2} = 99.261$$

$$\theta_{T0} = 99.1016 \ is \ selected$$

Step 10: Analyze the phase shift performance of the designed Parallel Loaded Line Digital Phase Shifter (PLL-DPS) using MatLab environment. For this purpose, we developed a main program called

"*Main_PLL_DPS_Example_3_3.m.*"

In this program, frequency-dependent center line loads are computed as

$$Y_L\left(\omega\right) = \omega C_D$$

$$\mu_T = tan\left(\omega\tau_{nT}\right)$$

$$\tau_{nT} = \frac{\theta_T(in \ radians)}{\omega_0}$$

$$Y_B\left(\omega\right) = \frac{Y_L + \mu_T}{1 - Y_L\mu_T}$$

$$Y_A\left(\omega\right) = -\frac{1}{\mu_T}$$

Execution of this program results in the electrical performances depicted in Figure 4.7b and Figure 4.7c.

Phase shifting performance of the designed PLL-DPS is shown in Figure 4.7b. At $f_{0a} = 3.041\ GHz$, the phase shift is about to 22.5. As we see from this figure, center frequency is shifted due to intrinsic mismatch of the topology. At $f_{1a} = 2.942\ GHz$, the phase shift is about 24.75°. At $f_{2a} = 3.1758\ GHz$, the phase shift is 20.25°. Therefore, 10% phase bandwidth is about $\Delta F = 942\ MHz$.

The loss curves of the phase shifter are shown in Figure 4.7c.

A close examination of Figure 4.7c reveals that loss performance is less than 0.1 dB over the 10% bandwidth, which is considered very good. We as also observe the intrinsic miss match at the design frequency, which is due to the selected topology.

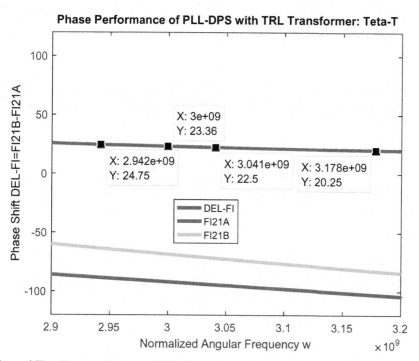

Figure 4.7b   Phase performance of PLL-DPS of Example 3.3 for $\Delta\theta_0 = -22.5°$ phase shift.

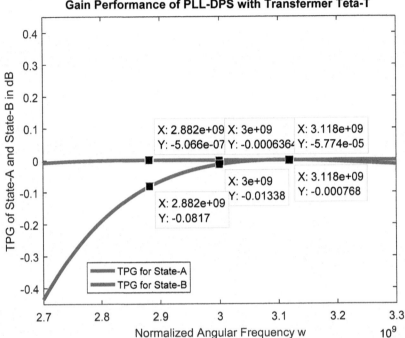

Figure 4.7c   Loss performance of PLL-DPS of Example 3.2A for $\Delta\theta_0 = 22.5°$ phase shift.

4.5 A Perfectly Matched PLL-DPS Loaded with Tuned Circuits

A perfectly matched, Parallel Loaded Line Digital Phase Shifter or in short PLL-DPS loaded with tuned parallel resonant circuits is shown in Figure 4.8a.

This figure includes two identical-shunt digital loads. In State-A, diodes are forwards biased and they ideally, exhibit short circuits. In this mode of operation, the effective admittance acts as a parallel capacitor C at the normalized angular frequency ω_0 such that

$$
Y_{nA}(j\omega_0) = \frac{1}{j\omega_0 L_A} + j\omega C_A = j\omega_0 C = -\frac{j\omega_0}{\omega_0^2 L_A} + j\omega_0 C_A
$$
$$
= j\omega_0 \frac{(\omega_0^2 L_A C_A - 1)}{\omega_0^2 L_A} \tag{4.44a}
$$

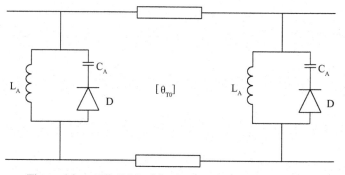

Figure 4.8a PLL-DPS with tuned parallel resonant circuits.

$$C = \frac{\omega_0^2 L_A C_A - 1}{\omega_0^2 L_A} \tag{4.44b}$$

In State-B, diodes are reverse biased with junction capacitor C_D. In this case, C_A in series with C_D forms an equivalent capacitor C_T such that

$$C_T = \frac{C_A C_D}{C_A + C_D} \tag{4.44c}$$

The reverse biased capacitor C_D is selected by the designer to optimize the performance of the phase shifter. It may be worthwhile to mention that C_D may be considered as a tuning capacitor that works on the capacitor C_A. In State-B, L_A resonates with C_T yielding

$$C_T = \frac{1}{L_A \omega_0^2} \tag{4.45a}$$

In this State, at ω_0, the equivalent shunt admittances are equal to zero (i.e. $Y_{nB}(j\omega_0) = jY_B$; $Y_B = 0$).
Employing (4.46b) in (4.45b), we end up with

$$C = C_A - C_T \tag{4.45b}$$

Inserting (4.46a) into (4.46c), we obtained a quadratic equation in terms of C_A such that

$$C_A^2 - (C) C_A - (CC_D) = 0 \tag{4.45c}$$

Thus, once C is determined, C_A is found as

$$C_{A(1,2)} = \frac{C \mp \sqrt{C^2 + 4CC_D}}{2} = \frac{C + \sqrt{C^2 + 4CC_D}}{2} \tag{4.45d}$$

In (4.46d), one must use "+" sign to determine C_A since C and C_D are positive quantities.

Let us further elaborate the design equations. In State-A, the effective shunt admittance is described as in (4.45a).

$$Y_{nA}(j\omega_0) = jY_A(\omega) = j\omega_0 C \qquad (4.46)$$

At the design frequency ω_0, the digital phase shift $\Delta\theta_0$ is given by

$$\Delta\theta_0 = [\varphi_{21B}(\omega_0) - \varphi_{21A}(\omega_0)]$$
$$= -2\left[\tan^{-1}\left(\frac{Y_B}{2}\right) - \tan^{-1}\left(\frac{Y_A}{2}\right)\right] \qquad (4.47a)$$

In the above equation, $Y_B = 0$, yielding the new major design parameter β_Y as

$$\beta_Y = \tan\left(\frac{\Delta\theta_0}{2}\right) = \frac{Y_A}{2} = \frac{\omega_0 C}{2} \qquad (4.47b)$$

or the effective capacitor C of State-B is directly computed from the digital phase shift $\Delta\theta_0$ or in terms of β_Y as

$$C = \frac{2|\beta_Y|}{\omega_0} \qquad (4.47c)$$

In (4.47d), the effective capacitor must be positive. Therefore, at the beginning of the design, digital phase shift must be selected as a negative quantity or we should take the absolute value of β_Y to end up with a positive value for C.

Finally, using the perfect transmission condition in State-A, we compute the center line length θ_{T0} from the perfect match condition of (4.39f) as

$$\theta_{T0} = tan^{-1}\left(\frac{2}{Y_A}\right) + k_{T0}\pi = tan^{-1}\left(\frac{2}{\omega_0 C}\right) + k_{T0}\pi$$
$$= tan^{-1}\left(\frac{1}{\beta_Y}\right) + k_{T0}\pi$$

or

$$\theta_{T0} = tan^{-1}\left(\frac{1}{\beta_Y}\right) + k_{T0}\pi \qquad (4.47d)$$

In (4.47e), we if the quantity $tan^{-1}\left(\frac{2}{\omega_0 C}\right)$ becomes too small, we can always extend the length of the central line by adding 180° to it selecting $k_{T0} = 1$.

All the above derivations can be gathered under a design algorithm.

Algorithm 5: Design of Capacitive PLL-DPS with Parallel Resonant Circuits

This algorithm designs a PLL-DPS with parallel resonant circuits. In this circuit, we employed PIN diodes as switching elements with reverse biased capacitors C_D. However, the same configuration can be implemented with MOS devices as MMIC.

Inputs:
f_{0a}: Actual center frequency (ACF) of the design
ω_0: Normalized angular frequency to design PLL-DPS with resonating loads
$\Delta\theta_0$: Digital phase shift at ω_0
C_{Da}: Actual switch capacitance

Computational Steps:

Step 1a: Compute the major design parameter

$$\beta_Y = tan\left(\frac{\Delta\theta_0}{2}\right)$$

Step 1b: Compute the normalized switch capacitor C_D

$$C_D = (2\pi f_{0a})(RC)$$

Step 2: Compute the effective capacitor C of State-A

$$C = \frac{2\beta_Y}{\omega_0}$$

Step 3: Compute normalized value of the auxiliary or tuning capacitor C_A

$$C_A = \frac{C + \sqrt{C^2 + 4C_D}}{2}$$

Step 4: Compute the actual value of C_A

$$C_{Aa} = \frac{C_A}{2\pi f_{0a}R}$$

Step 5: Compute the switch capacitor C_T

$$C_T = \frac{C_A C_D}{C_A + C_D}$$

Step 6a: Compute the normalized value of the shunt inductor L_A from the resonance condition of State-B

$$L_A = \frac{1}{\omega_0^2 C_T}$$

Step 6b: Compute the actual value of L_A

$$L_{Aa} = \frac{R L_A}{2\pi f_{0a}}$$

Step 7: Compute the length of the central line θ_{T0}

$$\theta_{T0} = tan^{-1}\left(\frac{2}{\omega_0 C}\right) + k_{T0} \times 180$$

Note: At this step, k_{T0} is set to 1 if the phase $tan^{-1}\left(\frac{2}{\omega_0 C}\right)$ is too small to be manufactured. Otherwise, it is set to zero.

Step 8: Analyze the electric performance of the PLL-DPS using Matlab.

During performance analysis, the following frequency-dependent admittances load the center line. Hence, In State-A, the central line is load is given by

$$Y_A(\omega) = \frac{\omega^2 L_A C_A - 1}{\omega^2 L_A}$$

In State-B, L_A resonates with the equivalent capacitor C_T yielding an open circuit at ω_0. Therefore, in this state, we have

$$Y_B(\omega) = \frac{\omega^2 L_A C_T - 1}{\omega^2 L_A}$$

In this case, we can write a Matlab program to generate the scattering parameters using the above loads.

It should be noted that at this step, we develop a Matlab program in which performance analysis is carried out using our function

$$[RO21, FI21, RO11, FI11] = SPAR\_Parallel\_Loaded\_TRL$$
$$(w,\ w0,\ Y,\ \tau_n)$$

This function generates the scattering parameters of PLL-DPS with identical center line loads Y. Therefore, it has to be called by the main program twice both for state A and B. In this function, normalized delay length is computed in radians such that, if θ_{T0} is specified in degree, then

$$\tau_n = \left(\frac{\pi}{180}\right)\left(\frac{\theta_{T0}}{\omega_0}\right)$$

Now let us run an example.

Example 4.4: Referring to Figure 4.8a, it is desired to design a parallel loaded line-digital phase shifter with parallel L//C load employing identical switching diodes.

Let the termination resistors, characteristic impedances of the transforming lines and the centerline be equal such that $R = Z_0 = Z_T = 50\,\Omega$.

Let the reverse biased capacitance of the switching PIN diodes be $C_{Da} = 100\,fF$.

Let the actual center frequency be $f_{0a} = 3\,GHz$.

At the center frequency $f_{0a} = 3\,GHz$, let the phase shift $\Delta\theta_0$ between the switching states be

$$\Delta\theta_0 = -22.5°$$

Determine the electrical lengths θ_T and θ_{T0} of the impedance transforming lines and the centerline, respectively.

Solution

Step 1a: Compute the major design parameter

$$\beta_Y = tan\left(\frac{\Delta\theta_0}{2}\right) = -0.1989$$

Step 1b: Compute the normalized switch capacitor C_D

$$C_D = (2\pi f_{0a})(RC) = 0.9481$$

Step 2: Compute the effective capacitor C of State-A

$$C = \frac{2\beta_Y}{\omega_0} = 0.3978$$

Step 3: Compute normalized value of the auxiliary or tuning capacitor C_A

$$C_A = \frac{C + \sqrt{C^2 + 4C_D}}{2} = 0.8445$$

Step 4: Compute the actual value of C_A

$$C_{Aa} = \frac{C_A}{2\pi f_{0a} R} = 8.9602e - 13 = 0.896\ pF$$

Step 5: Compute the switch capacitor C_T

$$C_T = \frac{C_A C_D}{C_A + C_D} = 0.4467$$

Step 6a: Compute the normalized value of the shunt inductor L_A from the resonance condition of State-B

$$L_A = \frac{1}{\omega_0^2 C_T} = 2.2389$$

Step 6b: Compute the actual value of L_A

$$L_{Aa} = \frac{R L_A}{2\pi f_{0a}} = 5.9388e - 09 = 5.93\ nH$$

Step 7: Compute the length of the central line θ_{T0}

$$\theta_{T0} = tan^{-1}\left(\frac{2}{\omega_0 C}\right) + k_{T0} \times 180 = 78.75\ degree$$

Note: At this step, k_{T0} is set to 1 if the phase $tan^{-1}\left(\frac{2}{\omega_0 C}\right)$ is too small to be manufactured. Otherwise, it is set to zero.

Step 8: Analyze the electric performance of the PLL-DPS using Matlab.

Analyze the phase shift performance of the designed Parallel Loaded Line Digital Phase Shifter (PLL-DPS) using MatLab environment. For this purpose, we developed a main program called

"*Main_PLL_DPS_Example_3_4.m*".

Execution of this program results in the electrical performances depicted in Figure 4.8b and Figure 4.8c.

Phase shifting performance of the designed PLL-DPS is shown in Figure 4.7b. At $f_{0a} = 2.172\ GHz$, the phase shift is about to 24.75. As we see from this figure, center frequency is shifted due to intrinsic mismatch of the topology. At $f_{1a} = 3.651\ GHz$, the phase shift is also 24.75°. Therefore, 10% phase bandwidth is about $\Delta F = 1480\ MHz$, which is much better than that of PLL-DPS designed with transformer lines. Depending on the application, this phase shifter can be even utilized over 1.7–5 GHz.

The loss curves of the phase shifter are shown in Figure 4.8c.

A close examination of Figure 4.8c reveals that loss performance is better than 0.35 dB over the 10% bandwidth which is considered very good. We as also observe the intrinsic miss-match at the design frequency, which is due to the selected topology. Moreover, from 2 to 4 GHz, loss curves are less than 0.8 dB, which is good. Hence, we can confidently state that, PLL-DPS configuration with perfect matched can be employed as a wideband phase shifter topology whenever is needed.

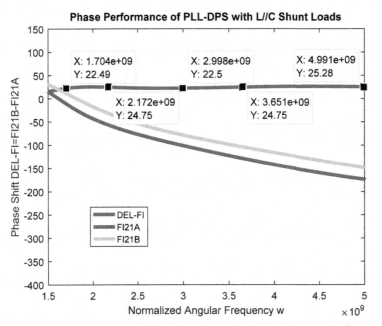

Figure 4.8b Phase performance of PLL-DPS of Example 3.3 for $\Delta \theta_0 = -22.5°$ phase shift.

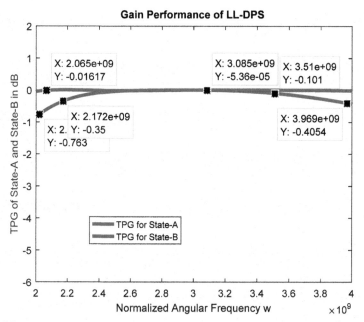

Figure 4.8c   Loss performance of PLL-DPS of Example 3.2A for $\Delta\theta_0 = 22.5°$ phase shift.

An alternative perfectly matched PLL-DPS can be designed with effective shunt inductor L of State-B as described in the following section.

4.6 Perfectly Matched PLL-DPS with Effective Inductor "*L*"

Perfectly Matched PLL-DPS with effective inductor (PLL-DPS-L) uses the same topology of Figure 4.8a with slightly different concept.

In State-A, when the switching element is forward biased (i.e. ideally, it is shorted), inductor L_A resonates with the capacitor C_A at the design frequency ω_0 yielding

$$\omega_0^2 = \frac{1}{L_A C_A} \tag{4.48a}$$

In this mode of operation, the load admittance $Y_{nA}(j\omega) = jY_A(\omega)$ is zero. Therefore, at ω_0, the digital phase shift $\Delta\theta_0$ is given by

$$\Delta\theta_0 = 2\left[\tan^{-1}\left(\frac{Y_B}{2}\right) - \tan^{-1}\left(\frac{Y_A}{2}\right)\right] = 2\tan^{-1}\left(\frac{Y_B}{2}\right) \tag{4.48b}$$

or

$$\beta_Y = \tan\left(\frac{\Delta\theta_0}{2}\right) = \left(\frac{Y_B}{2}\right) \tag{4.48c}$$

In state B, at the design frequency ω_0, the load admittance is set to an effective inductor L such that

$$Y_B(\omega_0) = -\frac{1}{\omega_0 L} = 2\beta_Y \tag{4.48d}$$

If we start with a negative digital phase shift $\Delta\theta_0$, then β_Y will be a negative quantity. In this case, we can confidently state that

$$L = \frac{1}{2\omega_0 |\beta_Y|} \tag{4.48e}$$

On the other hand, in State-B operation, the parallel load admittance $Y_{nB}(j\omega)$ is given by

$$Y_{nB}(j\omega) = \frac{1}{j\omega L_A} + j\omega C_T \tag{4.49a}$$

or at the design frequency ω_0, effective inductance L is given by

$$Y_{nB}(j\omega_0) = \frac{1}{j\omega_0 L_A} + j\omega_0 C_T = \frac{1}{j\omega_0 L} \tag{4.49b}$$

or

$$L = \frac{L_A}{1 - \omega_0^2 L_A C_T} \tag{4.49c}$$

By (4.48a), inductor L_A is

$$L_A = \frac{1}{\omega_0^2 C_A} \tag{4.49d}$$

Then, using (4.49d) in (4.49c), the effective inductor is found as

$$L = \frac{1}{\omega_0^2 (C_A - C_T)} \tag{4.49e}$$

or

$$(C_A - C_T) = \frac{1}{\omega_0^2 L} \tag{4.49f}$$

where C_T is given by

$$C_T = \frac{C_A C_D}{C_A + C_D} \tag{4.49g}$$

where C_D is the reverse biased switch capacitor as before. Using (4.49g) in (4.49f), one ends up with second order equation in C_A such that

$$C_A - \frac{C_A C_D}{C_A + C_D} = \frac{1}{\omega_0^2 L}$$

or

$$C_A^2 + C_D C_A - C_D C_A - \frac{(C_A + C_D)}{\omega_0^2 L} = 0$$

or

$$\left(\omega_0^2 L\right) C_A^2 - C_A - C_D = 0 \tag{4.49h}$$

Solution of (4.49h) is given by

$$C_A = \frac{1 + \sqrt{1 + 4\omega_0^2 L C_D}}{2\omega_0^2 L} \tag{4.49i}$$

Then, by (4.48a), the parallel inductor L_A is given by

$$L_A = \frac{1}{\omega_0^2 C_A} \tag{4.49j}$$

Finally, the center line length is computed from the perfect match condition of State-B as in (4.10) such that

$$\theta_{T0} = \tan^{-1}\left(\frac{2}{Y_B}\right) + k_{T0}\pi = -\tan^{-1}\left(\frac{2}{\frac{1}{\omega_0 L}}\right) + k_{T0}\pi$$

or

$$\theta_{T0} = -\tan^{-1}\left(2\omega_0 L\right) + k_{T0}\pi \tag{4.49k}$$

or

$$\theta_{T0} = -\tan^{-1}\left(\frac{1}{|\beta_Y|}\right) + k_{T0}\pi \tag{4.49l}$$

All the above derivations can be combined in an algorithm to design perfectly matched PLL-DPS with effective inductance L as follows.

Algorithm 6. Design of Perfectly Matched PLL-DPS with Effective Inductance L

In this algorithm, we design a PLL-DPS with effective inductor L. In this circuit, we employed PIN diodes as switching elements with reverse biased capacitors C_D. However, the same configuration can be implemented with MOS devices as MMIC.

Inputs:

f_{0a}: Actual center frequency (ACF) of the design

ω_0: Normalized angular frequency to design PLL-DPS with resonating loads

$\Delta\theta_0$: Digital phase shift at ω_0 (in this step, we can start with a negative value of $\Delta\theta_0$).

C_{Da}: Actual switch capacitance

Computational Steps:

Step 1a: Compute the major design parameter

$$\beta_Y = tan\left(\frac{\Delta\theta_0}{2}\right)$$

Step 1b: Compute the normalized switch capacitor C_D

$$C_D = (2\pi f_{0a})(RC)$$

Step 2: Compute the effective inductor L of State-B

$$L = \frac{1}{2\omega_0 |\beta_Y|}$$

Step 3: Compute normalized value of the auxiliary or tuning capacitor C_A

$$C_A = \frac{1 + \sqrt{1 + 4\omega_0^2 L C_D}}{2\omega_0^2 L}$$

Step 4: Compute the actual value of C_A

$$C_{Aa} = \frac{C_A}{2\pi f_{0a} R}$$

Step 6a: Compute the normalized value of the shunt inductor L_A from the resonance condition of State-B

$$L_A = \frac{1}{\omega_0^2 C_A}$$

Step 6b: Compute the actual value of L_A

$$L_{Aa} = \frac{R L_A}{2\pi f_{0a}}$$

Step 7: Compute the length of the central line θ_{T0}

$$\theta_{T0} = -tan^{-1}\left(\frac{1}{|\beta_Y|}\right) + k_{T0} \times 180$$

Note: At this step, k_{T0} is set to 1 if the phase $tan^{-1}\left(\frac{1}{|\beta_Y|}\right)$ is too small to be manufactured. Otherwise, it is set to zero.

Step 8: Analyze the electric performance of the PLL-DPS using Matlab.

During performance analysis, the following frequency-dependent admittances load the center line.

In State-A, the switches are forward biased. That is, they are ideally short. Therefore, in this mode, the central line is loaded by admittance

$$Y_A(\omega) = \frac{\omega^2 L_A C_A - 1}{\omega^2 L_A}$$

It should be also reminded that, at the design frequency ω_0, inductor L_A resonates with the capacitor C_A. Thus, in this mode of operation (State-A), the digital load acts like open circuits. This fact may be utilized to check the correctness of the performance analysis of the PLL-DPS.

In State-B, switches are reversed biased and they act like capacitors C_D. Therefore, in this state, we have the following load admittance loading the central line.

$$Y_B(\omega) = \frac{\omega^2 L_A C_T - 1}{\omega^2 L_A}$$

To analyze the performance of the PLL-DPS, we developed a Matlab program to generate the scattering parameters using the above generic forms of the loads. In this program, we employed the Matlab function

$$[RO21, FI21, RO11, FI11] = SPAR\_Parallel\_Loaded\_TRL$$
$$(w, \ w0, \ Y, \ \tau_n)$$

to generate the scattering parameters. In this function, normalized delay length τ_n is computed in radians such that if θ_{T0} is specified in degree, then,

$$\tau_n = \left(\frac{\pi}{180}\right)\left(\frac{\theta_{T0}}{\omega_0}\right)$$

Now let us run an example.

Example 4.5: Repeat Example 3.4 to design a PLL-DPS with effective inductive "L" load in State-B.

Answer: Following the computational steps of Algorithm 6, we have;

Inputs:

$$R = 50;$$
$$f0a = 3e9$$
$$w0 = 1$$
$$Del_{Teta} = -22.5 \ degree,$$
$$CDa = 100 \ fF,$$

Computational Steps:

Step 1a: Compute the major design parameter

$$\beta_Y = tan \left(\frac{\Delta\theta_0}{2} \right) = -0.1989$$

Step 1b: Compute the normalized switch capacitor C_D

$$C_D = (2\pi f_{0a}) (RC) = 0.9481$$

Step 2: Compute the effective inductor L of State-B

$$L = \frac{1}{2w_0\beta_Y} = 2.5137$$

Step 3: Compute normalized value of the auxiliary or tuning capacitor C_A

$$C_A = \frac{1 + \sqrt{1 + 4w_0^2 L C_D}}{2w_0^2 L} = 0.8445$$

Step 4: Compute the actual value of C_A

$$C_{Aa} = \frac{C_A}{2\pi f_{0a} R} = 8.9602e - 13 = 0.896 \ pF$$

Step 6a: Compute the normalized value of the shunt inductor L_A from the resonance condition of State-B

$$L_A = \frac{1}{w_0^2 C_A} = 1.1842$$

Step 6b: Compute the actual value of L_A

$$L_{Aa} = \frac{RL_A}{2\pi f_{0a}} = 3.1411e - 09 = 3.14\,nH$$

Step 7: Compute the length of the central line θ_{T0} at $f_{0a} = 3\,GHz$

$$\theta_{T0} = -tan^{-1}\left(\frac{1}{\beta_Y}\right) + 180 = 101.25\,degree$$

Step 8: Analyze the electric performance of the PLL-DPS using Matlab. For this purpose, we developed a MatLab program called

"*Main_PLL_DPS_Example_3_5.m*"

This program generates the element values of the PLL-DPS with inductive loads of State-B as computed above and then analyzes its electrical

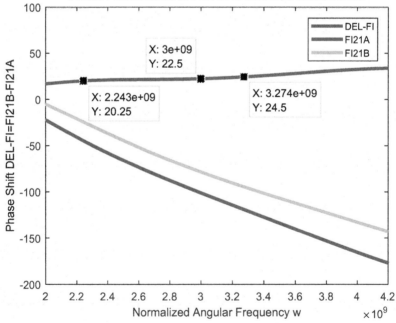

Figure 4.8d   Phase performance of PLL-DPS of Example 3.5 for $\Delta\theta_0 = -22.5°$ phase shift.

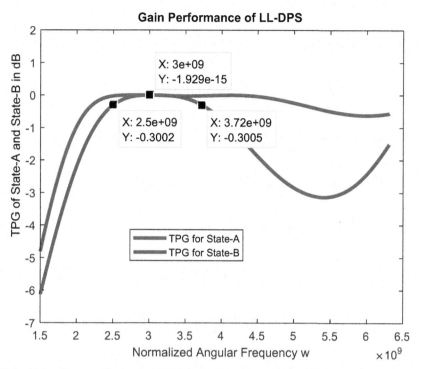

Figure 4.8e Loss performance of PLL-DPS of Example 3.5 for $\Delta\theta_0 = 22.5°$ phase shift.

performance using the ideal switching elements. Thus, digital phase and the loss performances are depicted in Figure 4.8d and Figure 4.8e, respectively.

A close examination of the above figure reveals that phase shifter bandwidth is about 1 GHz.

In Figure 4.8e, loss performance of the PLL-DPS with effective inductive load is depicted. As it is seen from the above figure, at $f_{0a} = 3\ GHz$, PLL-DPS is perfectly matched in both State-A and State-B. Loss performance over 2.5–3.5 GHz is better than 0.3 dB that is very good.

Perfectly matched PLL-DPS is capable of providing wide phase range up to 180° with the expense of sacrificing bandwidth. In the following example, we will investigate this situation for designing a PLL-DPS with $\Delta\theta_{T0} = 135°$ at $f_{0a} = 3\ GHz$.

Example 4.6: Repeat 4.5 for $\Delta\theta_{T0} = -135°$ at $f_{0a} = 3\ GHz$.

Solution

Computational Steps:

Step 1a: Compute the major design parameter

$$\beta_Y = tan\left(\frac{\Delta\theta_0}{2}\right) = -2.4142$$

Step 1b: Compute the normalized switch capacitor C_D

$$C_D = (2\pi f_{0a})(RC) = 0.9481$$

Step 2: Compute the effective inductor L of State-B

$$L = \frac{1}{2\omega_0\beta_Y} = 0.2071$$

Step 3: Compute normalized value of the auxiliary or tuning capacitor C_A

$$C_A = \frac{1 + \sqrt{1 + 4\omega_0^2 LC_D}}{2\omega_0^2 L} = 5.6401$$

Step 4: Compute the actual value of C_A

$$C_{Aa} = \frac{C_A}{2\pi f_{0a}R} = 5.9843e - 12 = 5.98\,pF$$

Step 6a: Compute the normalized value of the shunt inductor L_A from the resonance condition of State-B

$$L_A = \frac{1}{\omega_0^2 C_A} = 0.1773$$

Step 6b: Compute the actual value of L_A

$$L_{Aa} = \frac{RL_A}{2\pi f_{0a}} = 4.7031e - 10 = 0.47\,nH$$

Step 7: Compute the length of the central line θ_{T0} at $f_{0a} = 3\,GHz$

$$\theta_{T0} = -tan^{-1}\left(\frac{1}{\beta_Y}\right) + 180 = 157.5\,degree$$

Step 8: Analyze the electric performance of the PLL-DPS using Matlab.
For this purpose, we developed a MatLab program called

"Main_PLL_DPS_Example_4_6.m"

Electric performance of the perfectly matched PLL-DPS is shown in Figure 4.9a and Figure 4.9b.

As can be seen from the above figure, 10% width ($\delta_\theta = \mp 13.5°$) of the phase shifter runs from $f_{1a} = 2.973\ GHz$ to $f_{2a} = 3.029\ GHz$, which makes about $560\ MHz$ bandwidth. Over the bandwidth, loss performance is less 0.6 dB, which is reasonable.

As pointed earlier, bandwidth of the phase shifter is drastically reduced from $1\ GHz$ down to $560\ MHz$ when the electrical performances of Example 4.5 and 4.6 is compared.

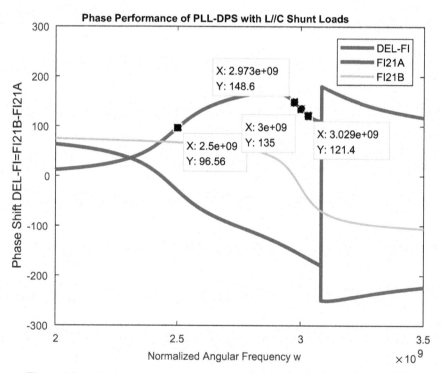

Figure 4.9a Digital phase shift performance of Example 4.6 with $\Delta\theta_0 = -135°$.

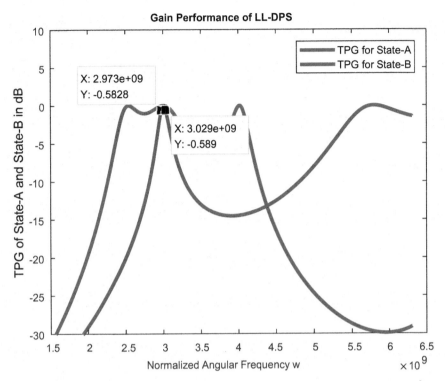

Figure 4.9b   Loss performance of the PLL-DPS of Example 4.6 with $\Delta\theta_0 = -135°$.

4.7 Reflection Phase Shifters

Reflection phase shifters are widely used in practice due to their large phase shift capabilities. A phase shift of $180°$ can be accomplished without any difficulty.

A reflection type phase shifter may be built using a branch line coupler.

A branch line coupler is a four-port device consist of four quarter wave length (i.e. $\lambda/4$) transmission lines designed at the specified operation frequency f_0. Lines are connected to form a rectangular loop as shown in Figure 4.10.

In a typical reflection phase shifter, Port-3 and Port-4 of the branch line coupler is terminated in symmetric reactive loads $Z_L(j\omega) = jX(\omega)$.

If one selects the characteristic impedances of the top and the bottom lines as Z_0, then the side lines characteristic impedances are fixed at $\frac{Z_0}{\sqrt{2}}$. In this case, the phase $\emptyset(\omega_0)$ of the load reflectance

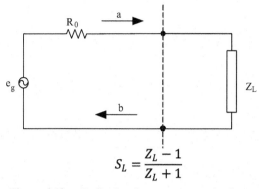

$$S_L = \frac{Z_L - 1}{Z_L + 1}$$

Figure 4.10a Reflection from a pure reactive load.

$\left\{ S_L \left(j\omega_0 \right) = \rho_L \left(\omega_0 \right) e^{j\varphi_L (\omega_0)} \text{ such that } \rho_L \left(\omega_0 \right) = 1, \quad \varphi_L \left(\omega_0 \right) = \emptyset \left(\omega_0 \right) \right\}$ becomes the phase shift between port-1 and port-2 and it is measured by the transfer scattering parameter $\left\{ S_{21} \left(j\omega_0 \right) = \rho_{21} \left(\omega_0 \right) e^{j\varphi_{21} (\omega_0)}; \quad \varphi_{21} \left(\omega_0 \right) = \emptyset \left(\omega_0 \right) \right\}$.

Referring to Figure 4.10a, when an incident wave "a" hits the lossless load (Z_L), it is completely reflected back with a certain amount of phase shift. At the load terminal plane, the phase shift between the incident wave "a" and the reflected wave "b" is given by the phase $\emptyset$ of the reflection coefficient S_L, which is normalized with respect to R_0 such that

$$S_L = \frac{b}{a} = 1e^{j\emptyset}$$
$$S_L = \frac{Z_L - 1}{Z_L + 1} = \frac{1 - Y_L}{1 + Y_L} \tag{4.49a}$$

where $Z_L = jX_L$ is the normalized reactive load impedance and $Y_L = 1/Z_L$ is the corresponding load admittance.

In order to make a physically realizable phase shifter, reflected wave "b" from the reactive termination should be received by a different port rather than input. Therefore, a simple reflection type time delay phase shifter must contain at least three ports, which is realized as a circulator. However, at high frequencies, physical implementation of the circulators becomes difficult. Instead, a lossless four-port branch line coupler is used with symmetric reactive loading to design a phase shifter. The phase shift between the input and output ports is equal to the phase $\emptyset(\omega)$ of the reflectance S_L of the load.

A typical reflection phase shifter is realized using a branch-line coupler as shown in Figure 4.10b.

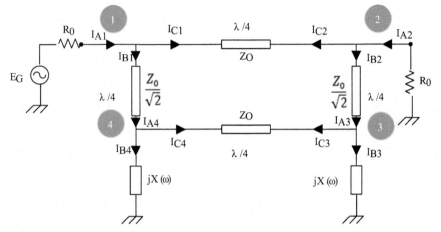

Figure 4.10b   Reflection-type phase shifter implemented using branch line coupler.

Appendix 4: MatLab Programs for Chapter 4

Program List 4.1a. MatLab Program developed for Example 4.1A

```
% Main_DPS_LoadedTRL_Example_4_1A.m
clear; clc; close all;
% This program designs a simple loaded transmission line digital phase
% shifter with PIN-Diodes using the design
   steps given by Algorithm 3.1
% This program is written by BS Yarman on May 20, 2018; Vanikoy.
% Inputs:
   % Desired phase shifts: DEL-FI
   % Actual Design Frequency f0 (Center Frequency)
   % Normalized angular Frequency w0
   % (it is usually selected as unity)
   % Port normalization number R
   % (It is usually selected as R=50 ohms)
   % z0: Normalized Characteristic impedance of the transmission line
   % (usually, z0=Z0/R
   % Note: Z0 is selected as the port normalization number R=50 ohms.
   % Then, normalized characteristic impedance of (4.5) is z0=1
   % vp: Propagation velocity of the transmission line medium.
   % ----------------------------------------------------------------
% Output:
% L=physical length of TRL
% CDa: Actual Diode capacitance
% ----------------------------------------------------------------
% Inputs:
        Del_Teta=-45
        f0a=3e9
        w0=1
```

```
R=50; Z0=R; z0=Z0/R;
eps_r=3.4
%  -----------------------------------------------------------
% Computational Steps:
% Step 1. Compute XB from (4.13) by setting XA=0
XB=2*tand(Del_Teta/2);
% Step 2. Employing (4.10), compute the physical length of the
transmission line.
Teta_T0=(atand(2/XB));
if Teta_T0<0
   Teta_T0=180+Teta_T0;
end
mu=tand(Teta_T0); % See equation (4.6b)
Teta_Rad=Teta_T0*pi/180
tau=Teta_Rad/2/pi/f0a, % actual delay length of TRL
%  -----------------------------------------------------------
% Compute the normalized delay length tau_n of TRL
tau_n=2*pi*f0a*tau/w0;
%  -----------------------------------------------------------
v0=3e8; % Free Space Velocity of Propagation
vp=v0/sqrt(eps_r)
length=tau*vp, % Actual physical length of TRL
% Note: If the transmission line is realized using microstrip line,
% then v_p is computed employing (4.16).
% Step 3. Compute the normalized value of CD.
CD=abs(1/w0/XB), % Normalized Capacitanceof the PIN-Diode [D]
% Step 4. Compute the actual value of CDa
CDa=CD/2/pi/f0a/R,
   % Actual value of the revere biased capacitance [D]
%  -----------------------------------------------------------
w1=0;   w2=2; N=10001;
DW=(w2-w1)/(N-1);
w=w1;
%  -----------------------------------------------------------
W=zeros(1,N); DEL_FIA=zeros(1,N);
RO21A_Ar=zeros(1,N); RO21B_Ar=zeros(1,N);
FI21A_Ar=zeros(1,N); FI21B_Ar=zeros(1,N);
for i=1:N
   W(i)=w*f0a;
% [ Zin ] =Zin_Loaded_TRL_X(w,z0,tau_n, XB);
[ RO21B,FI21B ] = SPAR_Loaded_TRL(w, w0, XB, tau_n);
[ RO21A,FI21A ] = SPAR_Loaded_TRL(w, w0, 0, tau_n);
FI21A_Ar(i)=FI21A; FI21B_Ar(i)=FI21B;
RO21A_Ar(i)=20*log10(RO21A); RO21B_Ar(i)=20*log10(RO21B);
DEL_FIA(i)=FI21B-FI21A;
%
   w=w+DW;
end
%
figure
plot(W,DEL_FIA,W,FI21A_Ar,W,FI21B_Ar)
xlabel('Normalized Angular Frequency w')
ylabel('Phase Shift DEL-FI=FI21B-FI21A')
```

```
title ('Phase Performance of Loaded TRL for DEL-FI=FI21B-FI21A')
legend('DEL-FI','FI21A','FI21B')
%
figure
plot(W,RO21A_Ar,W,RO21B_Ar)
xlabel('Normalized Angular Frequency w')
ylabel('TPG of State-A and State-B in dB')
title ('Gain Performance of LL-DPS ')
legend('TPG for State-A','TPG for State-B')
```

Program List 4.1b. MatLab Program developed for Example 4.1B

```
% Main_DPS_LoadedTRL_Example_3_1B.m
clear; clc; close all;
% This program designs a simple loaded transmission line digital phase
% shifter with PIN-Diodes using the designsteps given by Algorithm 3.1
% This program is written by BS Yarman on May 20, 2018; Vanikoy.
% Inputs:
    % Desired phase shifts: DEL-FI
    % Actual Design Frequency f0 (Center Frequency)
    % Normalized angular Frequency w0
    % (it is usually selected as unity)
    % Port normalization number R
    % (It is usually selected as R=50 ohms)
    % z0: Normalized Characteristic impedance of the transmission line
    % (usually, z0=Z0/R
    % Note: Z0 is selected as the port normalization number R=50 ohms.
    % Then, normalized characteristic impedance of (4.5) is z0=1
    % vp: Propagation velocity of the transmission line medium.
% ---------------------------------------------------------------------
% Output:
% L=physical length of TRL
% CDa: Actual Diode capacitance
% ---------------------------------------------------------------------
% Inputs:
        Del_Teta=-90
        f0a=3e9
        w0=1
        R=50; Z0=R; z0=Z0/R;
        eps_r=3.4
% ---------------------------------------------------------------------
% Computational Steps:
% Step 1. Compute XB from (4.13) by setting XA=0
XB=2*tand(Del_Teta/2);
% Step 2. Employing (4.10), compute thephysical length of the
transmission line.
Teta_T0=(atand(2/XB));
if Teta_T0<0
    Teta_T0=180+Teta_T0;
end
```

```
mu=tand(Teta_T0); % See equation (4.6b)
Teta_Rad=Teta_T0*pi/180
tau=Teta_Rad/2/pi/f0a, % actual delay length of TRL
%
% Compute the normalized delay length tau_n of TRL
tau_n=2*pi*f0a*tau/w0;
%
v0=3e8; % Free Space Velocity of Propagation
vp=v0/sqrt(eps_r)
length=tau*vp, % Actual physical length of TRL
% Note: If the transmission line is realized using microstrip line,
% then v_p is computed employing (4.16).
% Step 3. Compute the normalized value of CD.
CD=abs(1/w0/XB), % Normalized Capacitance of the PIN-Diode [D]
% Step 4. Compute the actual value of CDa
CDa=CD/2/pi/f0a/R, % Actual value of the revere biased capacitance [b]
%
w1=0; w2=2; N=10001;
DW=(w2-w1)/(N-1);
w=w1;
%
W=zeros(1,N); DEL_FIA=zeros(1,N);
RO21A_Ar=zeros(1,N); RO21B_Ar=zeros(1,N);
FI21A_Ar=zeros(1,N); FI21B_Ar=zeros(1,N);
for i=1:N
    W(i)=w*f0a;
% [ Zin ]=Zin_Loaded_TRL_X(w,z0,tau_n, XB);
[ RO21B,FI21B ] = SPAR_Loaded_TRL(w, w0, XB, tau_n);
[ RO21A,FI21A ] = SPAR_Loaded_TRL(w, w0, 0, tau_n);
FI21A_Ar(i)=FI21A; FI21B_Ar(i)=FI21B;
RO21A_Ar(i)=20*log10(RO21A); RO21B_Ar(i)=20*log10(RO21B);
DEL_FIA(i)=FI21B-FI21A;
%
    w=w+DW;
end
%
figure
plot(W,DEL_FIA,W,FI21A_Ar,W,FI21B_Ar)
xlabel('Normalized Angular Frequency w')
ylabel('Phase Shift DEL-FI=FI21B-FI21A')
title ('Phase Performance of Loaded TRL for DEL-FI=FI21B-FI21A')
legend('DEL-FI','FI21A','FI21B')
%
figure
plot(W,RO21A_Ar,W,RO21B_Ar)
xlabel('Normalized Angular Frequency w')
ylabel('TPG of State-A and State-B in dB')
title ('Gain Performance of LL-DPS ')
legend('TPG for State-A','TPG for State-B')
```

Program List 4.1C. MatLab Program developed for Example 4.1C

```matlab
%Main_ISLL_DPS_with_Inductors_Example_4_1C.m
% This program designs a simple loaded linephase shifter with series
% inductive load. In this configuration,when the PIN diodes are
on, they
% are shorted. Then, the line teta_T0 isloaded with the
%normalized inductor L.
% If the diodes are off, the diode reverse biased capacitor
CD resonates
% with the loading inductor L. In this case, XA=w*L-1/(w*CD)
 or at w=w0,
% CD=1/(w0^2*L).
clear; clc; close all;
% This program designs a simple loaded transmission line digital phase
% shifter with ideal switches using the design steps given
by Algorithm 3.1
% This program is written by BS Yarman on May 20, 2018; Vanikoy.
 % Inputs:
    % Desired phase shifts: DEL-FI
    % Actual Design Frequency f0 (Center Frequency)
    % Normalized angular Frequency w0
    % (it is usually selected as unity)
    % Port normalization number R
    % (It is usually selected as R=50 ohms)
    % z0: Normalized Characteristic impedance of the transmission line
    % (usually, z0=Z0/R
    % Note: Z0 is selected as the port normalization number R=50 ohms.
    % Then, normalized characteristic impedance of (4.5) is z0=1
    % vp: Propagation velocity of the transmission line medium.
% ------------------------------------------------------------
% Output:
% L=physical length of TRL
% CDa: Actual Diode capacitance
% ------------------------------------------------------------
% Inputs:
        Del_Teta=45
        f0a=3e9
        w0=1.
        R=50; Z0=R; z0=Z0/R;
        eps_r=3.4
% ------------------------------------------------------------
% Computational Steps:
% Step 1. Compute XB from (4.13) by setting XA=0
XB=2*tand(Del_Teta/2);
% Step 2. Employing (4.10), compute the physical length
of the transmission line.
Teta_T0=(atand(2/XB));
if Teta_T0<0
    Teta_T0=180+Teta_T0;
end
mu=tand(Teta_T0); % See equation (4.6b)
Teta_Rad=Teta_T0*pi/180
```

```
tau=Teta_Rad/2/pi/f0a, % actual delay length of TRL
%  ------------------------------------------------------------
% Compute the normalized delay length tau_n of TRL
tau_n=2*pi*f0a*tau/w0;
%  ------------------------------------------------------------
v0=3e8; % Free Space Velocity of Propagation
vp=v0/sqrt(eps_r)
length=tau*vp, % Actual physical length of TRL
% Note: If the transmission line is realized using microstrip line,
% then vp is computed employing (4.16).
% Step 3. Compute the normalized value of Inductor L:
% Note that XB=w0*L, L=XB/w0;
L=XB/w0, % Normalized value of the series load inductor
% Step 4. Compute the actual value La of the inductor
L (Normalized value)
La=L*R/2/pi/f0a, % Actual value of the series load inductor
CD=1/w0/w0/L,
    % Normalized value of the reverse biased diode capacitance
CDa=CD/R/2/pi/f0a,%Actual value of the reverse biased diode
capacitance
%  ------------------------------------------------------------

w1=0; w2=2; N=10001;
DW=(w2-w1)/(N-1);
w=w1;
%  ------------------------------------------------------------
W=zeros(1,N); DELFIA = zeros(1,N);
RO21A_Ar = zeros(1,N);RO21B_Ar=zeros(1,N);
FI21A_Ar=zeros(1,N); FI21B_Ar=zeros(1,N);
for i=1:N
    W(i)=w*f0a;
% [ Zin ] =Zin_Loaded_TRL_X(w,z0,tau_n, XB);
XA_w=w*L-1/w/CD;
XB_w=w*L;
[ RO21B,FI21B ] = SPAR_Loaded_TRL(w, w0, XB_w, tau_n);
[ RO21A,FI21A ] = SPAR_Loaded_TRL(w, w0, XA_w, tau_n);
FI21A_Ar(i)=FI21A; FI21B_Ar(i)=FI21B;
RO21A_Ar(i)=20*log10(RO21A); RO21B_Ar(i)=20*log10(RO21B);
DEL_FIA(i)=FI21B-FI21A;
%
    w=w+DW;
end
%
figure
plot(W,DEL_FIA,W,FI21A_Ar,W,FI21B_Ar)
xlabel('Normalized Angular Frequency w')
ylabel('Phase Shift DEL-FI=FI21B-FI21A')
title ('Phase Performance of Loaded TRL for DEL-FI=FI21B-FI21A')
legend('DEL-FI','FI21A','FI21B')
%
figure
plot(W,RO21A_Ar,W,RO21B_Ar)
xlabel('Normalized Angular Frequency w')
```

```
ylabel('TPG of State-A and State-B in dB')
title ('Gain Performance of LL-DPS')
legend('TPG for State-A','TPG for State-B')
```

Program List 4.2. Computation of Scattering Parameters for Loaded
Transmission Line

```
function [ RO21,FI21 ] = SPAR_Loaded_TRL(w, w0, X, tau_n)
% This function generates the scattering parameters of the loaded
% transmission line [X]-[TRL]-[X]
% Inputs:
% w: Normalized angular frequency
% w0: Normalized angular frequency
% X: Normalized series load of TRL
% tau_n: normalized length of TRL
% where Teta_T0 is the length of the TRL in degree.
% Output:
% S11: Input Reflection Coefficient of the loaded TRL: [X-TRL-X]
% S21: Transfer Scattering Parameter of the loaded TRL
% RO11=abs(S11)
% FI11=Phase of S11=atan2d(X11,R11)
% RO21=abs(S21)
% FI21=Phase of S21=atan2d(X21,R21)
% -------------------------------------------------------------
j=sqrt(-1);
% Step 1: Compute mu=tan(w0*tau_n)
mu=tan(w0*tau_n);
% Step 2: Compute length of TRL as Teta=w*tau_n
Teta=w*tau_n;
% Step 3: Compute S21T
S21T=cos(Teta)-j*sin(Teta);
% Step 4: Compute S11L and S21L
gL=2+j*X;
S11L=j*X/gL;
S21L=2/gL;
%-------------------------------------------------------------
\% Step 5: Compute S11 and S21 of X-TRL-X
\% S21=[S21L^2 ][S21T ]/(1-S21T^2 S11L^2)
%-------------------------------------------------------------
S11=j*(2*X-mu*X*X)/(2*(1-mu*X)+j*(2*mu+2*X-mu*X*X));
R11=real(S11); X11=imag(S11); FI11=atan2d(X11,R11);
S21=S21L*S21L*S21T/(1-S21T*S21T*S11L*S11L);
R21=real(S21); X21=imag(S21); FI21=atan2d(X21,R21);
%-------------------------------------------------------------
% Step 6: Generate {RO11, FI11},{RO21,FI21}
% RO11=abs(S11);
RO21=abs(S21);
% error1=RO21*RO21-(1-RO11*RO11);
% error2=2*FI21-(180+2*FI11);
end
```

Program List 4.3. Main program for Example 4.2

```
% Main_SLL_DPS_Example_4_2.m
clc
close all
clear
% This program designs a Series Loaded-Line DPS with Impedance
 Transforming
% Lines
% This program is developed by BS Yarman, Vanikoy, June 1, 2018.
% Inputs:
% R: Termination resistors
% and normalization numbers for the scattering parameters
% ZT: Characteristic impedance of the transforming line Z_T
% f0a: Actual Design Frequency
% DE1-Teta0: Desired Phase Shift between the switching states
% which is specified at the design frequency f0a
% w0: Normalized Angular Frequency at f0a
% CDa: Actual Reverse Biased Capacitance CDa of the switching Diodes
% Computational Steps
% Step 1a: Normalize ZT
% zT=ZT/R
% Step 1b: Normalize CDa
% CD=R*?0a*CDa=R X 2 X ? X f0a X CDa
% Step 1c: Compute the major design parameter ?
% beta=1/2 tan{(Del-Teta0)/2)
% Step 2: Compute the normalized load reactance XL
% XL=-1/(w0*CD)
% Step 3:Compute the coefficients {a,b,c} of the quadratic
 equation (4.24a)
% as in (4.26)
% a=(zT^2*beta+XL)
% b=-(4*beta*XL-zT*beta*XL+zT-zT^2)
% c=(4*zT*beta+zT*XL)
% Step 4:Compute the discriminant of the quadratic equation
 as in(4.25b)
% Discriminant=b^2-4ac>0
% Step 5:Check if Discriminant>0.If yes, continue with the following
 steps.
% If not, vary the design parameters w0 or CD until you end up with a
 % positive discriminant.
% Step 6: Compute the loading reactance XA of State-A
% XA(1,2)=(-b+/-sqrt(Discriminant))/2a
% In this step, prefer the positive value for XA to make
 the centerline as
% short as possible. If both values of XA(1,2) is negative,
 then prefer
% the smallest value of |X_A(1,2)| to minimize the length of the
 centerline.
% Step 7: Compute electrical length of the line transformer Teta_T
% Teta_T=atand|XA|+k1*pi
% In this step, if XA is positive set k1=0.
% If XA is negative, then set k1=1.
% Step 8: Compute the electrical lengths Teta_0A and Teta_0B
```

```
as in (4.29)
% Teta_(0A)=atand(2/XA)+k2*pi>0
% Teta_(0B)=atand(2/XB)+k3*pi>0
% Step 9: Compute the mismatched lengths ?T01 and ?T02 of the
  centerline
% as in (4.30)
% Teta_T0,1=sqrt(Teta_0A*Teta_0B)
% Teta_T0,2=(Teta_(0_A)+Teta_(0_B))/2
% Step 10: Analyze the phase shift performance of the designed Series
% Loaded-Line-Digital Phase Shifter (SLL-DPS) using MatLab both for
% Teta(T0,1) and Teta(T0,2), plot the results and identify the
  best value
% for Teta_T0 out of {Teta(T0,1),Teta(T0,2)}.


%-------------------------------------------------------------------
R=50, ZT=50, Z0=50,
f0a=3e9,
CDa=2.8e-12,
w0=1.
Del_Teta=22.5,
%-------------------------------------------------------------------
% Step 1a: Normalize ZT
zT=ZT/R,
% Step 1b: Normalize CDa
w0a=2*pi*f0a,
% C_D=R?_0a C_Da=RX2X?f_0aXC_Da
CD=R*w0a*CDa,
% Step 1c: Compute the major design parameter ?=1/2 tan((??_0)/2)
% beta=(1/2)* tan((Teta_0)/2)
beta=tand(Del_Teta/2)/2,
% Step 2: Compute the normalized load reactance XL
XL=-1/w0/CD,
% Step 3: Compute the coefficients {a,b,c} of the quadratic equation
  (4.24a) as in (4.26)
a=zT*zT*beta+XL,
b=-(4*beta*XL-zT*beta*XL+zT-zT*zT),
c=(4*zT*beta+zT*XL),
% Step 4: Compute the discriminant_a of the quadratic equation
  as in (4.25b)
% ?_a=b^2-4ac?0
Discriminant=b*b-4*a*c,
% Step 5: Check if ?_a>0.
%If yes, continue with the following steps.
% If not, vary the design parameters ?_0 orC_D until you end up with
  positive discriminant ?_a.
if Discriminant<0
    Attention='Negative Discriminator: Change CDa or w0'
end
if Discriminant>0
% Step 6a: Compute the loading reactance X_A of State-A
XA1=(-b-sqrt(Discriminant))/2/a,
XA2=(-b+sqrt(Discriminant))/2/a,
if XA1>0
```

```
      XA=XA1,
      ROA1=9999999;
      if XA2>0
            if XA1>XA2
                  XA=XA2,
                  ROA2=9999999;
            end
            if XA2<XA1
                  XA=XA1,
            end
      end
end
if XA1<0
      ROA1=abs(XA1);
end
if XA2<0
      ROA2=abs(XA2)
end
if ROA1>ROA2
      XA=XA2
end
if ROA2>ROA1
      XA=XA1
end
% Step 6b: Compute XB from XA
XB=(XA-4*beta)/(1+beta*XA),
% Step 7: Compute electrical length of the line transformer Teta_T
if XA>0
Teta_T=atand(XA)
end
if XA<0
      Teta_T=atand(XA)+180,
end
% Step 8: Compute the electrical lengths Teta_0A and Teta_0B
  as in (4.29)
if XA<0
Teta_0A=atand(XA)+180,
end
if XA>0
 Teta_0A=atand(XA),
end
% Teta_0A=atand(2/XA)+k2*pi>0
% Teta_0B=atand(2/XB)+k3*pi>0
if XB<0
Teta_0B=atand(XB)+180,
end
if XB>0
 Teta_0B=atand(XB),
end
%
% Step 9: Compute the mismatched lengths Teta_(T0,1) and Teta_(T0,2) of
%the centerline as in (4.30)
% Teta_T01=sqrt(Teta_0A*Teta_0B)
```

```
Teta_T01=sqrt(Teta_0A*Teta_0B),
Teta_T02=(Teta_0A+Teta_0B)/2,
% Select the short line length:
if Teta_T01>Teta_T02
    Teta_T0=Teta_T02
end
if Teta_T02>Teta_T01
    Teta_T0=Teta_T01
end
%------------------------------------------------
% In order to use our analysis tool, compute the normalized
  delay length
tau_n=Teta_T0/w0
%------------------------------------------------
end; % End of positive Discriminant loop
%------------------------------------------------
w1=0.9; w2=1.1; N=10001;
DW=(w2-w1)/(N-1);
w=w1;
%------------------------------------------------
W=zeros(1,N); DEL_FIA=zeros(1,N);
RO21A_Ar=zeros(1,N); RO21B_Ar=zeros(1,N);
FI21A_Ar=zeros(1,N); FI21B_Ar=zeros(1,N);
for i=1:N
    W(i)=w*f0a;
% [ Zin ] =Zin_Loaded_TRL_X(w,z0,tau_n, XB);
[ RO21B,FI21B ] = SPAR_Loaded_TRL(w, w0, XB, tau_n);
[ RO21A,FI21A ] = SPAR_Loaded_TRL(w, w0, XA, tau_n);
FI21A_Ar(i)=FI21A; FI21B_Ar(i)=FI21B;
RO21A_Ar(i)=20*log10(RO21A); RO21B_Ar(i)=20*log10(RO21B);
DEL_FIA(i)=FI21B-FI21A;
%

    w=w+DW;
end
%
figure
plot(W,DEL_FIA,W,FI21A_Ar,W,FI21B_Ar)
xlabel('Normalized Angular Frequency w')
ylabel('Phase Shift DEL-FI=FI21B-FI21A')
title ('Phase Performance of Loaded TRL for DEL-FI=FI21B-FI21A')
legend('DEL-FI','FI21A','FI21B')
%
figure
plot(W,RO21A_Ar,W,RO21B_Ar)
xlabel('Normalized Angular Frequency w')
ylabel('TPG of State-A and State-B in dB')
title ('Gain Performance of LL-DPS')
legend('TPG for State-A','TPG for State-B')
%------------------------------------------------
  if Discriminant<0
    Attention='Discriminant is negative change CDa or w0'
end
```

Program List 4.4. Computation of Scattering Parameters for PLL-DPS

```
function [ RO21,FI21,RO11,FI11 ] = SPAR_Parallel_Loaded_TRL(w, w0, Y,
    tau_n)
% This function generates the scattering parameters of the loaded
% transmission line [X]-[TRL]-[X].
% Developed by BS Yarman on June 10, 2018,Philadelphia,
                                Pennsylvania, USA
 % Inputs:
    % w: Normalized angular frequency
    % w0: Normalized angular frequency
    % Y: Normalized Parallel load of TRL
    % tau_n: normalized length of TRL
    % where Teta_T0 is the length of the TRL in degree.
 % Output:
    % S11: Input Reflection Coefficient of the loaded TRL: [X-TRL-X]
    % S21: Transfer Scattering Parameter of the loaded TRL
    % RO11=abs(S11)
    % FI11=Phase of S11=atan2d(X11,R11)
    % RO21=abs(S21)
    % FI21=Phase of S21=atan2d(X21,R21)
    % ------------------------------------------------------------
j=sqrt(-1);
% Step 1: Compute mu=tan(w0*tau_n)
mu=tan(w0*tau_n);
% Step 2: Compute length of TRL asTeta=w*tau_n
Teta=w*tau_n;
% Step 3: Compute S21T
S21T=cos(Teta)-j*sin(Teta);
% Step 4: Compute S11L and S21L
gL=2+j*Y;
S11L=(-1)*j*Y/gL;
S21L=2/gL;
%------------------------------------------------------------
% Step 5: Compute S11 and S21 of Y-TRL-Y
% S21=[S21L^2 ][S21T ]/(1-S21T^2S11L^2)
%------------------------------------------------------------
S11=(-1)*j*(2*Y-mu*Y*Y)/(2*(1-mu*Y)+j*(2*mu+2*Y-mu*Y*Y));
R11=real(S11); X11=imag(S11); FI11=atan2d(X11,R11);
S21=S21L*S21L*S21T/(1-S21T*S21T*S11L*S11L);
R21=real(S21); X21=imag(S21); FI21=atan2d(X21,R21);
%------------------------------------------------------------
% Step 6: Generate {RO11, FI11},{RO21,FI21}
RO11=abs(S11);
RO21=abs(S21);
% error1=RO21*RO21-(1-RO11*RO11);
% error2=2*FI21-(180+2*FI11);
   end
```

Program List 4.5. Main_PLL_DPS_Example_4_3.m

% Main_PLL_DPS_Example_4_3.m

```
clc
close all
clear
% This program designs a Parallel Loaded-Line DPS with
% Impedance Transforming Lines
% This program is developed by BS Yarman,Philladelphia,
  USA June 10, 2018.
% Inputs:
% R: Termination resistors
% and normalization numbers for the scattering parameters
% ZT: Characteristic impedance of the transforming line ZT
% f0a: Actual Design Frequency
% DEl-Teta0: Desired Phase Shift between the switching states
% which is specified at the design frequency f0a
% w0: Normalized Angular Frequency at f0a
% CDa: Actual Reverse Biased Capacitance CDa of the switching Diodes
% Computational Steps
% Step 1a: Normalize ZT
% zT=ZT/R
% Step 1b: Normalize CDa
% CD=R*w0a*CDa=RX2XpiXf0aXCDa
% Step 1c: Compute the major design parameter beta
% beta=1/2 tan((Del-Teta0)/2)
% Step 2: Compute the normalized load susceptance YL=w*CD
% YL=w*CD
% Step 3:Compute the coefficients {aY,bY,cY} as in (4.40g)
% aY=1+beta*YL;
% bY=3*beta;
% cY=4*beta*YL+1
% Step 4:DELY=bY*bY-4*aY*cY;
%
% Step 5:Check if DELY>0.If yes, continue with the following steps.
% If not, vary the design parameters w0 or CD until you end up with a
% positive discriminant.
% Step 6: Compute the loading reactance XA of State-A
% YA(1,2)=(-bY+/-sqrt(DELY))/2aY
% % Step 8: Compute the electrical lengths Teta_0A and Teta_0B
  as in (4.29)
% Teta_(0A)=atand(2/YA)+kA*pi>0
% Teta_(0B)=atand(2/YB)+kB*pi>0
% Step 9: Compute the mismatched lengths?T01 and Teta_T02 of the
  centerline
% as in (4.30)
% Teta_T0,1=sqrt(Teta_0A*Teta_0B)
% Teta_T0,2=(Teta_(0_A)+Teta_(0_B))/2
% Step 10: Analyze the phase shift performance of the designed Series
% Loaded-Line-Digital Phase Shifter (SLL-DPS) using MatLab both for
% Teta(T0,1) and Teta(T0,2), plot the results and identify the
  best value
% for Teta_T0 out of {Teta(T0,1),Teta(T0,2) }.
%-----------------------------------------------------------------
R=50, ZT=50, Z0=50,
f0a=3e9,
```

```
CDa=2.8e-12,
w0=1,
Del_Teta=-22.5,% in degree
%---------------------------------------------------------------
% Step 1a: Normalize ZT
zT=ZT/R,
% Step 1b: Normalize CDa
w0a=2*pi*f0a,
% CD=R*w0a*CDa=RX2XpiXf0aXCDa
CD=R*w0a*CDa,
% Step 1c: Compute the major design parameter
% beta=(1/2)* tan((Teta_0)/2)
beta=tand(Del_Teta/2)/2,
% Step 2: Compute the normalized load susceptance YL
YL=w0*CD,
% Step 3: Compute the coefficients{aY,bY,cY}
aY=1+beta*YL,
bY=+3*beta,
cY=4*beta*YL+1,
% Step 4: Compute the discriminant DELY
DELY=bY*bY-4*aY*cY,
% Step 5: Check if ?_a>0.
%If yes, continue with the following steps.
% If not, vary the design parameters CD until you end up with
                        positive discriminant DELY.
if DELY<0
    Attention='Negative Discriminator: ChangeCDa'
end
if DELY>0
% Step 6a: Compute the loading reactance YA of State-A
YA1=(-bY-sqrt(DELY))/2/aY,
YA2=(-bY+sqrt(DELY))/2/aY,
ROA2=abs(YA2),
ROA1=abs(YA1)
if YA1>0
    YA=YA1,
    ROA1=9999999;
    if YA2>0
        if YA1>YA2
            YA=YA2,
            ROA2=9999999;
        end
        if YA2<YA1
            YA=YA1,
        end
    end
end
if YA1<0
    ROA1=abs(YA1);
end
if YA2<0
    ROA2=abs(YA2)
end
```

```
if ROA1>ROA2
    YA=YA2
end
if ROA2>ROA1
    YA=YA1
end
% Step 6b: Compute YB from YA
YB=(YA+4*beta)/(1-beta*YA),
YB2=(YL*YA-1)/(YA+YL),
% Step 7: Compute electrical length of the line transformer Teta_T
if YA<0
Teta_T=-atand(1/YA)
end
if YA>0
    Teta_T=-atand(1/YA)+180,
end
% Step 8: Compute the electrical lengths Teta_0A and Teta_0B
                       as in (4.29)
if YA<0
Teta_0A=atand(2/YA)+180,
end
if YA>0
 Teta_0A=atand(2/YA),
end
\% Teta_0A=atand(2/YA)+kA*pi>0
\% Teta_0B=atand(2/YB)+kB*pi>0
if YB<0
Teta_0B=atand(2/YB)+180,
end
if YB>0
 Teta_0B=atand(1/YB),
end
%
% Step 9: Compute the mismatched lengths Teta_(T0,1) and Teta_(T0,2) of
%the centerline as in (4.30)
% Teta_T01=sqrt(Teta_0A*Teta_0B)
Teta_T01=sqrt(Teta_0A*Teta_0B),
Teta_T02=(Teta_0A+Teta_0B)/2,
% Select the short line length:
if Teta_T01>Teta_T02
    Teta_T0=Teta_T02
end
if Teta_T02>Teta_T01
    Teta_T0=Teta_T01
end
%------------------------------------------------------------
% In order to use our analysis tool, compute the normalized delay
                       lengths
% as
tau_n=pi*Teta_T0/w0/180
tau_nT=pi*Teta_T/w0/180
%------------------------------------------------------------

end; % End of positive Discriminant loop
```

```
%---------------------------------------------------------------
w1=0.9; w2=1.1; N=10001;
DW=(w2-w1)/(N-1);
w=w1;
%---------------------------------------------------------------
W=zeros(1,N); DEL_FIA=zeros(1,N);
RO21A_Ar=zeros(1,N); RO21B_Ar=zeros(1,N);
FI21A_Ar=zeros(1,N); FI21B_Ar=zeros(1,N);
for i=1:N
    W(i)=w*f0a;
    YL=w*CD;
    muT=tan(w*tau_nT);
    YB=(YL+muT)/(1-YL*muT);
    YA=-1/muT;
[ RO21B,FI21B,RO11B,FI11B ] = SPAR_Parallel_Loaded_TRL(w, w0, YB,
    tau_n);
[ RO21A,FI21A,RO11A,FI11A ] = SPAR_Parallel_Loaded_TRL(w, w0, YA,
    tau_n);
FI21A_Ar(i)=FI21A; FI21B_Ar(i)=FI21B;
RO21A_Ar(i)=20*log10(RO21A); RO21B_Ar(i)=20*log10(RO21B);
DEL_FIA(i)=FI21B-FI21A;
%
    w=w+DW;
end
%
figure
plot(W,DEL_FIA,W,FI21A_Ar,W,FI21B_Ar)
xlabel('Normalized Angular Frequency w')
ylabel('Phase Shift DEL-FI=FI21B-FI21A')
title ('Phase Performance of PLL-DPS with TRL Transformer: Teta-T')
legend('DEL-FI','FI21A','FI21B')
%
figure
plot(W,RO21A_Ar,W,RO21B_Ar)
xlabel('Normalized Angular Frequency w')
ylabel('TPG of State-A and State-B in dB')
title ('Gain Performance of PLL-DPS with Transfermer Teta-T')
legend('TPG for State-A','TPG for State-B')
%---------------------------------------------------------------
 if DELY<0
    Attention='Discriminant is negative change CDa or w0'
end
```

Program List 4.6. Main_PLL_DPS_Example_4_4.m

```
% Main_PLL_DPS_Example_4_4.m
clc
close all
clear
%---------------------------------------------------------------
%Design of PLL-DPS with Parallel Resonant Circuits
% This algorithm designs a PLL-DPS with parallel resonant circuits.
```

```
%In this circuit, we employed PIN diodes as switching elements
%with reverse biased capacitors CD.
% However, the same configuration can be implemented using MOS devices
% perhaps as MMIC.
% Inputs:
% f0a: Actual Center Frequency (ACF) of the designy
% w0: Normalized Angular Frequency (NAF) of the design
% Del_Teta: Digital Phase Shift at w0
% CD: Reverese biase capacitance
% Computational Steps:
% -----------------------------------------------------------------
% Inputs:
R=50;
f0a=3e9
w0=1
Del_Teta=-22.5,
CDa=1006e-15,
% Step 1a: Compute the major design parameter
beta_Y=tand(Del_Teta/2),
% Step 1b: Compute the normalized value of CD
w0a=2*pi*f0a,
CD=w0a*R*CDa,
% Step 2: Compute the effective capacitor C of State-A
C=-2*(beta_Y)/w0,
% Step 3: Compute Normalized value of the auxiliary or
%  tuning capacitor CA
 CA=(C+sqrt(C*C+4*CD*C))/2,
% Step 4: Compute the actual value of CA
 CAa=CA/2/pi/f0a/R,
% Step 5: Compute the switch capacitor C_T
 CT=CA*CD/(CA+CD),
% Step 6a: Compute the normalized value of the shunt inductor LA from
% the resonance condition of State-B
LA=1/w0/w0/CT,
% Step 6b: Compute the actual value of LALAa=R*LA/2/pi/f0a,
% Step 7: Compute the length of the centralline Teta_T0
 Teta_T0=atand(2/w0/C)
 if Teta_T0<0
     Teta_T0=Teta_T0+180
 end
%
% -----------------------------------------------------------------
w1=0.5; w2=2.1; N=10001;
DW=(w2-w1)/(N-1);
w=w1;
tau_n=pi*Teta_T0/w0/180
%-----------------------------------------------------------------
W=zeros(1,N); DEL_FIA=zeros(1,N);
RO21A_Ar=zeros(1,N); RO21B_Ar=zeros(1,N);
FI21A_Ar=zeros(1,N); FI21B_Ar=zeros(1,N);
for i=1:N
   W(i)=w*f0a;
YA=(w*w*LA*CA-1)/w/w/LA;
```

```
YB=(w*w*LA*CT-1)/w/w/LA;
[ RO21B,FI21B,RO11B,FI11B ] = SPAR_Parallel_Loaded_TRL(w, w0, YB,
    tau_n);
[ RO21A,FI21A,RO11A,FI11A ] = SPAR_Parallel_Loaded_TRL(w, w0, YA,
    tau_n);
FI21A_Ar(i)=FI21A; FI21B_Ar(i)=FI21B;
RO21A_Ar(i)=20*log10(RO21A); RO21B_Ar(i)=20*log10(RO21B);
DEL_FIA(i)=FI21B-FI21A;
%
    w=w+DW;
end
%
figure
plot(W,DEL_FIA,W,FI21A_Ar,W,FI21B_Ar)
xlabel('Normalized Angular Frequency w')
ylabel('Phase Shift DEL-FI=FI21B-FI21A')
title ('Phase Performance of PLL-DPS with L//C Shunt Loads')
legend('DEL-FI','FI21A','FI21B')
%
figure
plot(W,RO21A_Ar,W,RO21B_Ar)
xlabel('Normalized Angular Frequency w')
ylabel('TPG of State-A and State-B in dB')
title ('Gain Performance of LL-DPS')
legend ('TPG for State-A','TPG for State-B')
```

Program List 4.7. Main_PLL_DPS_Example_4_5.m

```
% Main_PLL_DPS_Example_4_5.m
clc
close all
clear
%-------------------------------------------------------------
%Design of Perfectly Matched PLL-DPS with effective loading inductor L
% in State-B
% This algorithm designs a Perfectly Matched PLL-DPS with loaded with
% effective inductor in State-B.
% In this circuit, we employed PIN diodes as switching elements
% with reverse biased capacitors CD.
% However, the same configuration can be implemented using MOS devices
% perhaps as MMIC.
% Inputs:
% f0a: Actual Center Frequency (ACF) of the design
% w0: Normalized Angular Frequency (NAF) of the design
% Del_Teta: Digital Phase Shift at w0 which must be a negative quantity
% CD: Reverese biase capacitance
% Computational Steps:
% -------------------------------------------------------------
% Inputs:
R=50;
f0a=3e9
```

```
w0=1
Del_Teta=-22.5,
CDa=1006e-15,
% Step 1a: Compute the major design parameter
beta_Y=tand(Del_Teta/2),
% Step 1b: Compute the normalized value of CD
w0a=2*pi*f0a,
CD=w0a*R*CDa,
% Step 2: Compute the effective inductor L of State-B
L=-1/beta_Y/w0/2,
if L<0
    L=-L
end
% Step 3: Compute Normalized value of the auxiliary or tuning
%   capacitor CA
Discriminant=1+4*w0*w0*L*CD
CA=(1+sqrt(Discriminant))/2/w0/w0/L,
% Step 4: Compute the actual value of CA
CAa=CA/2/pi/f0a/R,
% Step 5: Compute the switch capacitor C_T
CT=CA*CD/(CA+CD),
% Step 6a: Compute the normalized value of the shunt inductor LA from
% the resonance condition of State-B
LA=1/w0/w0/CA,
% Step 6b: Compute the actual value of LA
LAa=R*LA/2/pi/f0a,
% Step 7: Compute the length of the central line Teta_T0
% Teta_T0=-atand(1/beta_Y)
Teta_T0=-atand(2*w0*L)+180
% -------------------------------------------------------------------
w1=0.5; w2=2.1; N=10001;
DW=(w2-w1)/(N-1);
w=w1;
tau_n=pi*Teta_T0/w0/180
% -------------------------------------------------------------------
W=zeros(1,N); DEL_FIA=zeros(1,N);
RO21A_Ar=zeros(1,N); RO21B_Ar=zeros(1,N);
FI21A_Ar=zeros(1,N); FI21B_Ar=zeros(1,N);
for i=1:N
    W(i)=w*f0a;
YA=(w*w*LA*CA-1)/w/LA;
YB=(w*w*LA*CT-1)/w/LA;
[ RO21B,FI21B,RO11B,FI11B ] = SPAR_Parallel_Loaded_TRL(w, w0, YB,
    tau_n);
[ RO21A,FI21A,RO11A,FI11A ] = SPAR_Parallel_Loaded_TRL(w, w0, YA,
    tau_n);
FI21A_Ar(i)=FI21A; FI21B_Ar(i)=FI21B;
RO21A_Ar(i)=20*log10(RO21A); RO21B_Ar(i)=20*log10(RO21B);
DEL_FIA(i)=FI21B-FI21A;
%
    w=w+DW;
end
%
```

```
figure
plot(W,DEL_FIA,W,FI21A_Ar,W,FI21B_Ar)
xlabel('Normalized Angular Frequency w')
ylabel('Phase Shift DEL-FI=FI21B-FI21A')
title ('Phase Performance of PLL-DPS with L//C Shunt Loads')
legend('DEL-FI','FI21A','FI21B')

figure
plot(W,RO21A_Ar,W,RO21B_Ar)
xlabel('Normalized Angular Frequency w')
ylabel('TPG of State-A and State-B in dB')
title ('Gain Performance of LL-DPS')
legend('TPG for State-A','TPG for State-B')
```

References

[1] B. Yarman, Design of ultra wideband matching networks via simplified real frequency technique, New Jersey, USA: Springer, 2008.

[2] B. S. Yarman, "New Circuit Configurations for Designing 0–180 degree Digital Phase Shifters," IEE Proceeding H, Vol. 134, pp. 253–260, 1987.

[3] B. S. Yarman, Design of Ultra-Wideband Antenna Matching Networks via Simplified Real Frequency Technique, New Jersey: Springer, 2008.

[4] H. J. Carlin and P. Civaleri, Design of Broadband Networks, Taylor and Franshis CRC Press Inc., 1998.

5

Symmetric T/PI-Sections as Phase Shifters

Summary

Element values of a symmetric LC-Ladder can be arranged in such a way that at a specified single frequency f_0 or equivalent angular frequency $\omega_0 = 2\pi f_0$, it yields a desired phase shift $\theta = \varphi_{21}(\omega_0)$. In this representation, $\varphi_{21}(\omega_0)$ designates the phase of the transfer scattering parameter $S_{21}(j\omega_0)$.

In this section, the phase shifting properties of symmetric LC T- and PI-Sections are investigated.

First, the symmetric low-pass/high-pass T-Sections are considered as lossless two-ports and their real normalized scattering parameters are derived in Laplace domain $p = \sigma + j\omega$ and on the real frequency axis $p = j\omega$. Then, explicit equations are given to determine the component values of the symmetric T- and PI-Sections for a specified phase shift at a selected normalized angular frequency ω_0. Similarly, symmetric low-pass and high-pass PI sections are investigated as phase shifting cells.

Several examples are presented to exhibit the usage of the explicit design formulas. All the examples are developed on MatLab environment [1].

The content of this chapter is based on mostly our previous work published in the current literature. Some of these references are given in [2–6].

5.1 Scattering Parameters of a Symmetric T-Section

Referring to Figure 5.1, a symmetric T-Section is considered as a two-port with series arm impedance Z and a shunt arm admittance Y.

In order to drive scattering parameters of the Symmetric T-Section, first of all, we define voltage-based input and output incident a_j and reflected b_j wave parameters such that

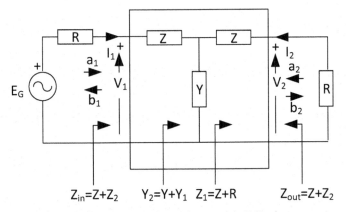

Figure 5.1 Generic form of symmetric T-Section.

$$a_j = \frac{1}{2}\left[\frac{V_j}{\sqrt{R}} + \sqrt{R}I_j\right]; \qquad j = 1, 2 \qquad (5.1a)$$

$$b_j = \frac{1}{2}\left[\frac{V_j}{\sqrt{R}} - \sqrt{R}I_j\right]; \qquad j = 1, 2 \qquad (5.1b)$$

where

$$v_j = \frac{V_j}{\sqrt{R}}; \qquad j = 1, 2 \qquad (5.1c)$$

$$i_j = \sqrt{R}I_j; \qquad j = 1, 2 \qquad (5.1d)$$

are the normalized voltage and currents, respectively.
or

$$a_j = \frac{1}{2}[v_j + i_j]; \qquad j = 1, 2 \qquad (5.2a)$$

$$b_j = \frac{1}{2}[v_j - i_j]; \qquad j = 1, 2 \qquad (5.2b)$$

Then, the scattering parameters of the symmetric T-Section are given by

$$S_{11} = \frac{b_1}{a_1}\Big|_{a_2=0} = \frac{v_1 - i_1}{v_1 + i_1} = \frac{Z_{in} - R}{Z_{in} + R} = \frac{z_{in} - 1}{z_{in} + 1} = \frac{z_{out} - 1}{z_{out} + 1} = S_{22}$$

$$S_{21} = \frac{b_2}{a_1}\Big|_{a_2=0} = \frac{v_2 - i_2}{v_1 + i_1}$$

where z_{in} is the normalized input impedance, z_{out} is the normalized output impedance, which are given by

$$z_{in} = \frac{Z_{in}}{R} = z_{out} = \frac{Z_{out}}{R}$$

In Figure 5.1, impedance Z_1 is given by

$$Z_1 = Z + R$$

The shunt admittance Y_2 is

$$Y_2 = Y + \frac{1}{Z_1} = Y + \frac{1}{Z + R} = \frac{ZY + RY + 1}{Z + R}$$

or

$$Z_2 = \frac{1}{Y_2} = \frac{Z + R}{ZY + RY + 1}$$

On the other hand,

$$Z_{in} = Z + Z_2 = Z + \frac{Z + R}{ZY + RY + 1}$$

or

$$Z_{in} = \frac{Z^2Y + RZY + 2Z + R}{ZY + RY + 1} = \frac{N}{D} \tag{5.3a}$$

and

$$Z_2 = \frac{1}{Y_2} = \frac{Z + R}{D} \tag{5.3b}$$

where

$$N = Z^2Y + RZY + 2Z + R \tag{5.3c}$$
$$D = ZY + RY + 1 \tag{5.3d}$$

Input reflectance S_{11} is given by

$$S_{11} = \frac{Z_{in} - R}{Z_{in} + R} = \frac{RD - N}{RD + N} = \frac{-Z^2Y - RZY - 2Z - R + RZY + R^2Y + R}{Z^2Y + RZY + 2Z + R + RZY + R^2Y + R}$$
$$= \frac{-Z^2Y - 2Z + R^2Y}{Z^2Y + 2RZY + 2Z + R^2Y + 2R}$$

In short,

$$S_{11} = \frac{\left(R^2 - Z^2\right) Y - 2Z}{Z^2 Y + 2RZY + 2Z + R^2 Y + 2R} \tag{5.4a}$$

As far as phase shifting property of the T-Section is concerned, at a specified frequency, input reflectance is forced to be zero. In this case, (5.3) demands that

$$\left(Z^2 - R^2\right) Y + 2Z = 0$$

or the perfect match condition of $S_{11} = 0$ yileds

$$Y = \frac{2Z}{R^2 - Z^2} \tag{5.4b}$$

Let us now derive S_{21}.

It is noted that impedance normalization is carried out with respect to termination resistors R. In this regard, normalized termination must be unity. Hence, setting $R = 1$, normalized driving excitation $e_g = \frac{E_G}{\sqrt{R}}$ is found as

$$e_g = v_1 + i_1 = 2a_1$$

or

$$a_1 = \frac{e_g}{2} \tag{5.5}$$

The normalized input current is derived as

$$i_1 = \frac{e_g}{1 + z_{in}}$$

where z_{in} is the normalized input impedance of port-1 and it is defined as $z_{in} = Z_{in}/R$.

Reflected wave of the output port is given by

$$b_2 = \frac{1}{2} \left[v_2 - i_2\right]$$

where the actual current $I_2 = -V_2/R$ or the normalized current $i_2 = -v_2$. Therefore, $b_2 = v_2$. In this case, transfer scattering parameter is given by

$$S_{21} = 2 \left(\frac{v_2}{e_g}\right) \tag{5.6}$$

Let the normalized voltage on Z_2 be v_y. Then,

$$v_y = z_2 i_1 = z_2 \left[\frac{e_g}{1 + z_{in}} \right]$$

where

$$z_{in} = \frac{Z_{in}}{R} = \frac{n}{d} \tag{5.7a}$$

and

$$z_{in} = \frac{Z_{in}}{R} = \frac{n}{d} \tag{5.7b}$$

are the normalized input impedance and equivalent shunt impedance, respectively. By (5.3b), z_2 is given as

$$z_2 = \frac{(z+1)}{(zy + y + 1)} = \frac{z+1}{d} \tag{5.7c}$$

In the above equations, numerator n, denonimator d, and their sum $n + d$ are given by

$$n = z^2 y + zy + 2z + 1 \tag{5.7d}$$
$$d = zy + y + 1 \tag{5.7e}$$
$$n + d = z^2 y + 2zy + 2z + y + 1 \tag{5.7f}$$

At the output port, normalized current i_2 is given by

$$-i_2 = \frac{v_y}{z+1} = \left[\frac{z_2}{(1 + z_{in})(z+1)} \right] e_g; \quad z = \frac{Z}{R}$$

In the above representations, the normalized shunt admittance y is specified by $y = YR$.

Based on the above nomenclature, shunt arm voltage v_y is found as

$$v_2 = \left(\frac{v_y}{z+1} \right) (1) = \left[\frac{z_2}{(1 + z_{in})(z+1)} \right] e_g = \left(\frac{z+1}{d} \right) \left(\frac{d}{d+n} \right) \left(\frac{1}{z+1} \right) e_g$$

or

$$v_2 = \frac{e_g}{d+n} \tag{5.8}$$

Hence, the transfer scattering parameter S_{21} is found as

$$S_{21} = 2\left(\frac{v_2}{e_g}\right) = \frac{2}{n+d} = \frac{2}{z^2 y + 2zy + 2z + y + 2} \qquad (5.9a)$$

Using normalized impedances and admittances, S_{11} of (5.4a) is simplified as

$$S_{11} = \frac{\left(1 - z^2\right) y - 2z}{z^2 y + 2zy + 2z + y + 2} \qquad (5.9b)$$

In terms of normalized impedance z and admittance y, the perfect transmission condition becomes

$$y = \frac{2z}{1 - z^2} \qquad (5.9c)$$

The above condition can only be satisfied at a single frequency using passive circuit components. For example, if $z = j\omega_0 L$, then $y = j\omega_0 C$.

Using (5.10c) in (5.10a), one can end up with the phase condition under perfect match.

In this case, at the specified frequency ω_0, the denominator term of (5.10a) becomes

$$D_{21} = z^2 y + 2zy + 2z + y + 2 = \frac{1}{1 - z^2}\{2z^3 + 4z^2 + 2z - 2z^3$$
$$+ 2z - 2z^2 + 2\}$$

or

$$D_{21} = \frac{1}{1 - z^2}\left(2z^2 + 4z + 2\right) = \frac{2}{1 - z^2}\left(z^2 + 2z + 1\right)$$
$$= \left(\frac{2}{1 - z^2}\right)(1 + z)^2$$

Then, at the given frequency ω_0, S_{21} of (5.10a) becomes

$$S_{21}(j\omega_0) = \frac{1 - z(j\omega_0)}{1 + z(j\omega_0)} \qquad (5.10)$$

5.2 A Low-pass Symmetric T-Section

A low-pass symmetric T-Section is obtained by setting normalized series arm impedances as $z = j\omega L$ and the shunt admittance as $y = j\omega C$ as shown in Figure 5.2.

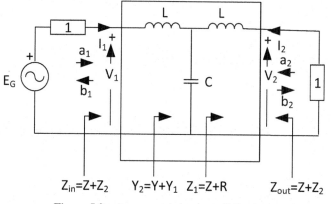

Figure 5.2 A symmetric low-pass T-Section.

Using the normalized inductor L and capacitor C, Equation (5.10) is modified.

Input reflectance $S_{11}(j\omega)$ becomes

$$S_{11} = -\frac{\left(1 + \omega^2 L^2\right) j\omega C - 2j\omega L}{2\left(1 - \omega^2 LC\right) + j\omega\left(2L + C - \omega^2 L^2 C\right)} \tag{5.11a}$$

Transfer scattering parameter S_{21} is

$$S_{21} = \frac{2}{2\left(1 - \omega^2 LC\right) + j\omega\left(2L + C - \omega^2 L^2 C\right)} = \frac{2}{d_r + jd_x} = \rho_{21} e^{j\varphi_{21}(\omega)} \tag{5.11b}$$

where

$$d_r = 2\left(1 - \omega^2 LC\right) \tag{5.11c}$$

and

$$d_x(\omega) = \omega\left(2L + C - \omega^2 L^2 C\right) \tag{5.11d}$$

$$\rho_{21} = \frac{2}{\sqrt{d_r^2 + d_x^2}}$$

$$\varphi_{21}(\omega) = -\tan^{-1}\left(\frac{d_x}{d_r}\right) = -\tan^{-1}\left[\frac{\omega\left(2L + C - \omega^2 L^2 C\right)}{2\left(1 - \omega^2 LC\right)}\right]$$

At a specified angular frequency, let $\theta = \varphi_{21}(\omega_0)$ be a desired phase. Then,

$$\mu = \tan(\theta) = -\frac{d_x}{d_R} = \frac{\omega\left(2L + C - \omega^2 L^2 C\right)}{2\left[\omega^2 LC - 1\right]}\Big|_{\omega=\omega_0} \quad (5.12)$$

At ω_0, perfect transmission condition of (5.9c) yields that

$$C = \frac{2L}{1 + \omega_0^2 L^2} \quad (5.13a)$$

In this case, μ is calculated in terms of the series arm inductor L as follows

$$
\begin{aligned}
d_x &= \omega\left(2L + \frac{2L}{1 + \omega_0^2 L^2} - \omega^2 L^2 \frac{2L}{1 + \omega_0^2 L^2}\right) \\
&= \frac{\omega}{1 + \omega_0^2 L^2}\left[2L + 2\omega_0^2 L^3 + 2L - 2\omega^2 L^3\right]
\end{aligned}
$$

or

$$d_x = \frac{4\omega L}{1 + \omega_0^2 L^2} \quad (5.13b)$$

Similarly,

$$d_r = 2\left(\frac{1 - \omega^2 L^2}{1 + \omega_0^2 L^2}\right) \quad (5.13c)$$

Then,

$$\mu = \tan(\theta) = \frac{2\omega_0 L}{\omega_0^2 L^2 - 1}$$

or

$$\mu \omega_0^2 L^2 - 2\omega_0 L - \mu = 0 \quad (5.13d)$$

Solution of (5.13d) results in L

$$L_{1,2} = \frac{\omega_0 \mp \sqrt{\omega_0^2 + \omega_0^2 \mu^2}}{\mu \omega_0^2} = \frac{1 \mp \sqrt{1 + \mu^2}}{\omega_0 \mu} \quad (5.13e)$$

$$= \frac{1}{\omega_0}\left(\frac{1}{\mu} \mp \sqrt{1 + \frac{1}{\mu^2}}\right) \quad (5.13f)$$

It should be noted that in the above formulation, for a "low-pass T-Section", the phase $\theta = \varphi_{21}(\omega_0)$ is a negative quantity and it varies between $0°$ and $-90°$. Then, the phase parameter μ is always negative and $\sqrt{1 + \mu^2}$ is always greater than 1. Therefore, in order to force the series inductor L to be positive, one must select the negative sign in (5.13e) or in (5.13f). Hence, the series arm inductor is given as

$$L = \frac{1 - \sqrt{1 + \mu^2}}{\omega_0 \mu} = \frac{1}{\omega_0} \left(\frac{1}{\mu} - \sqrt{1 + \frac{1}{\mu^2}} \right) > 0 \qquad ; -\infty \leq \mu \leq 0$$

(5.13g)

The phase parameter $1/\mu = 1/\tan(\theta)$ can be evaluated as follows. Let a new phase parameter be defined as

$$\eta = \tan(90° - \theta) \leq 0; \tag{5.14a}$$

$$\eta = \frac{\tan(90°) - \tan(\theta)}{1 + \tan(90°) . \tan(\theta)}$$

where $\tan(90°) = \infty$. Therefore, in the limit case,

$$\eta = \tan(90° - \theta) = \frac{1}{\tan(\theta)} = \frac{1}{\mu} \tag{5.14b}$$

Hence, the series arm inductance value is found as

$$L_{1,2} = \frac{1}{\omega_0} \left(\eta \mp \sqrt{\eta^2 + 1} \right) > 0 \tag{5.14c}$$

In (5.15c), series arm inductor L must be positive. Therefore, one must select the positive sign yielding a single value for the inductor

$$L = \frac{1}{\omega_0} \left(\eta + \sqrt{\eta^2 + 1} \right) = \left(\frac{1}{\omega_0} \right) \tan \left(\frac{|\theta|}{2} \right) > 0 \tag{5.14d}$$

Remark 1: For an LC low-pass symmetric T-section, the phase variation of $S_{21}(j\omega)$ is in the lower half of the phase plane.[1] Therefore, it may be appropriate to take the phase θ as a negative quantity.

Remark 2: Based on the above suggestion, for a typical low-pass LC–T section, phase varies between $0°$ and $-180°$ (i.e., $-180° < \theta < 0°$).

---

[1]In the context of phase shifter design, a phase plane with a $360°$ complete circle and with a diameter R is referred to as the "phase plane-PP". A central phase θ is measured in degrees or in radians, which is formed with selected points A and B on a piece of arc connected to the center point of the circle.

Remark 3: Depending on the value of the selected phase θ, $\mu = \tan(\theta)$ could be negative or positive. On the other hand, for $-90° < \theta < 0°$, phase-dependent parameter μ is always negative. In this case, the value of the series inductor must be selected as in (5.13g). However, if the phase varies as $-180° < \theta < -90°$, then phase-dependent parameter $\mu = \tan(\theta)$ becomes positive. In this case, inductor value L must be selected as

$$L = \frac{1 + \sqrt{1 + \mu^2}}{\omega_0 \mu} = \frac{1}{\omega_0}\left(\frac{1}{\mu} + \sqrt{1 + \frac{1}{\mu^2}}\right)$$

$$= \left(\frac{1}{\omega_0}\right)\tan\left(\frac{|\theta|}{2}\right) > 0; \quad -\infty < \mu < 0$$

Remark 4: if $\theta = -90°$, then inductor L must be set to $L = \frac{1}{\omega_0}$.

No matter, how we compute the series inductor L, perfect match condition ($S_{11} = 0$ or $|S_{21}| = 1$) requires the shunt capacitor C as specified by (5.13a) such that

$$C = \frac{2L}{1 + \omega_0^2 L^2}$$

Remark 5: Alternative definition of phase-dependent parameter $\eta = \tan(90° - \theta)$ simplifies the computations. Depending on the value of the phase θ, η could be positive or negative.

No matter what the value of η is, (5.14d) always results in positive value for the series inductor L such that

$$L = \frac{1}{\omega_0}\left(\eta + \sqrt{\eta^2 + 1}\right) > 0 \; \forall \eta$$

Therefore, it is handy to define the phase-dependent parameter η as $\eta = \tan(90° - \theta)$ and compute the series inductor L as in (5.14d) and generate the shunt capacitor as in (5.13a).

Hence, in summary, phase shifting design table for the low-pass symmetric T-Section (LPST) is given as in Table 5.1. Note that the phase shift θ must be specified between $-180°$ and $0°$ and it cannot take the exact values of $-180°$ and $0°$.

Let us now run some examples to utilize the above formulas.

Example 5.1:
(a) Using $\mu = \tan d(\theta)$-based formulas, design a low-pass symmetric T-Section that yields the phase $\theta = \varphi_{21}(\omega_0) = -45°$ at $\omega_0 = 1$ under perfect match condition (i.e., $S_{11}(\omega_0) = 0$ or equivalently $|S_{21}(j\omega_0)|_{\omega_0=1} = 1$).

Table 5.1 For a specified phase shift θ, design equations for a low-pass symmetric section (LPST)

$$\eta = \tan(90° - \theta)$$

$$L = \frac{1}{\omega_0}\left(\eta + \sqrt{\eta^2 + 1}\right) > 0$$

$$C = \frac{2L}{1 + \omega_0 L} > 0$$

Note that the phase shift θ must be specified between $-180°$ and $0°$ and it cannot take the exact values of $-180°$ and $0°$.

(b) In order to analyze the amplitude and phase variation of the transfer scattering parameter S_{21}, develop a MatLab program to generate the element values as well as the phase and amplitude variation of the transfer scattering parameters and plot the results.

Solution

(a) Using (5.13g), the series arm inductor L is computed as

$$L = \frac{1 - \sqrt{1 + \mu^2}}{\omega_0 \mu} = \frac{1 - \sqrt{1 + \tan^2(-45°)}}{1.\tan(-45°)} = \frac{1 - \sqrt{2}}{1.(-1)} = 0.4142$$

The shunt arm capacitor C is calculated using (5.13a)

$$C = \frac{2L}{1 + \omega_0^2 L^2} = 0.7071$$

(b) In order to examine the phase and the amplitude variations of the transfer scattering parameter S_{21}, we developed a Main MatLab program called

$$\textit{"Main\_Lowpass\_T\_with\_Classical\_Formulas.m"}$$

This program, first, evaluates (5.13g) and (5.13a) for a selected negative phase θ and angular frequency ω_0, and then evaluates the scattering parameters as they are defined by (5.9).

S-Parameters of the symmetric low-pass T is computed by the function

$$\textit{"[S11, S21, RO11, F11, RO21, F21]} = \textit{S\_Par\_Lowpass\_T}\,(w, L, C)\textit{"}$$

The phase variation of the low-pass symmetric-T is depicted in Figure 5.3a. The amplitude variation is shown in Figure 5.3b.

(a)   **Phase variation of a "Lowpass T-Section" with classical formulation**

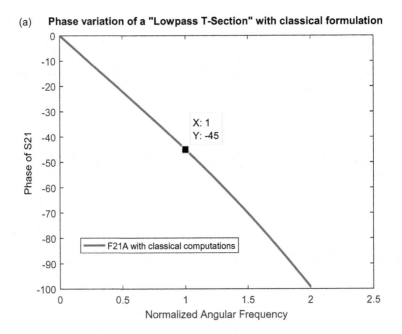

(b)   **Amplitude variation of a "Lowpass T-Section" with classical formulation**

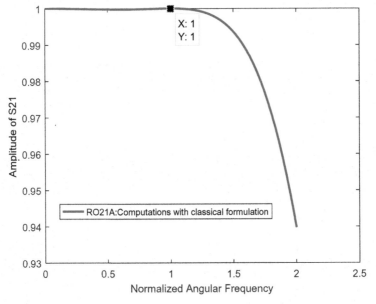

Figure 5.3   A symmetric low-pass T-section.

Program list of the main program and function S_Par is given as follows

Program List 5.1: Main_Lowpass_T_with_Classical_Formulas.m

```
% Main_Lowpass_T_with_Classical_Formulas.m
% December 7, 2018
% Developed by BS Yarman, Vanikoy, Istanbul, Turkey
% It should be noted that the formulas used in this program is valid
    for the values of teta between 0 and 90 degree.
% Exact 90 is not possible
clc, clear
close all
teta=input('Enter negative value for the phase
shift teta=')
while teta==0
 stop= 'Attention teta=0. Phase can never be zero degree. Therefore,
    change teta and re-run the program again.'
    break
end
 while teta==-180
    stop= 'Attention teta=-180. Phase can never be -180 degree.
    Therefore,  change teta and re-run the program again.'
    break
end
while teta>0
stop1='Attention: teta is positive. For a Lowpass Symmterric
T-Section  teta must be negative quantity.'
stop2='Please enter a negative value for teta and re-run the program'
    break
end
% Phase teta is proper. Then, start Computations
if teta>-180
if teta<0
w0=1;
mu=tand(teta)
eta=tand(90-teta)
[ L1,L2,L,C] = mu_Based_Components_of_Lowpass_T
( w0,teta )
%   L=(1-sqrt(1+mu*mu))/w0/mu;
%   C=2*L/(1+w0*w0*L*L);
% -------------------------------------------------
[ La,Ca ] = Components_of_Lowpass_T_Section
( w0, teta )
Error_L=La-L
Error_C=Ca-C
% -------------------------------------------------
w=0;N=1000;w1=0;w2=2;DW=(w2-w1)/N;
FRI(1:(N+1))=zeros;
for j=1:N+1
WA(j)=w;
% -------------------------------------------------
[ S11,S21,RO11,F11,RO21,F21 ] = S_Par_Lowpass_T
( w,La,Ca );
```

```
  F21A(j)=F21;
  RO21A(j)=RO21;
% ------------------------------------------------
    w=w+DW;
end
% ------------------------------------------------
% Phase of S21
figure
plot(WA,F21A)
title('Phase variation of a "Lowpass T-Section" with classical
    formulation')
legend('F21A with classical computations')
xlabel('Normalized Angular Frequency')
ylabel('Phase of S21')
% Amplitude of S21
figure
plot(WA,RO21A)
title('Amplitude variation of a "Lowpass T-Section" with classical
    formulation')
legend('RO21A:Computations with classical formulation')
xlabel('Normalized Angular Frequency')
ylabel('Amplitude of S21')
% ------------------------------------------------
end
end
```

Program List 5.2: function Components_of_Lowpass_T_Section

```
function [ L,C ] = Components_of_Lowpass_T_Section
( w0, teta )
% This function generates the component values for a lowpass T-Section
%   Developed by BS Yarman: December 6, 2018; Vanikoy, Istanbul, Turkey
% Developed by BS Yarman. December 7, 2018, Vanikoy, Istanbul, Turkey
% Inputs:
%       teta: Negative value of teta
%       w0: Normalized angular frequency
% Outputs:
%       L: Series arm inductor
%       C: Shunt arm capacitor
% ------------------------------------------------
eta=tand(90-teta);
L=(eta+sqrt(1+eta*eta))/w0;
C=2*L/(1+w0*w0*L*L);
End
```

Program List 5.3: function S_Par_Lowpass_T (w,L,C)

```
function [ S11,S21,RO11,F11,RO21,F21 ] = S_Par_
Lowpass_T ( w,L,C )
%This function generates the S-Parameters of Lowpass T Section from the
    computed
% series arm inductor L and the shunt capacitor C
%   Developed by BS Yarman: December 7, 2018, Vanikoy, Istanbul
```

```
%  -----------------------------------------------------
dr=2*(1-w*w*L*C);
dx=w*(2*L+C-w*w*L*L*C);
j=sqrt(-1);
D=dr+j*dx;
N11= j*((1+w*w*L*L)*w*C-2*w*L);
S11=N11/D; R11=real(S11); X11=imag(S11);F11=atan2d
(X11,R11);
S21=2/D;   R21=real(S21); X21=imag(S21);F21=atan2d
(X21,R21);
RO11=abs(S11);
RO21=abs(S21);
End
```

Program List 5.4: For a Lowpass Symmetrical-T Section $\mu = \tan(\theta)$
based component values

```
function [ L1,L2,La_mu,Ca_mu ] = mu_Based_
Components_of_Lowpass_T( w0,teta )
% In this function component values of a symmetric Lowpass T is
    computed
%   Input phase teta is defined as a negative angle mu=tand(teta);
% Solution of equation (5.13e) for inductor L(1,2)
L1=(1/w0)*(1/mu + sqrt(1+1/mu^2 ));
L2=(1/w0)*(1/mu - sqrt(1+1/mu^2 ));
%  -----------------------------------------------
if L1>0
     La_mu=L1;
end
if L2>0
     La_mu=L2;
end
if teta==-90
     L1=1/w0; L2=1/w0;
     La_mu=1/w0;
end
%  -----------------------------------------------
Ca_mu=2*La_mu/(1+w0*w0*La_mu*La_mu);
End
```

5.3 Alternative Way to Compute Series Arm Inductor *L*

At a specified angular frequency, under perfect transmission condition, $S_{21}(j\omega_0)$ is derived as in (5.10) such that

$$S_{21}(j\omega_0) = \frac{1 - z(j\omega_0)}{1 + z(j\omega_0)}$$

Setting $z = j\omega_0 L$,

$$S_{21}(j\omega_0) = \frac{1 - (j\omega_0 L)}{1 + (j\omega_0 L)} = 1.e^{j\theta}$$

or

$$\theta = -2\tan^{-1}(\omega_0 L)$$

Since the series arm inductor L must be always positive, it is computed as

$$L = \frac{1}{\omega_0} \tan\left(\frac{|\theta|}{2}\right) \qquad (5.15a)$$

The above formula is much simpler than (5.14). The shunt arm capacitor C is then computed as in (5.13a).

$$C = \frac{2L}{1 + \omega_0^2 L^2} \qquad (5.15b)$$

Eventually, scattering parameters can be derived using (5.11a) and (5.11b) such that

$$S_{11} = S_{22} = -\frac{\left(1 + \omega^2 L^2\right) j\omega C - 2j\omega L}{2\left(1 - \omega^2 LC\right) + j\omega\left(2L + C - \omega^2 L^2 C\right)}$$
$$= \rho_{11} e^{j\varphi_{11}(\omega)} = \rho_{22} e^{j\varphi_{22}(\omega)} \qquad (5.16a)$$

and

$$S_{21} = \frac{2}{2\left(1 - \omega^2 LC\right) + j\omega\left(2L + C - \omega^2 L^2 C\right)} = \frac{2}{d_r + jd_x} = \rho_{21} e^{j\varphi_{21}(\omega)}$$
$$\qquad (5.16b)$$

It should be noted that for a lossless two-port, the phase of the transfer scattering parameters φ_{21} is given by

$$\varphi_{21} = 90 - \frac{\varphi_{11} + \varphi_{22}}{2} \qquad (5.17a)$$

Since $S_{11} = S_{22}$, $\varphi_{11} = \varphi_{22}$ and $\rho_{11} = \rho_{22}$. In this case, (5.17a) takes the following form:

$$\varphi_{21} = 90 - \varphi_{11} \qquad (5.18a)$$
$$\varphi_{11} = 90 - \varphi_{21} \qquad (5.18b)$$

Let us now run an example to verify the above formulas.

Example 5.2:

(a) Run Example 5.1 for $\omega_0 = 0.5$ and $\theta = 75°$ and using the simplified formula given by (5.15a) and (5.15b).

(b) Write a MatLab program to generate phase and amplitude variation of transfer scattering parameter for the above-specified parameters.

Solution

(a) Using (5.15a), the shunt inductor L is given by

$$L = \frac{1}{\omega_0} \tan\left(\frac{|\theta|}{2}\right) = \frac{1}{0.5} \tan\left(\frac{75°}{2}\right) = 1.5347$$

The shunt arm capacitor is given by (5.15b) as

$$C = \frac{2L}{1 + \omega_0^2 L^2} = \frac{2 \times 1.5347}{1 + 0.25 \times (1.5347)^2} = 1.9319$$

(b) For this part of the example, we developed a MatLab program called

"*Main_Alternative_Lowpass_T.m*"

Execution of this program results in the following variations. In Figure 5.4a, phase variation is depicted.

In Figure 5.4b, the amplitude variation is shown.

Example 5.3: In this example, we will exhibit the usage of the design formulas for an extreme case.

Let $\theta = -179°$ at $\omega_0 = 1$. Compute the element values of the low-pass symmetric T-Section and plot the phase performance using the alternative formulas.

Solution

Employing Equation (5.15), the series inductor L and the shunt capacitor C are found as

$$L = \frac{1}{\omega_0} \tan\left(\frac{|\theta|}{2}\right) = 114.5887$$

$$C = \frac{2L}{1 + \omega_0^2 L^2} = 0.0175$$

The phase response of the phase shifter is depicted in Figure 5.5a. In this figure, it is interesting to observe that phase shift over $0.5 < \omega < 0.9$ is

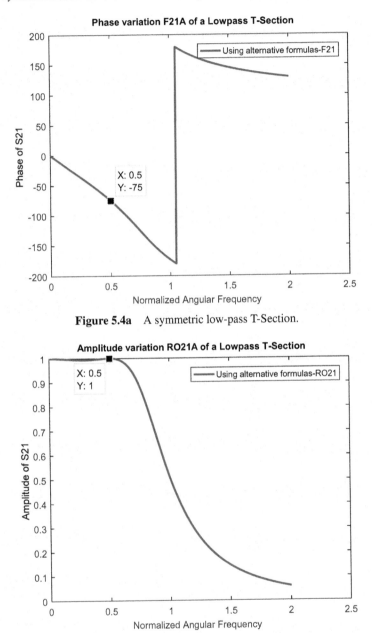

Figure 5.4a   A symmetric low-pass T-Section.

Figure 5.4b   Amplitude response of the symmetric low-pass T-Section of Example 5.2.

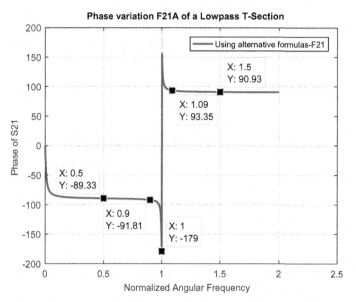

Figure 5.5a Phase response of the symmetric low-pass T-Section of Example 5.3.

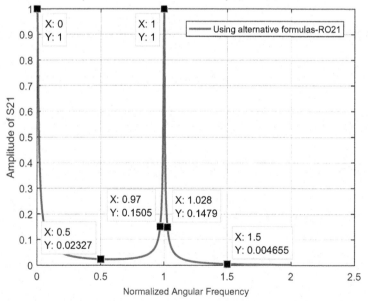

Figure 5.5b Amplitude response of the symmetric low-pass T-Section of Example 5.3.

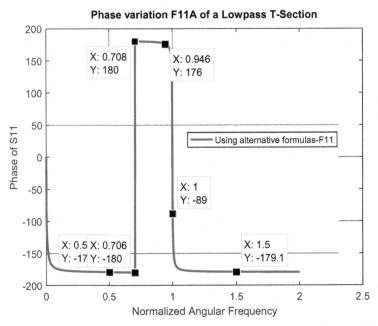

Figure 5.5c   Phase plot of S_{11} of the symmetric low-pass T-Section of Example 5.3.

approximately $-90°$ and almost flat. Similarly, over the frequency band of $1.09 < \omega < 1.5$, the phase is about $+90°$ and almost flat. In this regard, one may consider to utilize this phase shifter as a constant phase shifting unit at $90°$. Unfortunately, this is not the case since the amplitude response is very poor in both frequency intervals $0.5 < \omega < 0.9$ and $1.09 < \omega < 1.5$, as shown in Figure 5.5b.

The amplitude response of the phase shifter is depicted in Figure 5.5b. At the design frequency, $\omega_0 = 1$, $|S_{21}(j\omega)| = 1$. However, as mentioned above, it drops drastically, when we move a little bit away from the design frequency.

All the above results are produced using our program *"Main_Alter native_Lowpass_T.m"* developed on MatLab and it is given by Program List 5.5.

In the existing program, we also plot input reflection coefficient $S_{11} = S_{22}$.

In Figure 5.5c, phase plot of S_{11} is depicted.

A close examination of this figure reveals that $\varphi_{11} = 90 - \varphi_{21} = -89°$ as expected.

Amplitude plot of S_{11} is shown in Figure 5.5d.

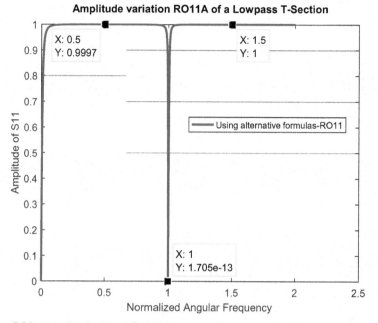

Figure 5.5d   Amplitude plot of S_{11} of the symmetric low-pass T-Section of Example 5.3.

A close examination of Figure 5.5d reveals that at $\omega=1$, $|S\_11\ (j\omega)|^2=1-|S\_21\ (j\omega)|^2=0$ as it should be. Then, it immediately jumps to 1 as expected. The MatLab program "Main_Alternative_Lowpass_T.m" is listed in Program List 5.5.

```
Program List 5.5: Main_Alternative_Lowpass_T.m
% Main_Alternative_Lowpass_T.m
% Developed by BS Yarman
% December 7, 2018, Vanikoy, Istanbul, Turkey
% Computations with alternative formulas
% December 7, 2018
% ----------------------------------------------
clc; close all
% Inputs:
teta=input('T-Section Phase Shift Teta in Degree=')
w0=0.5;
% Alternative way to generate inductor L
L=tand(teta/2)/w0;
C=2*L/(1+w0*w0*L*L);
% ----------------------------------------------
w=0;N=1000;w1=0;w2=2;DW=(w2-w1)/N;
FRI(1:(N+1))=zeros;
for j=1:N+1
    WA(j)=w;
% ----------------------------------------------
```

```
[S11,S21,RO11,F11,RO21,F21 ] = S_Par_Lowpass_T
( w,L,C );
 F21A(j)=F21;
 RO21A(j)=RO21;
% ------------------------------------------------
 F11A(j)=F11;
 RO11A(j)=RO11;
% ------------------------------------------------
 w=w+DW;
end
% ------------------------------------------------
% Phase of S21
figure
plot(WA,F21A)
title('Phase variation F21A of a Lowpass
T-Section')
legend('Using alternative formulas-F21')
xlabel('Normalized Angular Frequency')
ylabel('Phase of S21')
% Amplitude of S21
figure
plot(WA,RO21A)
title('Amplitude variation RO21A of a Lowpass
T-Section')
legend('Using alternative formulas-RO21')
xlabel('Normalized Angular Frequency')
ylabel('Amplitude of S21')
% ------------------------------------------------
figure
plot(WA,F11A)
title('Phase variation F11A of a Lowpass
T-Section')
legend('Using alternative formulas-F11')
xlabel('Normalized Angular Frequency')
ylabel('Phase of S11')
% Amplitude of S11
figure
plot(WA,RO11A)
 title('Amplitude variation RO11A of a Lowpass
 T-Section')
legend('Using alternative formulas-RO11')
xlabel('Normalized Angular Frequency')
ylabel('Amplitude of S11')
```

5.4 Programming with Positive Phase $|\theta| > 0$

In this approach, for a symmetric low-pass LC ladder, the negative quantity φ_{21} is expressed in terms of a positive phase $|\theta|$ such that

$$\varphi_{21} = -|\theta| \leq 0; \ \ 0 \leq \theta \leq 90° \tag{5.19a}$$

In this case, we set

$$\gamma = \tan |\theta| = -\mu \geq 0 \qquad (5.19b)$$

Then, the series arm inductor L is determined in terms of the positive phase parameter γ such that

$$L = \frac{\sqrt{1 + \gamma^2} - 1}{\omega_0 \gamma} = \frac{1}{\omega_0} \left[\sqrt{1 + \frac{1}{\gamma^2}} - \frac{1}{\gamma} \right] \geq 0 \qquad (5.19c)$$

Let $\lambda = \tan(90° - |\theta|)$ be a positive quantity. Then,

$$\lambda = \frac{1}{\gamma} \geq 0 \qquad (5.19d)$$

or

$$L = \frac{1}{\omega_0} \left[\sqrt{\lambda^2 + 1} - \lambda \right] \geq 0 \qquad (5.19e)$$

$$C = \frac{2L}{1 + \omega_0^2 L^2} \qquad (5.19f)$$

This section completes the phase shifting performance analysis of a symmetric low-pass T-Section. In the following section, we analyze the phase shifting performance of a symmetric high-pass T-Section.

5.5 A High-pass Symmetric T-Section

Derivation of the phase shifting formulas for a "High-pass Symmetric T-Section (HPST)" is rather simple using the impedance replacement technique.

In Figure 5.6, a typical high-pass symmetric T-Section is shown.

In the above figure, the series arm the normalized impedance z is defined as

$$z = \frac{1}{j\omega C} \qquad (5.20a)$$

Similarly, on the shunt arm, the normalized admittance y is given by

$$y = \frac{1}{j\omega L} \qquad (5.20b)$$

Hence, under the perfect transmission condition ($S_{11} = 0$) of (5.9c), normalized shunt arm admittance is given by

$$y = \frac{2z}{1 - z^2}$$

or

$$\frac{1}{j\omega_0 L} = \frac{2\left(\frac{1}{j\omega_0 C}\right)}{1 + \frac{1}{\omega_0^2 C^2}}$$

$$\frac{1}{L} = \frac{2\omega_0^2 C}{1 + (\omega_0 C)^2}$$

or the perfect transmission condition yields

$$L = \left(\frac{1}{\omega_0^2}\right)\left[\frac{1 + \omega_0^2 C^2}{2C}\right] \tag{5.20c}$$

On the other hand, transfer scattering parameter S_{21} is given by (5.9b)

$$S_{21} = \frac{2}{z^2 y + 2zy + 2z + y + 2}$$

For a high-pass T-Section, in complex Laplace variable $p = \sigma + j\omega$, the impedance z is given by

$$z = \frac{1}{pC}$$

and the admittance y is

$$y = \frac{1}{pL}$$

Then,

$$S_{21} = \frac{2}{\left(\frac{1}{pC}\right)^2 \left(\frac{1}{PL}\right) + 2\left(\frac{1}{pC}\right)\left(\frac{1}{PL}\right) + 2\left(\frac{1}{pC}\right) + \left(\frac{1}{pL}\right) + 2}$$

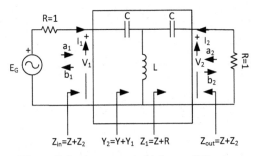

Figure 5.6   A symmetric high-pass T-Section.

or

$$S_{21} = \frac{2p^3 C^2 L}{1 + 2\,(pC) + 2\,(pC)\,(pL) + (pC)\,(pC) + 2\,(pC)\,(pC)\,(pL)}$$

or

$$S_{21} = \frac{2p^3 C^2 L}{2C^2 L p^3 + (2CL + C^2)\,p^2 + 2Cp + 1} \tag{5.21a}$$

On the real frequency axis $j\omega$, S_{21} becomes

$$S_{21} = \frac{-j\left(2\omega^3 C^2 L\right)}{-j\left(2\omega^3 C^2 L\right) - (2CL + C^2)\,\omega^2 + j\,(2\omega C) + 1} = \rho_{21}e^{j\varphi_{21}}$$

or

$$S_{21} = \frac{-j\left(2\omega^3 C^2 L\right)}{[1 - (2CL + C^2)\,\omega^2] + j\,[(2\omega C) - (2\omega^3 C^2 L)]} = \rho_{21}e^{j\varphi_{21}} \tag{5.21b}$$

Let

$$d_r = 1 - \left(2CL + C^2\right)\omega^2 \tag{5.21c}$$

and

$$d_x = 2\omega C\left[1 - \omega^2 CL\right] \tag{5.21d}$$

Let

$$D_{21} = d_r + jd_x = \left(\sqrt{d_r^2 + d_x^2}\right)e^{j\varphi_D} \tag{5.21e}$$

$$\varphi_D = \tan^{-1}\left(\frac{d_x}{d_r}\right) = \tan^{-1}\left(\frac{2\omega C\left[1 - \omega^2 CL\right]}{1 - (2CL + C^2)\,\omega^2}\right) \tag{5.21f}$$

and

$$N_{21} = -j\left(2\omega^3 C^2 L\right) = \left(2\omega^3 C^2 L\right)e^{-j90°} \tag{5.21g}$$

The phase φ_{21} is then given by

$$\varphi_{21} = -90° - \varphi_D = -90° - \tan^{-1}\left(\frac{2\omega C\left[1 - \omega^2 CL\right]}{1 - (2CL + C^2)\,\omega^2}\right) \geq 0$$

or

$$(\varphi_{21} + 90°) = -\tan^{-1}\left(\frac{2\omega C\left[1 - \omega^2 CL\right]}{1 - (2CL + C^2)\omega^2}\right)$$

or

$$-\tan(\varphi_{21} + 90°) = \frac{2\omega C\left(1 - \omega^2 CL\right)}{1 - (2CL + C^2)\omega^2} \geq 0$$

Now, let us define $\theta = \varphi_{21}(\omega_0) > 0$ and let η be

$$\eta = -\tan(90+\theta) = \frac{1}{\tan(\theta)} \geq 0; \quad 0 \leq \theta = \varphi_{21}(\omega_0) \leq 90° \quad (5.22a)$$

As we defined before, let $\mu = \tan(\theta) \geq 0; \quad 0 \leq \theta \leq 90°$
Then,

$$\eta = \frac{1}{\mu} = \frac{2\omega_0 C\left(1 - \omega_0^2 CL\right)}{1 - (2CL + C^2)\omega_0^2} = \frac{d_x}{d_r} \geq 0 \quad (5.22b)$$

where

$$L = \left(\frac{1}{\omega_0^2}\right)\left[\frac{1 + \omega_0^2 C^2}{2C}\right] \quad (5.22c)$$

$$d_x(\omega_0) = 2\omega_0 C\left(1 - \omega_0^2 CL\right) \quad (5.22d)$$

$$d_r(\omega_0) = 1 - \left(2CL + C^2\right)\omega_0^2 \quad (5.22e)$$

Using inductor L of (5.22c) in $d_x(\omega_0)$ of (5.22d), it is found that

$$d_x(\omega_0) = 2\omega_0 C\left(1 - \omega_0^2 C\left(\frac{1}{\omega_0^2}\right)\left[\frac{1 + \omega_0^2 C^2}{2C}\right]\right)$$

or

$$d_x(\omega_0) = \omega_0 C\left(1 - \omega_0^2 C^2\right)$$

Similarly, $d_r(\omega_0)$ is derived in terms of phase parameter η and series arm capacitor C as

$$d_r(\omega_0) = 1 - \left(2C\left(\frac{1}{\omega_0^2}\right)\left[\frac{1 + \omega_0^2 C^2}{2C}\right] + C^2\right)\omega_0^2$$

or

$$d_r(\omega_0) = -2\omega_0^2 C^2$$

Employing the above-simplified forms of $d_x(\omega_0)$ and $d_r(\omega_0)$ in (5.22c), we have

$$\frac{1}{\mu} = \frac{1 - \omega_0^2 C^2}{-2\omega_0 C} \geq 0 \qquad (5.22\text{f})$$

or for a specified phase parameter μ, the series arm capacitor C can be found as the solution of a second-order equation such that

$$\left(\mu\omega_0^2\right) C^2 - 2\left(\omega_0\right) C - \mu = 0 \qquad (5.22\text{g})$$

or

$$C = \left(\frac{1}{\omega_0\mu}\right)\left(1 + \sigma\sqrt{\mu^2 + 1}\right) > 0 \qquad (5.22\text{h})$$

where σ is a unitary sign constant in absolute value (i.e., $\sigma = \mp 1$).

It must be selected as $\sigma = +1$ if $0° < \theta < 90°$ or it is set to $\sigma = -1$ when $90° < \theta < 180°$ to make the series arm capacitor C strictly positive.

Eq (5.22h) can be re-arranged by setting $\eta = \tan(90° - \theta) = \frac{1}{\mu}$ such that

$$C = \left(\frac{1}{\omega_0}\right)\left(\eta + \sqrt{\eta^2 + 1}\right) > 0 ; \quad \forall\theta \qquad (5.22\text{i})$$

As in the LPST, in the following, an alternative way is presented to compute the capacitor C.

5.6 Alternative Way to Compute φ_{21} for a Symmetric LC High-pass T-Section

For perfect transmission, in terms of the series arm impedance z, generic form of S_{21} was computed by (5.10) as

$$S_{21}(j\omega_0) = \frac{1 - z(j\omega_0)}{1 + z(j\omega_0)} = \rho_{21}e^{j\varphi_{21}} = \frac{1 - \left(\frac{1}{j\omega_0 C}\right)}{1 + \left(\frac{1}{j\omega_0 C}\right)} = \frac{j\omega_0 C - 1}{j\omega_0 C + 1}$$

$$= -1.e^{j\left[2\tan^{-1}(\omega_0 C)\right]}$$

or phase φ_{21} is given by

$$\varphi_{21}(\omega_0) = 180 - 2\tan^{-1}\omega_0 C$$

Table 5.2 Design formulas for a high-pass symmetric T-Section (HPST) for a specified positive phase θ such that $0° < \theta < 180°$

Phase-dependent parameter-1: $\mu = \tan(\theta)$ Phase-dependent parameter-2: $\eta = \tan(90° - \theta) = \frac{1}{\mu}$	
Straightforward derivation of the component values	Alternative and simple way to determine the component values
$C = \left(\frac{1}{\omega_0 \mu}\right)\left(1 + \sigma\sqrt{\mu^2 + 1}\right); \begin{cases} \sigma = +1; 0° < \theta < 90° \\ \sigma = -1; 90° < \theta < 180° \end{cases}$ or in short $C = \frac{1}{\omega_0}\left(\eta + \sqrt{\eta^2 + 1}\right) > 0$	$C = \frac{1}{\omega_0}\tan\left(90 - \frac{\theta}{2}\right) > 0;$ $0° < \theta < 180°$
$L = \left(\frac{1}{\omega_0}\right)\left[\frac{1 + \omega_0 C}{2C}\right] > 0$	

Then, series arm capacitor C is found as

$$C = \frac{1}{\omega_0}\tan\left(90 - \frac{|\theta|}{2}\right) > 0; \qquad 0°0 < \theta < 180° \qquad (5.23)$$

It should be noted that for $0° < \theta < 180°$, the series arm capacitor of (5.23) is always positive. Therefore, (5.23) is always safe to use.

Henceforth, design equations for a high-pass symmetric T-Section is summarized in Table 5.2.

Let us run an example to show the utilization of the above formulas.

Example 5.4:

(a) Design an LC high-pass symmetric T-Section, which yields $\varphi_{21} = 65°$ at the specified normalized angular frequency $\omega_0 = 0.9$.
(b) Develop a MatLab program to analyze the phase performance of the high-pass T designed as above.

Solution

(a) Employing the design equations given by Table 5.2, the series arm capacitor C is found as

$$C = \frac{1}{\omega_0}\tan\left(\frac{\theta}{2}\right) = C = \left(\frac{1}{\omega_0 \mu}\right)\left(1 + \sqrt{\mu^2 + 1}\right) = \frac{1}{0.9}\tan\left(\frac{65}{2}\right) = 1.7441$$

and the shunt inductor is determined as below.

$$L = \left(\frac{1}{\omega_0^2}\right)\left[\frac{1 + \omega_0^2 C^2}{2C}\right] = \frac{1}{(0.9)^2}\left[\frac{1 + (0.9)^2(1.7441)^2}{2 \times 1.7441}\right] = 1.2260$$

The MatLab program called "$Main\_Highpass\_T.m$".

(b) As far as phase shifting performance analysis is concerned, execution of "*Main_Highpass_T.m*" reveals the phase and the amplitude response of the high-pass T-section. The phase performance is shown in Figure 5.7a.

The amplitude response of the high-pass T-Section is depicted in Figure 5.7b.

Let us run an extreme example.

Example 5.5:

(a) Design an LC high-pass symmetric T-Section which yields $\varphi_{21} = 179°$ at the specified normalized angular frequency $\omega_0 = 0.5$.

(b) Analyze the phase performance of the high-pass T designed above.

Solution

(a) Employing (5.23a) or (5.22h), the series arm capacitor C is found as

$$C = \frac{1}{\omega_0} \tan\left(\frac{\theta}{2}\right) = 0.0175$$

and by (5.20)

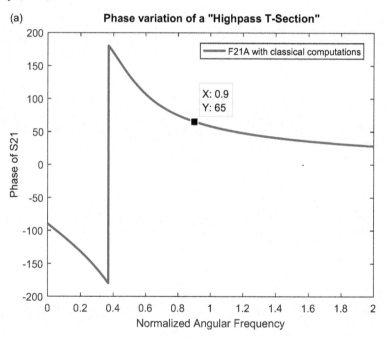

(a)

Phase variation of a "Highpass T-Section"

Figure 5.7   Phase performance of the high-pass T-Section of Example 5.4.

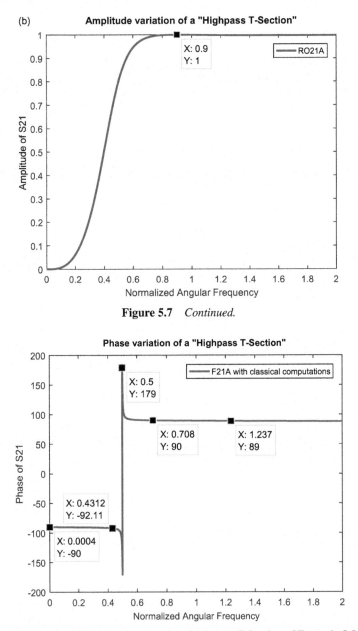

Figure 5.7 *Continued.*

Figure 5.8a Phase performance of the high-pass T-Section of Example 5.5.

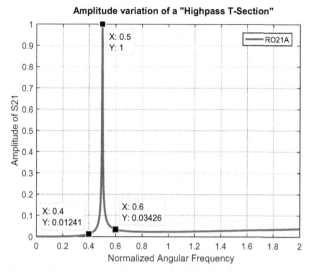

Figure 5.8b   Amplitude characteristic of HPST for Example 5.5.

$$L = \left(\frac{1}{\omega_0^2}\right)\left[\frac{1+\omega_0^2 C^2}{2C}\right] = 114.5974$$

(b) The phase characteristic of HPST is depicted in Figure 5.8a.

At $\omega = 0.5$, phase of HPST is $179°$ as it should be. As the frequency deviates even less than 5%, phase is almost constant at about $\mp 90°$, which was also the case for LPST.

The amplitude performance is shown in Figure 5.8b.

Program lists of the main program and its related functions are given as follows

Program List 5.5: Main_Highpass_T_with_Classical_Formulas.m

```
% Main_Highpass_T_with_Classical_Formulas.m
% January 6, 2019
% Developed by BS Yarman, Vanikoy, Istanbul, Turkey
clc, clear
close all
teta=input('Enter positive value for the phase shift
teta=')
w0=1;
[ L,C ] = AlternativeComponents_of_Highpass_T_
Section( w0, teta );
[ La,Ca,C1,C2 ] = mu_Based_Components_of_Highpass_
T( w0, teta )
Error_C=C-Ca
Error_L=L-La
```

```
% ----------------------------------------------
w=0;N=5000;w1=0;w2=2;DW=(w2-w1)/N;
FRI(1:(N+1))=zeros;
for j=1:N+1
    WA(j)=w;
% ----------------------------------------------
[ S21,RO21,F21 ] = S_Par_Highpass_T ( w,L,C );
    F21A(j)=F21;
    RO21A(j)=RO21;
% ----------------------------------------------

    w=w+DW;
end
% ----------------------------------------------
% Phase of S21
figure
plot(WA,F21A)
title('Phase variation of a "Highpass T-Section" ')
legend('F21A with classical computations')
xlabel('Normalized Angular Frequency')
ylabel('Phase of S21')
% Amplitude of S21
figure
plot(WA,RO21A)
title('Amplitude variation of a "Highpass
 T-Section" ')
legend('RO21A')
xlabel('Normalized Angular Frequency')
ylabel('Amplitude of S21')
```

Program List 5.6: function mu_Based_Components_of_Highpass_T

```
function [ La,Ca,C1,C2 ] = mu_Based_Components_of_
Highpass_T( w0, teta )
% This function generates the component values for
a lowpass T-Section
% Developed by BS Yarman: December 6, 2018;
  Vanikoy, Istanbul, Turkey
% Developed by BS Yarman. December 7, 2018, Vanikoy,
  Istanbul, Turkey
% Inputs:
%       teta: Positive angle at w0
%       w0: Normalized angular frequency
% Outputs:
%       C: Series arm Capacitor
%       L: Shunt arm Inductor
% ----------------------------------------------
mu=tand(teta);
if teta==90
    C1=1/w0;
    C2=1/w0;
    Ca=1/w0;
end
```

```
C1=(1/w0/mu)*(1+sqrt(mu*mu+1));
if C1>0
     Ca=C1;
end
C2=(1/w0/mu)*(1-sqrt(mu*mu+1));
if C2>0
     Ca=C2;
end
La=(1/w0/w0)*(1+w0*w0*Ca*Ca)/2/Ca;

End
```

Program List 5.7: AlternativeComponents_of_Highpass_T_Section

```
function [ L,C ] = AlternativeComponents_of_
Highpass_T_Section( w0, teta )
% This function generates the component
 values for a lowpass T-Section
% Developed by BS Yarman: December 6, 2018;
Vanikoy, Istanbul, Turkey
% Developed by BS Yarman. December 7, 2018,
Vanikoy, Istanbul, Turkey
% Inputs:
%        teta: Positive angle at w0
%        w0: Normalized angular frequency
% Outputs:
%        C: Series arm Capacitor
%        L: Shunt arm Inductor
% -------------------------------------------------
C=(1/w0)/tand(teta/2);
L=(1/w0/w0)*(1+w0*w0*C*C)/2/C;
end
```

5.7 Scattering Parameters of a Symmetric π-Section

Referring to Figure 5.9, a symmetric π-Section is considered as a two-port with a series-arm impedance Z and a shunt arm admittance Y.

Considering the normalized immittances, scattering parameters can be developed step by step as follows:

Step 1: Generate the first-level normalized admittance y_1 as

$$y_1 = y + 1$$

Step 2: Generate the second-level normalized impedance z_2 as

$$z_2 = z + \frac{1}{y+1} = \frac{zy + z + 1}{y+1}$$

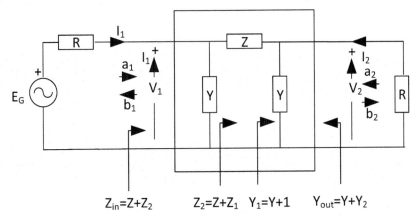

$Z_{in}=Z+Z_2$ $Z_2=Z+Z_1$ $Y_1=Y+1$ $Y_{out}=Y+Y_2$

Figure 5.9   Generic form of a symmetric π-Section.

or

$$y_2 = \frac{1}{z_2} = \frac{y+1}{zy+z+1}$$

Step 3: Compute the normalized input admittance

$$y_{in} = y_3 = y + y_2 = y + \frac{y+1}{zy+z+1} = \frac{zy^2+zy+2y+1}{zy+z+1}$$

or

$$z_{in} = \frac{zy+z+1}{zy^2+zy+2y+1} = \frac{N}{D}$$

where

$$N = zy+z+1$$
$$D = zy^2+zy+2y+1$$

Step 4: Compute the input reflection coefficient $S_{11} = \frac{z_{in}-1}{z_{in}+1}$. Then,

$$S_{11} = \frac{N-D}{N+D} = \frac{z\left(1-y^2\right)-2y}{zy^2+2zy+2y+z+2} \tag{5.24}$$

Step 5: By (5.6) $S_{21} = 2\left(\frac{v_2}{e_g}\right)$. Then,

Step 5a: The input current i_1 is given by

$$i_1 = \frac{e_g}{1+z_{in}}$$

Step 5b: The input voltage v_1 is given by

$$v_1 = z_{in} i_1 = \left(\frac{e_g}{1 + z_{in}} \right) (z_{in})$$

Step 5c: Compute the normalized current i_z on the series arm impedance z as

$$i_z = \frac{v_1}{z + z_1} = \frac{v_1}{z + \frac{1}{y+1}}$$

Replacing v_1 by $\left(\frac{e_g}{1+z_{in}} \right) (z_{in})$, the current i_z is drived as

$$i_z = \left(\frac{e_g}{1 + z_{in}} \right) \frac{(z_{in})(y + 1)}{(zy + z + 1)}$$

Step 5d: Output voltage v_2 is computed as

$$v_2 = \frac{i_z}{y + 1} = \left(\frac{e_g}{1 + z_{in}} \right) \left(\frac{z_{in}}{N} \right) = \left(\frac{e_g}{D + N} \right) \left(\frac{D}{N} \right) \left(\frac{N}{D} \right)$$

Hence,

$$S_{21} = 2 \left(\frac{v_2}{e_g} \right) = \frac{2}{zy^2 + 2zy + 2y + z + 2} \tag{5.25}$$

Step 6: By (5.24), at a specified normalized angular frequency ω_0, perfect match condition of $S_{11}(j\omega_0) = 0$ yields that

$$z = \frac{2y}{1 - y^2} \tag{5.26}$$

Step 7: Using (5.26) in (5.25), $S_{21}(j\omega_0)$ yields that

$$S_{21}(j\omega_0) = \frac{1 - y(j\omega_0)}{1 + y(j\omega_0)} = \rho_{21}(\omega_0) e^{j\varphi_{21}(\omega_0)} \tag{5.27}$$

In the following section, for the specified component values, we will drive the scattering parameters of the low-pass symmetric pi section (LPSπ).

5.8 Low-pass Symmetric π-Section (LPSP or LPSπ-Section)

In Figure 5.10, a low-pass symmetric π-section (LPSπ) is depicted.

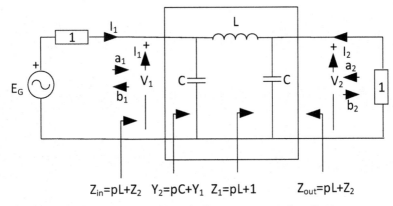

$Z_{in}=pL+Z_2$ $Y_2=pC+Y_1$ $Z_1=pL+1$ $Z_{out}=pL+Z_2$

Figure 5.10   Generic form of a low-pass symmetric π-Section.

In terms of the component values of Figure 5.10, S_{11} of (5.24) is found as

$$S_{11}(j\omega) = \frac{j\omega\left[L\left(1+\omega^2C^2\right)-2C\right]}{d_r + jd_x} = \rho_{11}(\omega)\,e^{j\varphi_{11}(\omega)} \qquad (5.28a)$$

where

$$\rho_{11}(\omega) = \frac{\omega\left[L\left(1+\omega^2C^2\right)-2C\right]}{\sqrt{d_r^2 + d_x^2}} \qquad (5.29a)$$

and

$$\varphi_{11}(\omega) = 90° - \tan^{-1}\left(\frac{d_x}{d_r}\right) \qquad (5.30a)$$

Similarly, S_{21} of (5.25) is given by

$$S_{21}(j\omega) = \frac{2}{d_r + jd_x} = \rho_{21}(\omega)\,e^{j\varphi_{21}(\omega)} \qquad (5.30b)$$

where

$$d_r = 2\left(1-\omega^2 LC\right) \qquad (5.30c)$$

and

$$d_x = \omega\left(2C + L - \omega^2 LC^2\right) \qquad (5.30d)$$

and

$$\rho_{21}\left(\omega\right) = \sqrt{d_r^2 + d_x^2} \tag{5.30e}$$

and

$$\varphi_{21}\left(\omega\right) = -\tan^{-1}\left(\frac{d_x}{d_r}\right) = -\tan^{-1}\left[\frac{\omega\left(2C + L - \omega^2 LC^2\right)}{2\left(1 - \omega^2 LC\right)}\right] \tag{5.30f}$$

At this point, we can compute the desired phase of $S_{21}\left(j\omega\right)$ at a specified normalized angular frequency ω_0 as in the following section.

5.9 Design of a Low-pass Symmetric PI Section for a Specified Phase $\varphi_{21}\left(\omega_0\right)$

A low-pass Symmetric PI-Section can be designed as a phase shifting unit. The phase φ_{21} is a negative quantity between $0°$ and $-180°$.

A phase shifter must satisfy two major conditions as follows.

The first one is the phase shift condition at the normalized angular frequency ω_0 such that

$$\theta = \varphi_{21}\left(\omega_0\right) < 0$$

The above condition is derived from (5.27). Hence, by setting $z = pL$ and $y = pC$, we have

$$S_{21}\left(j\omega_0\right) = \frac{1 - y\left(j\omega_0\right)}{1 + y\left(j\omega_0\right)} = \frac{1 - j\omega_0 C}{1 + j\omega_0 C} = 1.e^{-j2\tan^{-1}\left(\omega_0 C\right)}$$

Thus,

$$\theta = -2\tan^{-1}\left(\omega_0 C\right) < 0$$

or

$$C = \frac{1}{\omega_0}\tan\left(\frac{|\theta|}{2}\right) > 0; \quad 0° < |\theta| < 180° \tag{5.31a}$$

The second one is the perfect match condition at the operating frequency. This condition is satisfied employing (5.26). Thus, the series arm inductor L is computed as

$$L = \frac{2C}{1 + \omega_0^2 C^2} > 0 \tag{5.31b}$$

Equations (5.31a) and (5.31b) complete the design.

On the other hand, we can compute the shunt capacitor value using (5.31b) in (5.30f). At this point, it may be appropriate to define phase parameter μ as follows.

$$\mu = \tan(\theta) \qquad (5.31c)$$

Then, by (5.30f), we have

$$\mu = \left[\frac{\omega\left(2C + L - \omega^2 LC^2\right)}{2\left(\omega^2 LC - 1\right)} \right] \qquad (5.32a)$$

Using (5.31b) in (5.32) one ends up with the following quadratic equation to solve the shunt capacitor C. In other words,

$$\mu\omega_0^2 C^2 - 2\omega_0 C - \mu = 0 \qquad (5.32b)$$

or

$$C = \frac{1}{\omega_0}\left(\frac{1}{\mu} \mp \sqrt{1 + \frac{1}{\mu^2}}\right) \qquad (5.32c)$$

At this point, let us define the alternative phase parameter η as

$$\eta = \tan(90-\theta) = \frac{1}{\mu} \qquad (5.32d)$$

In this case, quadratic equation (5.32b) becomes

$$\omega_0^2 C^2 - \frac{2\omega_0}{\mu}C - 1 = 0$$

or

$$\omega_0^2 C^2 - 2\omega_0\eta C - 1 = 0 \qquad (5.32e)$$

or the shunt capacitor C becomes

$$C = \frac{1}{\omega_0}\left(\eta + \sqrt{\eta^2 + 1}\right) > 0; \quad -180 < \theta < 0 \qquad (5.32f)$$

Now, let us run an example to design LPSπ-Section.

Example 5.6:
(a) Design a low-pass symmetric π-Section which yields $\varphi_{21} = 67°$ at the specified normalized angular frequency $\omega_0 = 0.75$.

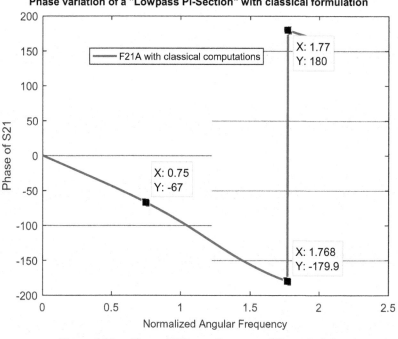

Figure 5.11a Phase shifting performance of Example 5.6.

(b) Analyze the phase performance of the low-pass π designed above.

Solution

(a) Employing (5.31a) or (5.32f), for $\theta = -67°$, the series arm capacitor C is found as

$$C = \frac{1}{\omega_0} \tan\left(\frac{|\theta|}{2}\right) = \frac{1}{\omega_0}\left(\eta + \sqrt{\eta^2 + 1}\right) = 0.8825$$

and by (5.20)

$$L = \left(\frac{1}{\omega_0^2}\right)\left[\frac{1 + \omega_0^2 C^2}{2C}\right] = 1.2273$$

(b) Phase shifting performance of the LPSP-Section under consideration is analyzed employing the MatLab program "*Main_Lowpass_PI_
Section.m*". In Figure 5.11a, phase variation of LPSP-Section is depicted.
A close examination of Figure 5.11a reveals that $\theta = -67°$ at $\omega = 0.75$ as
desired.

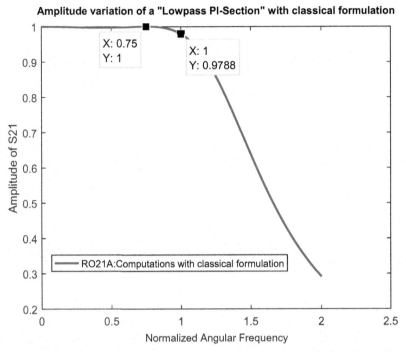

Figure 5.11b   Gain variation of the LPSP-Section of Example 5.6.

MatLab program $Main\_Lowpass\_PI\_Section.m$ calls two functions. Component values are determined function "$eta\_Based\_Components\_of$ $\_Lowpass_PI$", which programs Equation (5.31) and generates the component values C and L for specified phase θ and normalized angular frequency ω_0. Similarly, function "$Spar_{Lowpass_PI}$" generates the scattering parameters for the computed computed shunt capacitors C and the series inductor L at a given normalized angular frequency ω.

Program lists of the above-mentioned MatLab files are given in Program Lists 5.9, 5.10, and 5.11, respectively.

In Figure 5.11b, gain characteristic of the phase shifter is shown.

As expected, phase shifter is perfectly matched at $\omega = 0.75$. Then, the gain drops gradually.

In the following, MatLab programs are included for the program under consideration.

Program List 5.8: Main_Lowpass_PI_Section.m

```
% Main_Lowpass_PI_Section.m
% January 12, 2019
% Developed by BS Yarman, Vanikoy, Istanbul, Turkey
% It should be noted that the formulas used in this program is valid
  for the values of teta between 0 and -180 degree.
% Exact 90 is possible
clc, clear
close all
teta=input('Enter negative value for the phase
 shift teta=')
while teta==0
 stop= 'Attention teta=0. Phase can never be zero
 degree. Therefore, change teta and re-run the
  program again.'
 break

end
while teta==-180
    stop= 'Attention teta=-180. Phase can never be
    -180 degree. Therefore, change teta and re-run
    the program again.'
  break
end
while teta>0
stop1='Attention: teta is positive. For a Lowpass
 Symmetric PI-Section teta must be negative
  quantity.'
stop2='Please enter a negative value for teta and
 re-run the program'
    break
end
% Phase teta is proper. Then, start Computations
if teta>-180
if teta<0
w0=input('Enter the normalized angular frequency
 w0=')
 mu=tand(teta)
 eta=tand(90-teta)
% --------------------------------------------------
[ Ca,La,error ] = eta_Based_Components_of_Lowpass_
PI( w0,teta )
% --------------------------------------------------
w=0;N=1000;w1=0;w2=2;DW=(w2-w1)/N;
FRI(1:(N+1))=zeros;
for j=1:N+1
    WA(j)=w;
% --------------------------------------------------
[ S11,S21,RO11,F11,RO21,F21 ] = S_Par_Lowpass_PI
( w,La,Ca );
 F21A(j)=F21;
 RO21A(j)=RO21;
```

```
% ------------------------------------------------
  w=w+DW;
end
% ------------------------------------------------
% Phase of S21
figure
plot(WA,F21A)
title('Phase variation of a "Lowpass PI-Section"
 with classical formulation')
legend('F21A with classical computations')
xlabel('Normalized Angular Frequency')
ylabel('Phase of S21')
% Amplitude of S21
figure
plot(WA,RO21A)
title('Amplitude variation of a "Lowpass PI-
Section" with classical formulation')
legend('RO21A:Computations with classical
 formulation')
xlabel('Normalized Angular Frequency')
ylabel('Amplitude of S21')
end
end
```

Program List 5.9: function eta_Based_Components_of_

```
Lowpass_PI
function [ C,L,error ] = eta_Based_Components_of_
Lowpass_PI( w0,teta )
% In this function teta must be a negative quantity
% In this function component values of a symmetric
 Lowpass pi is computed
%   Input phase teta is defined as a negative
 quantity
if teta>=0
    stop='Attention: teta is positive. It must be a
     negative quantity'
    C='teta is positive. It must be negative'
    L='teta is positive. It must be negative'
    error='teta is positive. It must be negative'
end
if teta<0
eta=tand(90-teta);
% ------------------------------
C=(1/w0)*(eta + sqrt(1+eta*eta ));
L=2*C/(1+w0*w0*C*C);
C1=1/w0*tand(abs(teta/2));
error=norm(C-C1);
end
% ------------------------------------------------
end
```

Program List 5.10: function S_Par_Lowpass_PI

```
( w,L,C )
function [ S11,S21,RO11,F11,RO21,F21 ] = S_Par_
Lowpass_PI( w,L,C )
% This function generates the S-Parameters of
 Lowpass T Section from the
% computed series arm inductor L and the shunt
 capacitor C
%  Developed by BS Yarman: December 7, 2018,
 Vanikoy, Istanbul
% S11=-((1+w0^2 L^2 )jw0C-2jw0L)/(2(1-w0^2 LC)+jw0
(2L+C-w0^2 L^2 C) )
% S_21=2/(2(1-w0^2 LC)+jw0(2L+C-w0^2 L^2 C) )=2/
(d_r+jd_x )=ρ_21 e^(jθ_21 (w0))
% --------------------------------------------------
dr=2*(1-w*w*L*C);
dx=w*(2*C+L-w*w*C*C*L);
j=sqrt(-1);
D=dr+j*dx;
N11= j*((1+w*w*C*C)*w*L-2*w*C);
S11=N11/D; R11=real(S11); X11=imag(S11);F11=atan2d
(X11,R11);
S21=2/D;   R21=real(S21); X21=imag(S21);F21=atan2d
(X21,R21);
RO11=abs(S11);
RO21=abs(S21);
   end
```

5.10 High-pass Symmetric π-Section (HPSP or HPSπ-Section)

In Figure 5.12, a high-pass Symmetric π-Section (LPSπ) is depicted.

By setting $z = \frac{1}{j\omega C}$ and $y = \frac{1}{j\omega C}$, in terms of the component values of Figure 5.12, S_{11} of (5.24) is found as

$$S_{11}(p) = \frac{\left[p^2L^2 - 1 - 2p^2LC\right]/\left[p^3L^2C\right]}{\left[(1 + 2p^2LC + p^2L^2) + 2pL\left(1 + p^2LC\right)\right]/\left[p^3L^2C\right]}$$

or for $p = j\omega$,

$$S_{11}(j\omega) = -\frac{1 + \omega^2L(L - 2C)}{d_r + jd_x} = \rho_{11}(\omega)\,e^{j\varphi_{11}(\omega)} \qquad (5.33a)$$

where

$$\rho_{11}(\omega) = \frac{1 + \omega^2L(L - 2C)}{\sqrt{d_r^2 + d_x^2}} \qquad (5.33b)$$

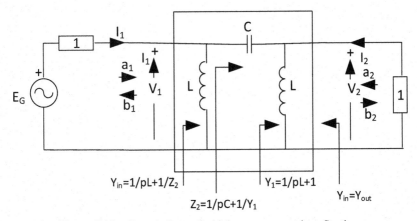

Figure 5.12   Generic form of a high-pass symmetric π-Section.

and

$$\varphi_{11}(\omega) = 180° - \tan^{-1}\left(\frac{d_x}{d_r}\right) \tag{5.33c}$$

Similarly, S_{21} of (5.25) is given by

$$S_{21} = \frac{-j\left(2\omega^3 L^2 C\right)}{-j\left(2\omega^3 L^2 C\right) - \left(2CL + L^2\right)\omega^2 + j\left(2\omega L\right) + 1} = \rho_{21}e^{j\varphi_{21}}$$

$$S_{21}(j\omega) = -j\frac{2\omega^3 L^2 C}{d_r + jd_x} = \rho_{21}(\omega)\,e^{j\varphi_{21}(\omega)} \tag{5.33d}$$

where

$$d_r = 1 - \omega^2\left(2LC + L^2\right) \tag{5.33e}$$

and

$$d_x = 2\omega L\left(1 - \omega^2 LC\right) \tag{5.33f}$$

and

$$\rho_{21}(\omega) = \sqrt{d_r^2 + d_x^2} \tag{5.33g}$$

and

$$\varphi_{21}\left(\omega\right) = -90^\circ - \tan^{-1}\left(\frac{d_x}{d_r}\right) = -90^\circ - \tan^{-1}\left[\frac{2\omega L\left(1 - \omega^2 LC\right)}{1 - \omega^2\left(2LC + L^2\right)}\right] > 0$$

or

$$\varphi_{21}\left(\omega\right) + 90^\circ = -\tan^{-1}\left(\frac{d_x}{d_r}\right) \tag{5.33h}$$

Let

$$\theta = \varphi_{21}\left(\omega_0\right) \tag{5.33i}$$

and

$$\mu = \tan\left(\theta\right) \tag{5.33j}$$

and let

$$\eta = -\tan\left(90^\circ + \theta\right) = \frac{1}{\mu} = \left[\frac{2\omega L\left(1 - \omega^2 LC\right)}{1 - \omega^2\left(2LC + L^2\right)}\right] > 0 \tag{5.33k}$$

At this point, we can compute the desired phase of $S_{21}\left(j\omega\right)$ at a specified normalized angular frequency ω_0 as in the following section.

5.11 Design of a High-pass Symmetric PI Section for a Specified Phase $\varphi_{21}\left(\omega_0\right)$

A high-pass symmetric PI-Section can be designed as a phase shifting unit. The phase φ_{21} is a positive quantity between 0° and 180°.

The high-pass T phase shifter must satisfy two major conditions. The first one is the phase shift condition at the normalized angular frequency ω_0 such that

$$\theta = \varphi_{21}\left(\omega_0\right) > 0$$

The above condition is derived from (5.27). Hence, by setting $z = 1/pC$ and $y = 1/pL$, and $p = j\omega$, we have

$$S_{21}\left(j\omega_0\right) = \frac{1 - y\left(j\omega_0\right)}{1 + y\left(j\omega_0\right)} = \frac{1 - 1/j\omega_0 L}{1 + 1/j\omega_0 L} = \frac{1 + j\frac{1}{\omega_0 L}}{1 - j\frac{1}{\omega_0 L}} = 1.e^{j2\tan^{-1}\left(\frac{1}{\omega_0 L}\right)}$$

Thus,

$$\theta = 2 \tan^{-1} \left(\frac{1}{\omega_0 L} \right) > 0$$

or

$$L = \left(\frac{1}{\omega_0} \right) \left[\frac{1}{\tan \left(\frac{\theta}{2} \right)} \right] = \left(\frac{1}{\omega_0} \right) \cotan \left(\frac{\theta}{2} \right) > 0; \quad 0° < \theta < 180° \quad (5.34a)$$

The second one is the perfect match condition at the operating frequency. This condition is satisfied employing (5.26). Thus, the series arm capacitor C is computed as

$$C = \left(\frac{1}{\omega_0^2} \right) \left[\frac{1 + \omega_0^2 L^2}{2L} \right] \quad (5.34b)$$

Equations (5.34a) and (5.34b) complete the design.

On the other hand, we can compute the shunt inductor value L of (5.34a) from (5.33f) by inserting (5.34b) in (5.33f).

Note that if we define a new phase shifting parameter

$$\eta = -\tan(90+\theta) = \frac{1}{\mu} = \frac{2\omega_0 L \left(1 - \omega_0^2 LC \right)}{1 - \omega_0^2 \left(2LC + L^2 \right)} = \frac{d_x}{d_r} \quad (5.34c)$$

Using (5.34b) in (5.34c) and after some algebraic manipulations, we have

$$\omega_0^2 L^2 - \frac{2\omega_0}{\mu} L - 1 = 0$$

or

$$\omega_0^2 L^2 - 2\omega_0 \eta L - 1 = 0 \quad (5.34d)$$

or the shunt inductor L becomes

$$L = \frac{1}{\omega_0} \left(\eta + \sqrt{\eta^2 + 1} \right) > 0; \quad 180 > \theta > 0 \quad (5.34e)$$

Now, let us run an example to design HPSπ-Section.

Example 5.7:
(a) Design a high-pass symmetric PI Section for $\theta = 75°$ and $\omega_0 = 0.75$.
(b) Analyze its performance by developing a MatLab program.

Solution

(a) First of all, let us compute the phase-dependent parameters μ, $\tan\left(\frac{\theta}{2}\right)$, and η as follows:

$$\mu = \tan(\theta) = 3.7321$$

$$\tan\left(\frac{\theta}{2}\right) = 0.7673$$

$$\eta = -\tan(90° + \theta) = \frac{1}{\mu} = 0.2679$$

Then, by (5.34a)

$$L = \left(\frac{1}{\omega_0}\right)\left[\frac{1}{\tan\left(\frac{\theta}{2}\right)}\right] = \left(\frac{1}{\omega_0}\right)\cotan\left(\frac{\theta}{2}\right) = 1.7376$$

By (5.34c)

$$\eta = -\tan(90+\theta) = 0.2679$$

By (5.34f)

$$L = \frac{1}{\omega_0}\left(\eta + \sqrt{\eta^2 + 1}\right) = 1.7376$$

By (5.34b)

$$C = \left(\frac{1}{\omega_0^2}\right)\left[\frac{1 + \omega_0^2 L^2}{2L}\right] = 1.3804$$

(b) Performances Analysis

Design equations of the high-pass PI-Section are programmed under the main program "$Main\_Highpass\_PI\_Section.m$". This program calls the function "eta_Based_Components_of_Highpass_PI" to generate the components values for L and C. Eventually, function "S_Par_Highpass_PI" generates phase shifting performance of the designed phase shifter.

Phase performance of the example under consideration is plotted in Figure 5.13a.

Loss performance of the high-pass symmetric PI-Section is depicted in Figure 5.13b.

A close examination of Figure 5.13 reveals that at $\omega_0 = 0.75$, phase $\varphi_{21}(\omega_0) = 75°$ with $|S_{21}(j\omega_0)|^2 = 1$ as desired.

(a)   **Phase variation of a "highpass PI-Section" with classical formulation**

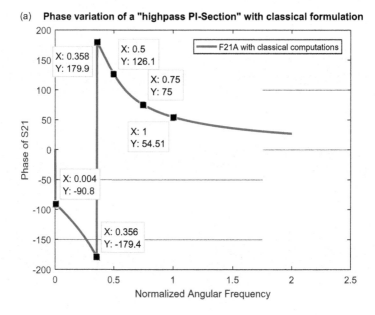

(b)   **Amplitude variation of a "Highpass PI-Section" with classical formulation**

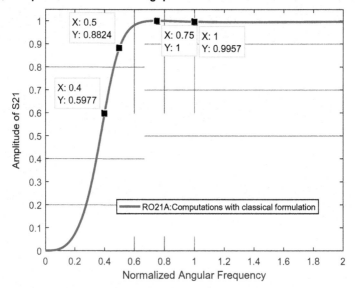

Figure 5.13   Phase performance of Example 5.7.

Program List 5.11: Main_Highpass_PI_Section.m

```
% Main_Highpass_PI_Section.m
% January 18, 2019
% Developed by BS Yarman, Vanikoy, Istanbul, Turkey
% It should be noted that the formulas used in this
program is valid for
% the values of teta between 0 and +180 degree.
% Exact 90 is possible
clc, clear
close all
teta=input('Enter positive value for the phase
shift teta=')
while teta==0
 stop= 'Attention teta=0. Phase can never be zero
 degree. Therefore, change teta and re-run the
 program again.'
    break
 end
  while teta==180
    stop= 'Attention teta=180. Phase can never be
    -180 degree. Therefore, change teta and re-run
    the program again.'
    break
 end
while teta<0
stop1='Attention: teta is negative. For a Highpass
Symmetric PI-Section teta must be positive
quantity.'
stop2='Please enter a positive value for teta and
re-run the program'
      break
end
% Phase teta is proper. Then, start Computations
if teta<180
if teta>0
w0=input('Enter the normalized angular frequency
w0=')
mu=tand(teta)
eta=tand(90-teta)
% -------------------------------------------------
[ Ca,La,error ] = eta_Based_Components_of_Highpass_
PI( w0,teta )
% -------------------------------------------------
w=0;N=1000;w1=0;w2=2;DW=(w2-w1)/N;
FRI(1:(N+1))=zeros;
for j=1:N+1
    WA(j)=w;
% -------------------------------------------------
[ S11,S21,RO11,F11,RO21,F21 ] = S_Par_Highpass_PI
(w,La,Ca );
    F21A(j)=F21;
    RO21A(j)=RO21;
```

```
% ----------------------------------------------------
   w=w+DW;
end
% ----------------------------------------------------
% Phase of S21
figure
plot(WA,F21A)
title('Phase variation of a "highpass PI-Section"
 with classical formulation')
legend('F21A with classical computations')
xlabel('Normalized Angular Frequency')
ylabel('Phase of S21')
% Amplitude of S21
figure
plot(WA,RO21A)
title('Amplitude variation of a "Highpass
PI-Section" with classical formulation')
legend('RO21A:Computations with classical
formulation')
xlabel('Normalized Angular Frequency')
ylabel('Amplitude of S21')
% ----------------------------------------------------
end
end
```

Program List 5.12: Function eta_Based_Components_of_Highpass_PI

```
function [ C,L,error ] = eta_Based_Components_of_
Highpass_PI( w0,teta )
% In this function teta must be a positive quantity
% In this function component values of a symmetric
% Highpass pi is computed
% Input phase teta is defined as a positive quantity
if teta<=0
   stop='Attention: teta is positive. It must be a
   negative quantity'
   C='teta is positive. It must be positive'
   L='teta is positive. It must be positive'
   error='teta is positive. It must be negative'
end
if teta>0
eta=-tand(90+teta);
% ------------------------------
L=(1/w0)*(1/tand(teta/2));
C=(1/w0/w0)*(1+w0*w0*L*L)/2/L;
L1=(1/w0)*(eta + sqrt(1+eta*eta ));
error=norm(L-L1);
end
% ----------------------------------------------------
End
```

Program List 5.13: Function S_Par_Highpass_PI

```
function [ S11,S21,RO11,F11,RO21,F21 ] = S_Par_Highpass_PI( w,L,C )
% This function generates the S-Parameters of
% Lowpass T Section from the
% computed series arm inductor L and the shunt capacitor C
% Developed by BS Yarman: December 7, 2018, Vanikoy, Istanbul
% S11=-((1+w0^2 L^{}2 )jw0C-2j?L)/(2(1-w0^2 LC)+jw0
(2L+C-w0^2 L^2 C) )
% S_21=2/(2(1-w0^2 LC)+jw0(2L+C-w0^2 L^2 C) )=2/
(d_r+jd_x )=ρ_21 e^(jθ_21 (w0))
% ----------------------------------------------------
dr=1-w*w*(2*L*C+L*L);
dx=2*w*L*(1-w*w*L*C);
j=sqrt(-1);
D=dr+j*dx;
N11=-(1+w*w*L*(L-2*C));
S11=N11/D; R11=real(S11); X11=imag(S11);F11=atan2d
(X11,R11);
S21=-j*2*w*w*w*L*L*C/D;   R21=real(S21); X21=imag
(S21);F21=atan2d(X21,R21);
RO11=abs(S11);
RO21=abs(S21);
```

End

References

[1] MatLab, "MatLab S/W Package," The MathWorks Inc., Natick, Massachusetts, USA, 2018.

[2] Binboga S. Yarman, "π-section digital phase shifter apparatus," U.S. Patent: 4604593, Washington DC, USA, August 5, 1986.

[3] B. S. Yarman, "New Circuit Configurations for Designing 0-180 degree Digital Phase Shifters," IEE Proceeding H, Vol. 134, pp. 253–260, 1987.

[4] B. Yarman, "T Section Digital Phase Shifter Apparatus". USA Patent 4.603.310, 29 July 1986.

[5] B. Yarman, A.Rosen and P.Stabile, "Lowloss EHF Digital Phase Shifters Suitable for Monolithic Implementation," in IEEE ISCAS, Montreal, 1984.

[6] B. S. Yarman, "Novel circuit configurations to design loss balanced 0°–360° digital phase shifters," AEU, vol. 45, no. 2, pp. 96–104, 1991.

[7] B. S. Yarman, "Low pass T-section digital phase shifter apparatus," U.S. Patent: 4630010, Washington DC, USA, December 16, 1986.

[8] B. S. Yarman, "Lowpass PI Section Digital Phase Shifter Apparatus". USA Patent 4.603.921, 30 September 1986.

[9] B. S. Yarman, "Design of Digital Phase Shifters Suitable for Monolithic Implementations," Bull. of Technical University of Istanbul, pp. 185–205, 1985.

[10] I. Bahl, Lumped Elements for RF Microwave Circuits, Artech House, 2003.

[11] I. Bahl and P. Bhartia, Microwave Solide State Circuit Design, Hoboken, N.J., USA: Wiley Interscience, 2003.

[12] K. Chang, Handbook of RF Microwave Components and Engineering, New York, USA, Chechester, UK: Wiley, 2003.

[13] B. Constantine A., Antenna Theory: Analysis and Design, pp. 302–303., ISBN 1119178983: 4th Ed. John Wiley & Sons., 2015.

[14] D. W. Kang, H. D. Lee, C. H. Kim and S. Hong, "Ku-band MMIC phase shifter using a parallel resonator with 0.18 m CMOS technology," IEEE Trans. on MTT, Vol. 54, No. 1, pp. 294–301, 2006.

[15] H. Fang, T. Xinyi, K. Mouthaan and R. Guinvarch, "Two Octave Digital All-Pass Phase Shifters for Phase Array Applications," IEEE Radio and Wireless Symposium (RWS), pp. 169–171, 2013.

[16] R. Garver, Microwave Doide Control Devices, Dedham, MA: Artech Inc., 1976.

[17] D. Kang and S. Hong, "A 4-bit CMOS Phase Shifter Using Distributed Active Switches," IEEE Trans on MTT, Vol. 55, No. 7, pp. 1476–1483, 2007.

[18] S. Lucyszyn and L. Roberson, "Synthesis Techniques for High Performance Octave Bandwidth 180 degree Analog Phase Shifters," IEEE Trans on MTT, Vol. 40, No. 4, pp. 731–740, 1992.

[19] K. J. Koh and G. M. Rebeiz, "0.13- m CMOS phase shifters for and K-band phased arrays," IEEE Trans. on MTT, Vol. 42, No. 11, p. 2535–2546, November 2007.

[20] R. Melik and H. V. Demir, "Implementation of High Qualityfactor, on-chip Tuned Microwave Resonators at 7 GHz," Microwave and Optical Technology Letters, vol. 51, no. 2, pp. 497–501, February 2009.

[21] R. S. Pengelley, S. M. Wood, S. T. Milligan, W. L. Sheppard and W. L. Pribble, "A rewiew of GaN on SiC High Electron Mobility Power Transistors and MMICs," IEEE Trans. MTT, Part I, vol. 60, no. 6, pp. 1764–1783, 2008.

[22] J. Quirarte and J. Straski, "Novel Shiffman Phase Shifter," IEEE Trans. on MTT Vo. 41, No.1, pp. 9–14, 1993.

[23] T. A. Milligan, Modern Antenna Design, ISBN 0471720607: John Wiley & Sons., 2005.

[24] H. Watson, Microwave Semiconductor Devices and thier applications, New York, NY: McGraw Hill, 1976.

[25] B. S. Yarman, "Lowpass T Section Digital Phase Shiter Apparatus". USA Patent 4600010, 16 December 1986.

[26] M. I. L. S.W. Amy, "A Simple Microstrip Phase Shifter," Journal of Physics, No.45, pp. 105–114, 1992.

[27] Pat Hindle, Editor, "The State of RF and Microwave Switches," Microwave Journal, vol. 53, no. 11, p. 20, November 2010.

[28] L. Yun-Wei, H. Yi-Chieh and C. Chi-Yang, "A Balanced Digital Phase Shifter by a Novel Switching-Mode Topology," IEEE Trans on MTT, pp. 2361–2370, 2013.

6

180° Low-pass-based T-Section Digital Phase Shifter Topology (LPT-DPS)

Summary

In this chapter, a symmetric 3-element LC low-pass-based T-Section digital phase shifter configuration is presented. The proposed phase shifter topology is compact and it is suitable to manufacture with discrete components or as microwave monolithic integrated circuits (MMIC).

The content of this chapter is developed on our research work already shared in the open literature [1–4].

6.1 Solid-State Microwave Switches

(a) Diode Switches

Digital phase shifters are realized using microwave switches [5–7]. A simple microwave switch can be realized using a simple solid-state diode as shown in Figure 6.1.

Diodes can be manufactured as PIN diodes, bipolar transistors, CMOS, etc. depending on the application. A simple model of a diode can be built as shown in Figure 6.1. When the diode D is reverse-biased, it may show a capacitance C_D in series with a small resistance R_r. The reverse-biased resistance R_r may vary from a few milliohms to a few ohms. Similarly, when the diode is forward biased, it acts like a small resistor R_f in the range of R_r or bigger.

In Table 6.1, for a typical millimeter wave PIN diode (operating at f_0 = 44.5 GHz), forward-biased resistance and reverse-biased capacitance are given[1]. For the same diode, the reverse-biased resistance is negligibly small. Therefore, it is omitted in the table.

---

[1]This PIN diode is manufactured by Dr. Arye Rosen and his team of Sarnoff Research Center.

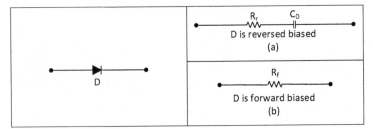

Figure 6.1   A typical diode employed as a microwave switch.

Table 6.1   Typical PIN diode switch parameters

Forward-Biased Current (mA)	$R_f(\Omega)$	Reverse-Biased Voltage (V) $R_r < 10^-\ \Omega$	Reverse-Biased Capacitance $C_D(pF)$
20	1.16	0	0.609
50	0.778	2	0.283
100	0.578	4	0.235
200	0.435	10	0.173
250	0.400	80	0.168

(b) MOS Switches

The metal oxide semiconductor field effect transistors (MOSFET) are widely used in RF and microwave integrated circuit switch applications. Transistor frequency performance is associated with its transit frequency f_T.

Here, f_T is the frequency upon which the current gain is unity. f_T values for different CMOS processes such as 0.25 µm, 0.18 µm, 0.13 µ, and 90 nm are 30, 50, 75, and 110 GHz, respectively.

In this chapter, we have the freedom to use TSMC's 0.18 µm NMOS transistors [8] as switching elements[2]. ON and OFF modes of an NMOS 0.18 µm transistor can be modeled as shown in Figure 6.2.

In the "ON" state ($V_{GS} = 1.8\ V$), most effective component of the model is the channel resistor R_{on}. For the process under consideration, selected R_{on} varies from 7.5 Ω to 60 Ω. In the "OFF" state ($V_{GS} = 0\ V$), the transistor simply exhibits a capacitor $C_{DS} = C_{off}$.

For the selected gate length $L = 180$ nm, the product $R_{on} \times C_{off}$ of the 0.18 um TSMC process is constant and it is given by

$$R_{on}\ (\Omega) \times C_{off}\ (Farad) = 672 \times 10^{-15} \tag{6.1}$$

---

[2]https://www.tsmc.com/english/dedicatedFoundry/technology/index.htm

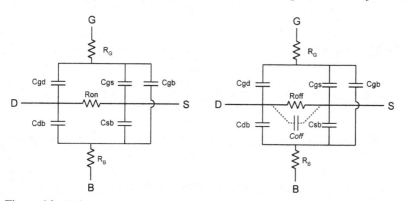

Figure 6.2 MOS switch equivalent circuits of ON state (left) and OFF state (right).

where R_{on} is the ON state resistance and C_{off} is the OFF state capacitor of the NMOS transistor.

For most practical cases, reverse-biased resistor R_{off} of the MOS switch is negligible. Therefore, the switch model of Figure 6.2 converges to Figure 6.1.

It is noted that for the switch under consideration, as the revere-biased capacitance C_{off} becomes smaller, ON state resistance R_{on} increases, which in turn increases the switch loss. Therefore, there is a trade-off between R_{on} and C_{off}.

Details of switches are out –of the scope of this chapter. Here, in this chapter, the focus is more on the design aspect of the digital phase shifter topologies with proper element values. Once, the designs are completed, the user can select a proper solid-state switch to implement the circuits.

6.2 Low-pass-based Symmetric T-Section Digital Phase Shifter

In Figure 6.3a, a symmetric low-pass-based, T-Section digital phase shifter $(LPT - DPS)$ is shown.

This configuration includes three switching diodes, namely in the left and right series arms, we have two identical diodes D_1, and in the shunt arm D_2.

In one state, say, State-A, when the diodes D_1s are forward-biased, the symmetric series arm impedances (Z) act as inductors L_{LT}, as in Figure 5.2. At the operating frequency f_0, in the shunt arm, D_2 is also forward-biased, yielding an equivalent shunt admittance $Y = (2\pi f_0)C_{LT}$.

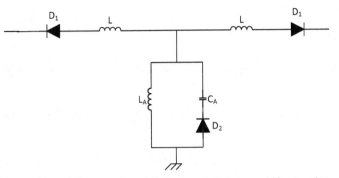

Figure 6.3a   A low-pass-based T-Section digital phase shifter topology.

In this mode of operation, ideally (i.e., $R_f = 0$), at a selected operation angular frequency ω_0 and phase $\varphi_A(\omega_0)$, element values of Figure 6.3a are specified as in Table 5.1 as follows.

$$L = \left(\frac{1}{\omega_0}\right) \tan\left(\frac{|\varphi_{21A}|}{2}\right) > 0 \tag{6.2a}$$

$$C = \frac{2L}{1 + \omega_0^2 L^2} > 0 \tag{6.2b}$$

In this state, shunt arm admittance is given by

$$Y(j\omega_0) = \frac{1}{j\omega_0 L_A} + j\omega_0 C_A = j\omega_0 C$$

or

$$C_A = C + \frac{1}{\omega_0^2 L_A} \tag{6.2c}$$

In State-B, all the diodes are reverse-biased. In this case, reverse-biased diode capacitance C_{D1} resonates with the series arm inductor L of (6.2a). Hence, at the given operating frequency ω_0, C_{D1} is specified by

$$C_{D1} = \frac{1}{\omega_0^2 L} \tag{6.2d}$$

Similarly, in the shunt arm, inductor L_A resonates with the equivalent capacitor C_T, which is given by

$$C_T = \frac{C_A C_{D2}}{C_A + C_2} \tag{6.2e}$$

At the resonance, capacitor C_T is given by

$$C_T = \frac{1}{w_0^2 L_A} = \frac{C_A C_{D2}}{C_A + C_2} \qquad (6.2f)$$

Using (6.2f) in (6.2c), we have

$$C_A = C + \frac{C_A C_{D2}}{C_A + C_{D2}}$$

Once C_{D2} is selected, the capacitor C_A can be determined by solving the following second-order equation

$$C_A^2 - (C)\, C_A - (C C_{D2}) = 0$$

or

$$C_{A(1,2)} = \left(\frac{C}{2}\right) \mp \frac{1}{2}\sqrt{C^2 + 4 C C_{D2}} > 0$$

Obviously, positive value for the capacitor C_A can only be obtained by selecting "+" sign in the above equation.

Hence,

$$C_A = \left(\frac{C}{2}\right) + \frac{1}{2}\sqrt{C^2 + 4 C C_{D2}} > 0 \qquad (6.2g)$$

Then, we can generate L_A as

$$L_A = \frac{1}{w_0^2 C_T} \qquad (6.2h)$$

6.3 Concept of Digital Phase Shift and Design Algorithm

The phase shifter topology introduced in Figure 6.3a describes two distinct operational states, namely State-A and State-B. The phase of each state is determined based on the position of the switching diodes. For each diode, forward-biased situation may be named as "switch is ON" or "switch is closed", and this position may be designated by the logic-state "1". When a diode is reversed-biased, it may be referred to as "switch is OFF" or "switch is open". This situation may be designated by the logic state "0".

"The phase $\theta(\omega)$ of a lossless two-port" is defined as the phase $\varphi_{21}(\omega)$ of the transfer scattering parameter $S_{21}(j\omega) = \rho(\omega)\, e^{j\varphi_{21}(\omega)}$. At the operating frequency, the objective is to make $\rho(\omega_0) = 1$ while reaching the desired phase $\theta(\omega_0) = \varphi_{21}(\omega_0)$.

Table 6.2   Description of switching states of LPT-DPS topology

	D1	D2	Phase θ of the Two-Port
State-A	1	1	$\theta_A = \varphi_{21B}(\omega_0)$
State-B	0	0	$\theta_B = \varphi_{21B}(\omega_0)$
Net Phase Shift Between States			$\Delta\theta = \theta_A - \theta_B$

For a two-port shown in Figure 6.3a, where the phase $\theta(\omega)$ varies depending on the position of the switches.

As described above, at a specified angular frequency ω_0, State-A refers to the phase $\theta_A = \varphi_{21A}(\omega_0)$ when all the switches are closed. In State-B, all the switches are opened and the two-port possesses the phase $\theta_B = \varphi_{21B}(\omega_0)$.

All the above explanations suggest the following truth table to define the phase-switching states of the topology shown in Figure 6.3a.

In view of Table 6.2, Figure 6.3a is named as low-pass-based T-Section digital phase shifter. It possesses two phase switching states, namely State-A and State-B. The net phase shift between the states is given by

$$\Delta\theta = \theta_A - \theta_B \tag{6.3}$$

However, it should be noted that:

(a) By its nature, in State-B, at the angular frequency ω_0, θ_B is forced to be zero
    and

(b) Range of θ_A must be less than 180° since $L = \frac{1}{\omega_0}\tan\left(\frac{|\varphi_{21A}(\omega_0)|}{2}\right)$ goes to infinity when $\varphi_{21A}(\omega_0) = 180°$.

In Figure 6.3b, equivalent immittance states of Figure 6.3a is depicted. Thus, we propose the following design algorithm to construct LPT-DPS

6.4 Algorithm to Design LPT-DPS for the Phase Range $180° < \Delta\theta = \theta_A < 0°$

In this algorithm, we present the design procedure to construct a low-pass T-Section Digital Phase Shifter, which can only cover the phase range of $0°–180°$. Using this phase shifter topology, it is not possible to obtain exact 180° phase shifting unit.

Inputs:

f_{1a}: Lower end of the passband of interest
f_{2a}: Upper end of the passband of interest

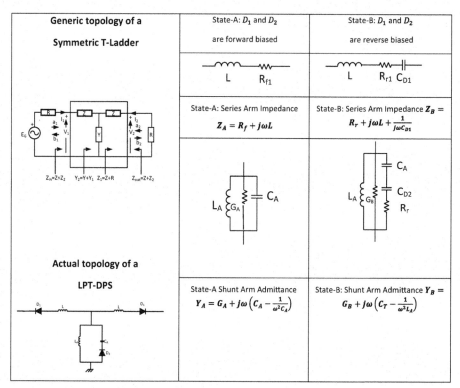

Figure 6.3b Immittance states of the low-pass-based T-Section digital phase shifter topology.

f_{0a}: Actual operating frequency such that $f_{1a} \leq f_{0a} \leq f_{2a}$

R: Normalization resistance. For most practical situations, it is selected as the termination resistors, which is perhaps 50Ω.

$\Delta\theta = \varphi_{21A}(\omega_0)$: Desired phase shift

G_A: Overall estimated normalized conductive loss of the shunt arm with respect to R in State-A.

G_B: Overall estimated normalized conductive loss of the shunt arm with respect to R in State-B.

Computational Steps:

Step 1: Select a normalized angular frequency ω_0 such that it is confined within

$$f_{1a}/f_{0a} \leq \omega_0 \leq f_{2a}/f_{0a}$$

It is useful to know that for many cases, ω_0 is set to unity (i.e., $\omega_0 = 1$)

Step 2: Compute the normalized series arm inductor $L = \frac{1}{\omega_0}\tan$ $\left(\frac{|\varphi_{21A}(\omega_0)|}{2}\right)$ of the ideal symmetric low-pass T-Section of Figure 5.2.

Step 3: Compute the normalized shunt arm capacitor C of the ideal symmetric low-pass T-Section of Figure 5.2 as

$$C = \frac{2L}{1 + \omega_0^2 L^2} > 0$$

Step 4: Compute the normalized value of the series arm, reverse-biased diode capacitor C_{D1} such that

$$C_{D1} = \frac{1}{\omega_0^2 L}$$

It is noted that, if C_{D1} is too small or too big to be manufactured, then one can select a reasonable diode capacitor value and adjust its value by connecting either a parallel or series capacitor to D_1, respectively.

Step 5a: Select a proper normalized value for C_{D2} and compute C_A as

$$C_A = \left(\frac{C}{2}\right) + \frac{1}{2}\sqrt{C^2 + 4CC_{D2}} > 0$$

At this point, the designer may wish to select $C_{D2} = C_{D1}$ if appropriate.

Step 5b: Generate C_T as in (6.2e)

$$C_T = \frac{C_A C_{D2}}{C_A + C_2}$$

Step 5c: Generate L_A as in (6.2h)

$$L_A = \frac{1}{\omega_0^2 C_T}$$

Step 6: Computation of S_{21A} of State-A

Step 6a: Obtain the forward bias resistance R_{f1} and reverse bias resistance R_{r1} of D_1, which may be derived from C_{D1} either by computations based on geometric size of the diode measurement or from the manufacturer and normalize then with respect to R. For example, if one employs an MOS diode, then $R_r \cong 0$ and for 180 μm Si process R_{f1} can be computed using (6.1) as

$$R_{f1}\,(\Omega) \times C_{D1}\,(Farad) = 672 \times 10^{-15}$$

Normalized forward-biased resistance is defined as $r_{f1} = R_{f1}/R$

Step 6b: Generate the normalized series arm impedance $z_a(j\omega) = r_a + j\omega L$, where ω is the normalized angular frequency. It may run from zero to perhaps up to $\omega = 5$. Resistance r_a designates the overall estimated loss of the series arms, which includes the forward bias resistance of diodes D_1 loss of the inductors L as well as the resistive loss of the connections. For many practical cases, $r_a = r_{f1}$ is selected.

Step 6c: Generate the normalized shunt arm admittance $y_a = G_A + j\omega \left(C_A - \frac{1}{\omega^2 L_A}\right)$ of State-A, where G_A is the overall estimated shunt conductive loss of the parallel combination inductor L_A and capacitor C_A. It is noted that, usually, G_A is negligible as compared to the loss effect of r_{f1}. In other words, if noticeable, r_{f1} is more dominant loss factor over G_A. Therefore, G_A may be set to zero in the course of performance computations.

Step 6d: Compute $S_{21A}(j\omega) = \rho_{21A}(\omega) e^{j\varphi_{21A}(\omega)}$ and $S_{11A}(j\omega) = \rho_{11A}(\omega) e^{j\varphi_{11A}(\omega)}$ as in (5.9a) and (5.9b), respectively, as

$$S_{21} = \frac{2}{z^2 y + 2zy + 2z + y + 2}; \quad \left\{ \begin{array}{l} z = z_a \\ y = y_a \end{array} \right\}$$

$$S_{11} = \frac{\left(1 - z^2\right) y - 2z}{z^2 y + 2zy + 2z + y + 2}; \quad \left\{ \begin{array}{l} z = z_b \\ y = y_b \end{array} \right\}$$

Step 7: Computation of S_{21B} and S_{11B} of State-B

Step 7a: Generate the normalized series arm impedance $z_b(j\omega) = r_b + j\omega L + \frac{1}{j\omega C_{D1}}$, where ω is the normalized angular frequency. It may run from zero to perhaps up to $\omega = 5$. The resistance r_b is associated overall loss of the series arms. For many practical cases, r_b is selected as the reverse bias resistance of the diodes D_1.

Step 7b: Generate the normalized shunt arm admittance $y_b = G_B + j\omega \left(C_T - \frac{1}{\omega^2 L_A}\right)$, where G_B is the estimated overall conductive loss of the shunt arm in State-B.

Step 7c: Compute $S_{21B}(j\omega) = \rho_{21B}(\omega) e^{j\varphi_{21B}(\omega)}$ and $S_{11B}(j\omega) = \rho_{11B}(\omega) e^{j\varphi_{11B}(\omega)}$ as in (5.9a) and (5.9b), respectively, as

$$S_{21B} = \frac{2}{z^2y + 2zy + 2z + y + 2}\Big|_{\substack{z = z_b; \\ y = y_b}}$$

$$S_{11B} = \frac{(1 - z^2)\, y - 2z}{z^2y + 2zy + 2z + y + 2}\Big|_{\substack{z = z_b; \\ y = y_b}}$$

Step 7d: Plot the results

Now, let us run an example to implement the above design algorithm.

Example 6.1: Let $\omega_0 = 1,\ R = 1,\ r_a = r_b = 0,\ G_A = G_B = 0$.

(a) Design a narrow bandwidth, $0° - -180°$ Phase Range LPT-DPG for the net-phase shift $\Delta\theta = 45°$.

(b) Investigate the effect of the 10% phase deviation on the frequency band.

(c) Investigate the gain characteristic of the LPT-DPS.

Solution

Let us follow the computational steps of the design algorithm

Step 1: Normalized angular frequency is already provided, which is $\omega_0 = 1$.

Step 2: For $\varphi_{21A}(\omega_0) = 45°$ normalized series arm inductor L is given by

$$L = \frac{1}{\omega_0}\tan\left(\frac{|\varphi_{21A}(\omega_0)|}{2}\right) = 0.4142$$

Step 3: Normalized ideal shunt arm capacitor C is

$$C = \frac{2L}{1 + \omega_0^2 L^2} = 0.7071$$

Step 4: Normalized series arm reverse bias diode capacitance is given by

$$C_{D1} = \frac{1}{\omega_0^2 L} = 2.4142$$

Step 5a: By selecting $C_{D2} = C_{D1} = 2.4142$, normalized shunt arm capacitor C_A is found as

$$C_A = \left(\frac{C}{2}\right) + \frac{1}{2}\sqrt{C^2 + 4CC_{D2}} = 1.7071$$

Step 5b: Compute C_T

$$C_T = \frac{C_A C_{D2}}{C_A + C_2} = 1$$

Step 5c: Compute L_A

$$L_A = \frac{1}{\omega_0^2 C_T} = 1$$

Steps 6 and 7 is implemented under a MatLab program called

"Main_Lowpass_TSection_DPS.m"

This program computes the scattering parameters of the $180°$ phase range LPT-DPS topology for both State-A and State-B and prints the performance of the digital phase sifter under consideration.

In Figure 6.4a, the phase variations of S_{21A} and S_{21B} are depicted.

A close examination of Figure 6.4a reveals that at $\omega_0 = 1$, $\Delta\theta = 45°$ as it should be. Furthermore, $\varphi_{21B}(1) = 0$ as expected.

In Figure 6.4b, gain performances of State-A and State-B are shown.

(b) Close examination of Figure 6.4a indicates that 10% phase fluctuation about $\omega_0 = 1$ results in phase-shift band over $\omega_1 = 0.849$ and $\omega_2 = 1.27$. Actually, the bandwidth is approximately 20%.

(c) Referring to Figure 6.4b, for 10% fluctuations of the phase shift performance, gain performance is better than that of 0.3 dB. Therefore, we can

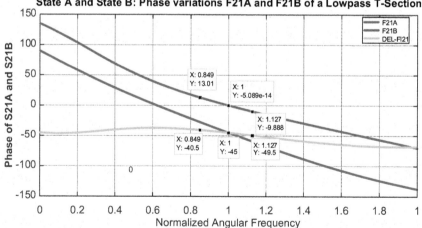

Figure 6.4a Phase variations of LPT-DPS for State-A, State-B, and $\Delta\theta = \varphi_{21A} - \varphi_{21B}$ of Example 6.1.

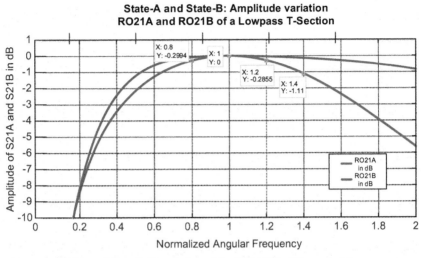

Figure 6.4b   Gain variation of the LPT-DPS of Example 6.1.

confidently state that LPT-DPS provides about 20% bandwidth for 10% phase fluctuation when $\Delta\theta\left(1\right) = 45°$.

Example 6.2: Repeat Example 6.1 for $\Delta\theta = 90°$.

Solution

In Figure 6.5a, phase characteristics of LPT-DPS are shown.
For the topology under consideration, the following component values are found.

Inputs:

$$\omega_0 = 1 \; (Given)$$

$$r_{f1} = r_{r2} = 0 \; (Given)$$

$$G_A = G_B = 0 \; (Given)$$

$$L = \frac{1}{\omega_0}\tan\left(\frac{|\varphi_{21A}\left(\omega_0\right)|}{2}\right) = 1$$

Computational steps:

$$\omega_0 = 1 \; (It \; is \; given \; as \; normalized)$$

$$C = \frac{2L}{1 + \omega_0^2 L^2} = 1$$

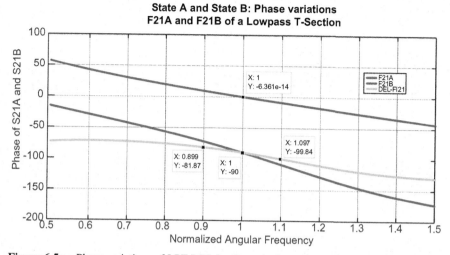

Figure 6.5a Phase variations of LPT-DPS for State-A, State-B and $\Delta\theta = \varphi_{21A} - \varphi_{21B}$ of Example 6.2.

$$C_{D1} = \frac{1}{\omega_0^2 L} = 1$$

$$C_{D2} = C_{D1} \; (our \; selection)$$

$$C_A = \left(\frac{C}{2}\right) + \frac{1}{2}\sqrt{C^2 + 4CC_{D2}} = 1.6180$$

$$C_T = \frac{C_A C_{D2}}{C_A + C_2} = 0.6180$$

$$L_A = \frac{1}{\omega_0^2 C_T} = 1.6180$$

The phase characteristic of the 90° LPT-DPS is depicted in Figure 5a.

A close examination of Figure 6.5a reveals that 10% phase shift fluctuations about $\omega_0 = 1$ shrinks the bandwidth approximately 10%. As compared to Example 6.1, we can confidently state that "larger the phase shift is smaller the bandwidth".

In Figure 6.5b, gain variation of the LPT-DPS is depicted. Over the 10% phase bandwidth, gain is above −0.3 dB. Therefore, we say that "gain characteristics is more comfortable than that of phase the characteristics of LPT-DPS".

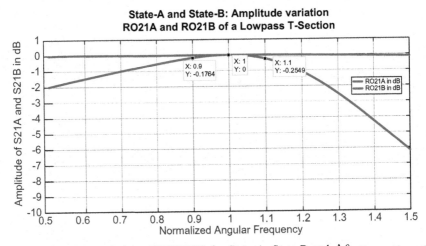

Figure 6.5b Gain variations of LPT-DPS for State-A, State-B and $\Delta\theta=\varphi_{21A}-\varphi_{21B}$ of Example 6.2.

6.5 Effect of Circuit Component Losses on the Electric Performance of the LPT-DPS

In this section, we investigate the phase and the gain performance of LPT-DPS with respect to diode and other component lossless. For this purpose, we presume that in State-A, the series arm impedance is approximated as

$$z_a\left(j\omega\right) \; = \; r_a+j\omega L \tag{6.4a}$$

where r_a includes forward bias diode resistance and other related component and connection lossless in the series arm.

Similarly, the shunt arm admittance of State-A is specified by

$$y_a \; = \; G_A+\frac{1}{j\omega L_A}+j\omega C_A \tag{6.4b}$$

where G_A is the overall estimated conductive losses of the circuit components and connections.

In State-B, the symmetric series arm impedance may be approximated as

$$z_b\left(j\omega\right) \; = \; r_b+j\omega L+\frac{1}{j\omega C_{D1}} \tag{6.4c}$$

and the shunt arm impedance is estimated as

$$y_b = G_B + \frac{1}{j\omega L_A} + j\omega C_T \tag{6.4d}$$

In a similar manner to State-A, r_b and G_B are associated with the resistive and conductive overall losses of the series and shunt arms, respectively.

The phase shift and the gain performance of LPT-DPS can be generated using (5.9) as follows

$$S_{11(A,B)} = \frac{(1-z^2)\, y - 2z}{z^2 y + 2zy + 2z + y + 2}; \quad \left\{ \begin{array}{l} \text{In State A} : z = z_a,\ y = y_a \\ \text{In State B} : z = z_b,\ y = y_b \end{array} \right\} \tag{6.4e}$$

and

$$S_{21(A,B)} = \frac{2}{z^2 y + 2zy + 2z + y + 2}; \quad \left\{ \begin{array}{l} \text{In State A} : z = z_a,\ y = y_a \\ \text{In State B} : z = z_b,\ y = y_b \end{array} \right\} \tag{6.4f}$$

Now let us estimate r_a, r_b, G_A, and G_B as follows.

In practice, a single inductor L includes a series resistive loss r_L. At the center frequency ω_0, the value of the loss resistance is proportional to inductance impedance $\omega_0 L$.

Let δ_{IND} be the multiplying constant of $\omega_0 L$ such that

$$r_L = (\omega_0 L)\, \delta_{IND} \tag{6.4g}$$

The size of δ_{IND} depends on the manufacturing technology of the inductor. For example, if one employs discrete component manufacturing technology, δ_{IND} could be much less than 10%. If the MMIC technology is used, then, it goes beyond 12.5%. Notice that quality factor Q_L of the inductor is given by

$$\delta_{IND} = \frac{1}{Q_L} \tag{6.4h}$$

In a similar manner, a capacitor C includes a a shunt conductive loss G_C. At the center frequency ω_0, G_C is proportional to the admittance $\omega_0 C$. In this case, one can find a proper multiplying constant δ_{CAP}, which estimates the conductive loss of a capacitor such that

$$G_C = (\omega_0 C)\, \delta_{CAP} \tag{6.4i}$$

For discrete capacitors, δ_{CAP} is much less than 10%. However, for MMIC implementation, δ_{CAP} value may even go beyond 15%. If Q_C is the quality factor of a capacitor, then,

$$\delta_{CAP} = \frac{1}{Q_C} \tag{6.4j}$$

In State-A, when the diode D_1 is forward biased, the series arm impedance becomes

$$z_a = (r_L + r_{f1} + r_s) + j\omega L = r_a + j\omega L$$

In this case, the equivalent series arm resistive loss is estimated as

$$r_a \cong r_L + r_{f1} + r_s \qquad (6.5a)$$

where r_s is the estimated resistive loss of all the connections. In (6.5a), the dominant term may be r_{f1}.

Therefore, r_s can be neglected as compared to r_{f1} and r_s.

In the shunt arm, the State-A inductor impedance is given by $r_{LA} + j\omega L_A$. Capacitor C_A is connected to forward-biased resistance r_{f2} of D_2. In this case, let the conductive loss of C_A be specified as G_{CA}, and the resulting admittance of the shunt capacitive arm is expressed as

$$Y_{CA} = G_{CA} + \frac{j\omega C_A}{1 + j\omega C_A r_{f2}}$$

In this case, equivalent admittance y_a is given by

$$\begin{aligned}
y_a &= \frac{1}{r_{LA} + j\omega L_A} + G_{CA} + \frac{j\omega C_A}{1 + j\omega C_A r_{f2}} \\
&= \frac{r_{LA}}{r_{LA}^2 + \omega^2 L_A^2} - j\frac{\omega L_A}{r_{LA}^2 + \omega^2 L_A^2} + G_{CA} \\
&\quad + \frac{\omega^2 r_{f2} C_A^2}{1 + \omega^2 r_{f2}^2 C_A^2} + \frac{j\omega C_A}{1 + \omega^2 r_{f2}^2 C_A^2}
\end{aligned}$$

where r_{f2} is the forward bias loss resistance of the diode D_2. At this point, let us note the following practical issues.

(a) Here, we assumed that r_{LA} also includes the series connection resistance r_{s2} of the shunt arm.

(b) The term $\omega_0^2 r_{f2}^2 C_A^2$ must be small enough as compare to unity due to negligible size of r_{f2}.

In the light of the above comments, y_a may be expressed in the neighborhood of ω_0 as

$$y_a = \left(G_{CA} + \frac{r_{LA}}{r_{LA}^2 + \omega_0^2 L_A^2} + \frac{\omega_0^2 r_{f2} C_A^2}{1 + \omega_0^2 r_{f2}^2 C_A^2} \right)$$

$$+\frac{j\omega C_A}{1+\omega^2 r_{f2}^2 C_A^2}-j\frac{\omega L_A}{r_{LA}^2+\omega^2 L_A^2}$$

$$=G_A+j\omega C_A+\frac{1}{j\omega L_A}$$

where the equivalent shunt conductive loss is approximated as

$$G_A \cong G_{CA}+\frac{r_{LA}}{r_{LA}^2+\omega_0^2 L_A^2}+\frac{\omega_0^2 r_{f2}C_A^2}{1+\omega_0^2 r_{f2}^2 C_A^2} \tag{6.5b}$$

Similarly, in State-B, the series arm loss r_b is estimated as

$$r_b \cong r_L+r_{r1}+r_s \tag{6.5c}$$

and G_B is approximated as

$$G_B \cong \frac{r_{LA}}{r_{LA}^2+\omega_0^2 L_A^2}+\frac{\omega_0^2 r_{r2}C_T^2}{1+\omega_0^2 r_{r2}^2 C_T^2}+G_{CA} \tag{6.5d}$$

where r_{r2} is the revere biase rsistive loss of the diode D_2.

It is noted that reverse bias diode loss resistance r_r is usually very small. It may be estimated by

$$r_{ri} = (\omega_0 C_{Di})\,\delta_D; \quad \{i = 1,2\} \tag{6.5e}$$

where δ_D is much less than 10%.

Let us propose the following algorithm to develop component losses introduced above.

6.6 Algorithm to Compute Component Lossless of LPT-DPS

This algorithm generates the component losses of LPT-DPS, which in turn yields series arm and shunt arm immittances in both State-A and State-B. Here, we presume that LPT-DPS is realized as MMIC on a silicon substrate using the 0.18u technology of TMSC.

Now let us investigate the possible losses of Example 6.1.

Inputs:

f_{0a}: Actual center frequency
ω_0: Normalized angular frequency

R: Normalization resistor

δ_{IND}: Multiplying factor to determine the series loss resistance r_L of an inductor L

δ_{CAP}: Multiplying factor to determine the shunt loss conductance G_C of a capacitor C

δ_D: Multiplying factor to determine the series loss resistance r_r of a reverse bias diode D

Computational Steps:

Step 1: Compute

$$L = \frac{1}{\omega_0} \tan\left(\frac{|\varphi_{21A}(\omega_0)|}{2} \right)$$

Step 2: Compute

$$C = \frac{2L}{1 + \omega_0^2 L^2}$$

Step 3: Compute C_{D1}

$$C_{D1} = \frac{1}{\omega_0^2 L}$$

Step 4: Select C_{D2} and compute

$$C_A = \left(\frac{C}{2} \right) + \frac{1}{2}\sqrt{C^2 + 4CC_{D2}}$$

Note: One may wish to set $C_{D2} = C_{D2}$

Step 5: Compute

$$C_T = \frac{C_A C_{D2}}{C_A + C_2}$$

Step 6: Compute

$$L_A = \frac{1}{\omega_0^2 C_T}$$

Step 7: Generate actual values for C_{D1} and C_{D2}

$$C_{D1a} = \frac{C_{D1}}{(2\pi f_{0a})R}$$

$$C_{D2a} = \frac{C_{D2}}{(2\pi f_{0a})R}$$

Step 8: Generate normalized values for r_{f1} and r_{f2}

$$R_{f1}\,(\Omega) \times C_{D1a}\,(Farad) = 672 \times 10^{-15}$$
$$R_{f2}\,(\Omega) \times C_{D2a}\,(Farad) = 672 \times 10^{-15}$$
$$r_{f1} = \frac{R_{f1}}{R}$$
$$r_{f2} = \frac{R_{f2}}{R}$$

Step 9: Generate r_{r1} and r_{r2}

$$r_{r1} = (\omega_0 C_{D1})\,\delta_D$$
$$r_{r2} = (\omega_0 C_{D2})\,\delta_D$$

Step 10: Generate r_{LA} and G_{CA}

$$r_{LA} = (\omega_0 L_A)\,\delta_L$$
$$G_{CA} = (\omega_0 C_A)\,\delta_C$$

Step 11: Generate r_L, r_a and G_A of State-A

$$r_L = (\omega_0 L)\,\delta_L$$
$$r_a = r_L + r_{f1}$$
$$G_A = G_{CA} + \frac{r_{LA}}{r_{LA}^2 + \omega_0^2 L_A^2} + \frac{\omega_0^2 r_{f2} C_A^2}{1 + \omega_0^2 r_{f2}^2 C_A^2}$$

Step 12: Generate r_b and G_B of State-B

$$r_b = r_L + r_{r1}$$
$$G_B = G_{CA} + \frac{r_{LA}}{r_{LA}^2 + \omega_0^2 L_A^2} + \frac{\omega_0^2 r_{r2} C_T^2}{1 + \omega_0^2 r_{r2}^2 C_T^2}$$

Step 13: Generate z_a and y_a of State-A

$$z_a\,(j\omega) = r_a + j\omega L$$
$$y_a = G_A + \frac{1}{j\omega L_A} + j\omega C_A$$

Step 14: Generate z_b and y_b of State-B

$$z_b(j\omega) = r_b + j\omega L + \frac{1}{j\omega C_{D1}}$$

$$y_b = G_B + \frac{1}{j\omega L_A} + j\omega C_T$$

Step 15: Generate the phase and the gain performance of LPT-DPS for State-A and State-B

$$S_{21(A,B)} = \frac{2}{z^2 y + 2zy + 2z + y + 2}; \quad \left\{ \begin{array}{l} \text{In State A}: z = z_a,\ y = y_a \\ \text{In State B}: z = z_b,\ y = y_b \end{array} \right\}$$

Step 16: Plot the phase and gain performance
   Now, let us run an example to evaluate the performance of an LPT-DPS constructed with lossy elements.

Example 6.3: Referring to Example 6.1, let the switching diodes are made of 0.180 um CMOS process of TSMC. Let the actual design frequency be $f_{0a} = 5\,GHz = 5 \times 10^9$ Hz.
   Let $\delta_{IND} = 10$, $\delta_{CAP} = 100$, $\delta_D = 100$,
   (a) Compute the ideal element values of the PT-DPS.
   (b) Compute the resistive and conductive losses associated with the ideal elements
   (c) Evaluate the phase shifting performance of the phase shifter constructed with lossy elements.
   (d) Compute the actual component values

Solution

(a) Ideal element values of the LPT-DPS under consideration are given as in Example 6.1.

$$L = 0.4142, C = 0.7071, CD1 = 2.4142, CD1 = CD2, CA = 1.7071,$$
$$CT = 1, LA = 1,$$

(b) Losses associated with component values:
Let us follow steps of the algorithm.

Step 8: Generate normalized values for r_{f1} and r_{f2}

$$R_{f1}(\Omega) \times C_{D1a}(Farad) = 672 \times 10^{-15}; \quad R_{f1} = 0.4372\,\Omega$$

$$R_{f2}\left(\Omega\right) \times C_{D2a}\left(Farad\right) = 672 \times 10^{-15}; \quad R_{f2} = 0.4372\Omega$$

$$r_{f1} = \frac{R_{f1}}{R} = 0.0087$$

$$r_{f2} = \frac{R_{f2}}{R} = 0.0087$$

Step 9: Generate r_{r1} and r_{r2}

$$r_{r1} = \left(\omega_0 C_{D1}\right)\delta_D = 0.0041$$

$$r_{r2} = \left(\omega_0 C_{D2}\right)\delta_D = 0.0041$$

Step 10: Generate r_{LA} and G_{CA}

$$r_{LA} = \left(\omega_0 L_A\right)\delta_L = 0.1000$$

$$G_{CA} = \left(\omega_0 C_A\right)\delta_C = 0.0171$$

Step 11: Generate r_L, r_a, and G_A of State-A

$$r_L = \left(\omega_0 L\right)\delta_L = 0.0414$$

$$r_a = r_L + r_{f1} = 0.0543$$

$$G_A = G_{CA} + \frac{r_{LA}}{r_{LA}^2 + \omega_0^2 L_A^2} + \frac{\omega_0^2 r_{f2} C_A^2}{1 + \omega_0^2 r_{f2}^2 C_A^2} = 0.1416$$

Step 12: Generate r_b and G_B of State-B

$$r_b = r_L + r_{r1} = 0.0497$$

$$G_B = G_{CA} + \frac{r_{LA}}{r_{LA}^2 + \omega_0^2 L_A^2} + \frac{\omega_0^2 r_{r2} C_T^2}{1 + \omega_0^2 r_{r2}^2 C_T^2} = 0.1248$$

Steps 13–15: All the above and rest of the steps are programmed under the MatLab Program called

$$"Main\_Example\_6\_3.m"$$

Execution of "$Main\_Example\_6\_3.m$ reveals the phase performance of the LPT-DPS under consideration as depicted in Figure 6.6a.

A close examination of Figure 6.6a reveals that, at $\omega_0 = 1$, the phase shift between the switching states is –44.87°. So, introduction of lossy elements perturbed the phase performance of the phase shifter about 0.2889%, which

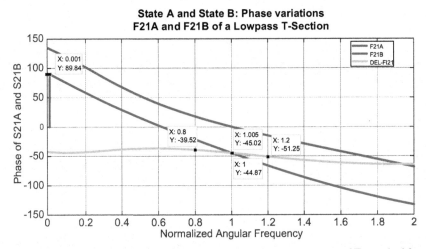

Figure 6.6a   Phase performance of LPT-DPS with lossy components of Example 6.3.

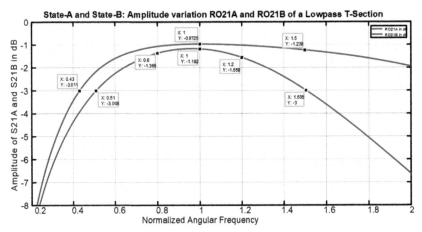

Figure 6.6b   Gain performance of LPT-DPS with lossy components of Example 6.3.

is tolerable. The lossy components did not affect the bandwidth of the phase shifter. It is still $\mp20\%$ about the center frequency.

Gain performance of the phase shifter topology is shown in Figure 6.6b. Over the $\mp20\%$ bandwidth, gain is better than -1.6 dB, which is acceptable. As far as gain curves are concerned, over the angular frequency interval of $0.5 < \omega < 1.5$, the gain is about -3dB.

MatLab codes of this Example 6.3 are given under Program Lists.

(d) Computation of actual element values
Actual element values for the capacitors are given as

$$C_{actual} = \frac{C_{normalized}}{2\pi f_{0a} R}$$

Similarly, actual element values for the inductors are computed as

$$L_{actual} = \frac{R L_{normalized}}{2\pi f_{0a}}$$

Hence, based on the above equations, actual diode capacitances are given as

$$C_{D1a} = C_{D2a} = 1.5369e - 12 = 1.53 \, pF$$

Actual series arm inductor L_a is

$$L_a = 6.5924e - 10 = 65.92 \, nH$$

In the shunt arm, shunt inductor L_{Aa} is given by

$$L_{Aa} = 1.5915e - 09 = 1.591 \, nH$$

And finally, the shunt arm capacitor is

$$C_{Aa} = 1.0868e - 12 = 1.086 \, pF$$

All the above values are realizable.
In the next example, we will investigate effect of the large phase shift on the phase and gain characteristics.

Example 6.4: Using MatLab program "$Main\_Example\_6\_3.m$", repeat Example 6.3 for $\theta = 120°$ and report the results.

Solution

Execution of "$Main\_Example\_6\_3.m$" results in the following actual element values

$$CD1a = 0.3676 \, pF; \ CAa = 0.8 \, pF;$$
$$LAa = 4.017 \, nH; \ La = 2.766 \, nH$$

Actual values of r_a and r_b are given by

$$r_{a-actual} = 11.3546\Omega$$

$$r_{b-actual} = 10.3923\Omega$$

And actual values of G_A and G_B are computed as

$$G_{A-actual} = 0.0022 \; siemens \; (or \; shunt \; resistor$$

$$R_{A-actual} = \frac{1}{G_{A-actual}} = 455\Omega)$$

$$G_{B-actual} = 0.0012 \; siemens \; (or \; shunt \; resistor$$

$$R_{B-actual} = \frac{1}{G_{B-actual}} = 868\Omega)$$

The phase performance of the LPT-DPS is depicted in Figure 6.7a.

A close examination of Figure 6.7a reveals that the 10% phase variation is confined over the bandwidth of $0.907 \leq \omega \leq 1.124$. In this interval, phase changes from $108°$ to $132°$. Target phase of $120°$ shift occurs at $\omega_0 = 1.019$. So, there is a small drift of $\Delta\omega = 0.019$ from the center frequency $\omega_0 = 1$. This shift is tolerable.

Gain curves of both State-A and State-B are given in Figure 6.7b. For $\Delta\theta = 120°$, component losses heavily penalize the gain below half power or $-3dB$. At the center frequency, gain is about -3.47 dB, which is not acceptable.

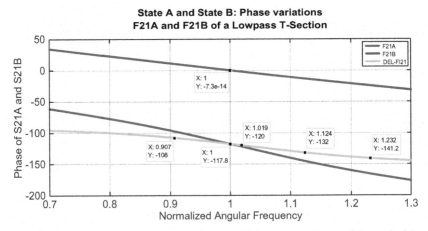

Figure 6.7a   Phase performance of LPT-DPS with lossy components of Example 6.4.

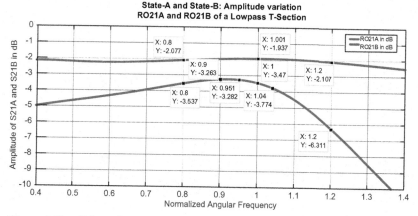

Figure 6.7b Gain performance of LPT-DPS with lossy components of Example 6.4.

6.7 General Comments and Conclusion

The symmetric 3-element LCL-type low-pass-based T-Section digital phase shifter topology (LPT-DPS) is easy to implement as an MMIC circuit. It is compact and safely provides 10% bandwidth with reasonable loss. On the other hand, LPT-DPS is not suitable to provide 180° phase shift. Nevertheless, two units of 90° LPT-DPS can be cascaded to design a 180° phase shifting cell.

Examples presented in this chapter verifies the below statements.

1. LPT-DPS is good for narrow phase range and small bandwidth requirements. It offers reliable phase shift up 90°
2. Bandwidth of the LPT-DPS decreases while phase shift ranges increases.
3. Component losses do not much effect the phase shifting performance of the LPT-DPS topology.
4. LPT-DPS topology is suitable for MMIC implementation.

Appendix 6: MatLab Programs for Chapter 6

Program List 6.1: Main_Lowpass_TSection_DPS.m

```
% Main Program: Main_Lowpass_TSection_DPS.m
% February 20, 2019
% Developed by BS Yarman, Vanikoy, Istanbul
% Enter Positive value
%
% --------------------------------------------------------
clc; close all
% Inputs:
```

```
teta=input('Alternative Formulas to Design Lowpass T Section. Enter
positive value for Teta in Degree=')
w0=1;
% Alternative way to generate inductor L
L=tand(teta/2)/w0;\% C=
2*L/(1+w0*w0*L*L);
%-----------------------------------------------------------------
% Component Values
CD1=1/(w0^2*L);
CD2=CD1;
DEN=2;
Delta=sqrt(C*C+4*C*CD2);
CA=(C)/DEN+Delta/DEN;
CT=CA*CD2/(CA+CD2);
LA=1/w0/w0/CT;
j=sqrt(-1);
w=0;N=1000;w1=0;w2=2;DW=(w2-w1)/N;
FRI(1:(N+1))=zeros;
for i=1:N+1
    WA(i)=w;
%-------------------
% State A:
%    --------------
za=j*w*L;
ya=1/j/w/LA+j*w*CA;
  [ S11a,S21a,RO11a,F11a,RO21a,F21a ] = S_Par_LPT_DPS (za,ya);
    F21A(i)=F21a;
    RO21A(i)=RO21a;
    F11A(i)=F11a;
    RO11A(i)=RO11a;
  %  --------------------
% State-B
%   -------------------
  zb=j*w*L+1/j/w0/CD1;
  yb=1/j/w/LA+j*w*CT;
  [ S11b,S21b,RO11b,F11b,RO21b,F21b ] = S_Par_LPT_DPS (zb,yb);
    F21B(i)=F21b;
    RO21B(i)=RO21b;
    F11B(i)=F11b;
    RO11B(i)=RO11b;
%------------------------------------------------------------------
  w=w+DW;
end
%-----------------------------------------------------------------
Plot_State_AB_LPT_DPS(WA,F21A,RO21A, F11A,RO11A,F21B,RO21B, F11B,RO11B)
```

Program List 6.2: S_Par_LPT_DPS

```
function [ S11,S21,RO11,F11,RO21,F21 ] = S_Par_LPT_DPS (z,y)
%This function generates the S-Parameters of a Lowpass T Section
% Phase Shifter from the series arm impedance Z(jw) and
% the shunt arm admittance Y(jw)
% computed series arm inductor L and the shunt capacitor C
```

```
% Developed by BS Yarman: Feb 20, 2019, Vanikoy, Istanbul
% See Equations (5.9)
%  ---------------------------------------------------------
D=z*z*y+2*z*y+2*z+y+2;
S11=((1-z*z)*y-2*z)/D;
S21=2/D;
R11=real(S11); X11=imag(S11);F11=atan2d(X11,R11);
R21=real(S21); X21=imag(S21);F21=atan2d(X21,R21);
RO11=abs(S11);
RO21=abs(S21);
end
```

Program List 6.3: function Plot_State_AB_LPT_DPS

```
function Plot_State_AB_LPT_DPS(WA,F21A,RO21A, F11A,RO11A,
F21B,RO21B, F11B,RO11B)
figure
plot(WA,F21A,WA,F21B)
title('State A and State B: Phase variations F21A and F21B of
a Lowpass T-Section')
legend('F21A','F21B')
xlabel('Normalized Angular Frequency')
ylabel('Phase of S21 and S21B')
% Amplitude of S21
figure
plot(WA,RO21A,WA,RO21B)
title('State-A and State-B: Amplitude variation RO21A and RO21B of a
Lowpass T-Section')
legend('RO21A','RO21B')
xlabel('Normalized Angular Frequency')
ylabel('Amplitude of S21A and S21B')
%---------------------------------------------------------
figure
plot(WA,F11A, WA, F11B)
title('State-A and State-B: Phase variation F11A and F11B of a
Lowpass T-Section')
legend('F11A')
xlabel('Normalized Angular Frequency')
ylabel('Phase of S11A and S11B')
% Amplitude of S11
figure
plot(WA,RO11A, WA, RO11B)
title('State-A and State-B: Amplitude variation RO11A and RO11B of a
Lowpass T-Section')
legend('RO11A','RO11B')
xlabel('Normalized Angular Frequency')
ylabel('Amplitude of S11A and S11B')

end
```

Program List 6.4: Main Program: Main_Example_6_3.m

```
% Main Program: Main_Example_6_3.m
```

```
% February 22, 2019
% Developed by BS Yarman, Vanikoy, Istanbul
% This program evaluates the lossy performance of a LPT-DPS for a
specified
% center actual frequency f0 in Hz.
%-----------------------------------------------------------------
clc; close all
% Inputs:
teta=input('Alternative Formulas to Design Lowpass T Section. Enter
positive value for Teta in Degree=')
% Inputs
f0a=input('Enter the actual center frequency in Hz f0a =')
w0=input('At f0a, enter the normalized angular frequency w0 =')
R=input('Enter the normaliziation Resistor R =')
%-----------------------------------------------------------------
% Compute the normalized element values of LPT-DPS
% Alternative way to generate inductor L
L=tand(abs(teta)/2)/w0;
C=2*L/(1+w0*w0*L*L);
%-----------------------------------------------------------------
% Component Values and their related resistive losses:
CD1=1/(w0^2*L); % See equation (6.2d) of Chapter 6
% ASSUMPTION 1:
% It is assumed that the series loss Ron of a forward biased diode is
% equal to the on channel resistor of an CMOS Switch
% [see Equation (6.1) of Chapter 6].
[ RF1,rf1 ] =Channel_Resistance_of_a_CMOS(CD1,R,f0a);
% ASSUMPTION 2:
% Reverse biased resistive loss of a diode is the "Percent_RVS"
amount of
% its reverse baised impedance at w0
Percent_RVS=100;
rr1=1/w0/CD1/Percent_RVS;
CD2=CD1;
 [ RF2,rf2 ] =Channel_Resistance_of_a_CMOS(CD2,R,f0a);
rr2=1/w0/CD2/Percent_RVS;
%
DEN=2;
Delta=sqrt(C*C+4*C*CD2);
CA=(C)/DEN+Delta/DEN;
CT=CA*CD2/(CA+CD2);
LA=1/w0/w0/CT;
%-----------------------------------------------------------------
% Loss Computations for both State-A and State-B:
% Assumption 3: Loss of an inductor is Percent_L amount of its impedance
% value at w0.
% Assumption 4: Connectivity loss of an inductor is "Percent_S"amount of
% its impedance value at w0.
% Assumption 5: Conductive loss of a Capacitor is "Percent_C" amount of
its
% admittance value at w0.
Percent_S=100;
Percent_L=10;
```

```
Percent_C=100;
% Resistive loss of the series arms in State-A:
rL=w0*L/Percent_L;
rs=w0*L/Percent_S;
ra=rL+rf1+rs;

% Conductive loss of the Shunt arm in State-A:
rLA=w0*LA/Percent_L;
GCA=w0*CA/Percent_C;
GA=GCA+rLA/(rLA*rLA+w0*w0*LA*LA)+(w0*w0*rf2*CA*CA)
/(1+w0*w0*rf2*rf2*CA*CA);

% Resistive loss of the series arms in State-B:
rb=rL+rr1+rs;

% Conductive loss of the Shunt arm in State-:

GB=GCA+rLA/(rLA*rLA+w0*w0*LA*LA)+(w0*w0*rf2*CT*CT)
/(1+w0*w0*rr2*rr2*CT*CT);

j=sqrt(-1);
w=0;N=2000;w1=0;w2=2;DW=(w2-w1)/N;
FRI(1:(N+1))=zeros;
for i=1:N+1
    WA(i)=w;
% --------------------
% State A:
% --------------------
za=ra+j*w*L;
ya=GA+1/j/w/LA+j*w*CA;
  [ S11a,S21a,RO11a,F11a,RO21a,F21a ] = S_Par_LPT_DPS (za,ya);
    F21A(i)=F21a;
    RO21A(i)=20*log10(RO21a);
    F11A(i)=F11a;
    RO11A(i)=20*log10(RO11a);
% --------------------
% State-B
% --------------------
  zb=rb+j*w*L+1/j/w0/CD1;
  yb=GB+1/j/w/LA+j*w*CT;
  [ S11b,S21b,RO11b,F11b,RO21b,F21b ] = S_Par_LPT_DPS (zb,yb);
    F21B(i)=F21b;
    RO21B(i)=20*log10(RO21b);
    F11B(i)=F11b;
    RO11B(i)=20*log10(RO11b);
DEL_FI21(i)=F21A(i)-F21B(i);
    w=w+DW;
 end

Plot_State_AB_LPT_DPS(WA,F21A,RO21A, F11A,RO11A,F21B,
RO21B,F11B,RO11B,DEL_FI21)
```

Program List 6.5: Channel_Resistance_of_a_CMOS

```
function [ Ron,ron ] = Channel_Resistance_of_a_CMOS(Coff,R,f0)
% This function generates the on forward biased channel resistance of a
% 0.180um CMOS switch manufactured by TSMC
% February 22, 2019, Vanikoy, Istanbul, Turkey
% Developed by BS Yarman
% Inputs:
% Coff: Normalized value of the Inductor L
% f0=Actual operating frequency (Normalization frequency)
% R: Actual terminatons resistors (Normalization Resistor)
% Output:
% Ron: Actual value of the channel resistor
% ron: Normalized Value of the channel resistor
%-----------------------------------------------------------------
[ Coff_Actual ] =ActualValues_of_a_Capacitor(Coff,R,f0);
% Ron (?) x C_D1 (Farad)=672 x 10^(-15)
Ron=672e-15/Coff_Actual;
ron=Ron/R;

end
```

Program List 6.6: function ActualValues_of_a_Capacitor

```
function [ C_actual ] = ActualValues_of_a_Capacitor(C,R,f0)
% February 22, 2019, Vanikoy, Istanbul, Turkey
% Developed by BS Yarman
% Inputs:
% C: Normalized value of the capacitor C
% f0=Actual operating frequency (Normalization frequency)
% R: Actual terminatons resistors (Normalization Resistor)
% Output:
% C_actual: Actual value of the capacitor
%-----------------------------------------------------------------
C_actual=C/2/pi/f0/R;
end
```

Program List 6.7: function ActualValues_of_an_Inductor

```
function [ L_actual ] =ActualValues_of_an_Inductor(L,R,f0)
% February 22, 2019, Vanikoy, Istanbul, Turkey
% Developed by BS Yarman
% Inputs:
% L: Normalized value of the Inductor L
% f0=Actual operating frequency (Normalization frequency)
% R: Actual terminatons resistors (Normalization Resistor)
% Output:
% L_actual: Actual value of the Inductor
%-----------------------------------------------------------------
L_actual=L*R/2/pi/f0;

end
```

References

[1] B. S. Yarman, "Low pass T-section digital phase shifter apparatus," U.S. Patent: 4630010, Washington DC, USA, December 16, 1986.

[2] B. Yarman, A. Rosen and P. Stabile, "Lowloss EHF Digital Phase Shifters Suitable for Monolithic Implementation," in IEEE ISCAS, Montreal, 1984.

[3] B. S. Yarman, "Design of Digital Phase Shifters Suitable for Monolithic Implementations," Bull. of Technical University of Istanbul, pp. 185–205, 1985.

[4] B. S. Yarman, "New Circuit Configurations for Designing 0–180 degree Digital Phase Shifters," IEE Proceeding H, Vol. 134, pp. 253–260, 1987.

[5] J. White, Microwave Semiconductor Devices, New York: Van Nostrand Reinhold Co., 1982.

[6] H. Watson, Microwave Semiconductor Devices and their applications, New York, NY: McGraw Hill, 1976.

[7] R. Garver, Microwave Doide Control Devices, Dedham, MA: Artech Inc., 1976.

[8] TSMC, "Taiwanese Semiconductor Manufacturing Company," (https://www.tsmc.com/english/aboutTSMC/company_profile.htm), Taiwan, 2019.

[9] B. S. Yarman, "Novel circuit configurations to design loss balanced 0°–360° digital phase shifters," AEU, vol. 45, no. 2, pp. 96–104, 1991.

[10] Binboga S. Yarman, "π-section digital phase shifter apparatus," U.S. Patent: 4604593, Washington DC, USA, August 5, 1986.

[11] H. M. Greenhouse, "Design of planar rectangular microelectronic inductors," IEEE Transactions on Parts, Hybrids, and Packaging, vol. 10, pp. 101–109, 1974.

[12] K. Chang, Handbook of RF Microwave Components and Engineering, New York, USA, Chechester, UK: Wiley, 2003.

[13] I. Bahl and B. Prakash, Microwave Solid State Circuit Design, 2. (1. John Wiley and Sons, Ed., 2003).

7

180° Low-pass-based PI-Section Digital Phase Shifter Topology (LPI-DPS)

Summary

In this chapter, a symmetric 3-element LC Low-pass-based PI-Section Digital Phase Shifter configuration is presented. The proposed phase shifter topology is compact and it is suitable to manufacture with discrete components, or as microwave monolithic integrated circuits (MMIC).

The content of this chapter is based on our previously completed research projects published in the open literature [1–5].

7.1 Low-pass-based Symmetric PI-Section Digital Phase Shifter

In Figure 7.1 a symmetric low-pass-based, PI-Section, Digital Phase Shifter ($LPI - DPS$) is shown.

This configuration includes three switching diodes, namely in the mid-series arm, we have diode D_1, and in the shunt arms, diodes D_2s, respectively.

In one state, say, State-A, when all diodes are forward-biased, the mid-series arm impedance Z acts as an inductor L, and at the operating frequency f_0, the shunt arms behave like a capacitor C yielding the admittance $Y = (2\pi f_0)C$, as shown in Figure 5.2.

In this mode of operation, ideally (i.e., $R_f = 0$), at a selected operation angular frequency w_0, and phase $\varphi_A(w_0)$, element values of Figure 7.1 are specified as in (5.31a) and (5.31b) as follows.

$$C = \left(\frac{1}{w_0}\right) \tan\left(\frac{|\theta|}{2}\right) > 0 \tag{7.1a}$$

$$L = \frac{2C}{1 + w_0^2 C^2} > 0 \tag{7.1b}$$

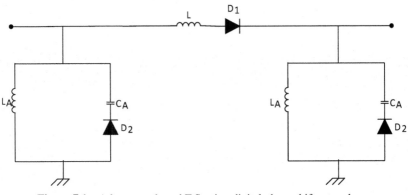

Figure 7.1   A low-pass-based T-Section digital phase shifter topology.

In this state, the shunt arm admittances are given by

$$Y(j\omega_0) = \frac{1}{j\omega_0 L_A} + j\omega_0 C_A = j\omega_0 C$$

or

$$C_A = C + \frac{1}{\omega_0^2 L_A} \tag{7.1c}$$

In State-B, all the diodes are reversed-biased. In this case, reverse-biased diode capacitance C_{D1} resonates with the mid-series arm inductor L of (7.1b). Hence, at the given operating frequency ω_0, C_{D1} is specified by

$$C_{D1} = \frac{1}{\omega_0^2 L} \tag{7.1d}$$

Similarly, in the shunt arms, inductor L_A resonates with the equivalent capacitor C_T, which is given by

$$C_T = \frac{C_A C_{D2}}{C_A + C_2} \tag{7.1e}$$

At the resonance, capacitor C_T is given by

$$C_T = \frac{1}{\omega_0^2 L_A} = \frac{C_A C_{D2}}{C_A + C_2} \tag{7.1f}$$

Using (7.1f) in (7.1c), we have

$$C_A = C + \frac{C_A C_{D2}}{C_A + C_{D2}}$$

Once C_{D2} is selected, the capacitor C_A can be determined by solving the following second-order equation.

$$C_A^2 - (C)\,C_A - (CC_{D2}) = 0$$

or

$$C_{A(1,2)} = \left(\frac{C}{2}\right) \mp \frac{1}{2}\sqrt{C^2 + 4CC_{D2}} > 0$$

Obviously, positive value for the capacitor C_A can only be obtained by selecting "+" sign in the above equation.

Hence,

$$C_A = \left(\frac{C}{2}\right) + \frac{1}{2}\sqrt{C^2 + 4CC_{D2}} > 0 \tag{7.1g}$$

Then, we can generate L_A as

$$L_A = \frac{1}{\omega_0^2 C_T} \tag{7.1h}$$

7.2 Algorithm to Design a Lowpass Based *PI* – Section Digital Phase Shifter

The phase shifter topology introduced in Figure 7.1 describes two distinct operational states, namely State-A and State-B. The phase of each state is determined based on the position of the switching diodes. For each diode, forward-biased situation may be named as "switch is ON" or "switch is closed"; and this position may be designated by the logic-state "1". When a diode is reverse-biased, it may be referred to as "switch is OFF" or "switch is open". This situation may be designated by the logic state "0".

"The phase $\theta(\omega)$ of a lossless two-port" is defined as the phase $\varphi_{21}(\omega)$ of the transfer scattering parameter $S_{21}\,(j\omega) = \rho\,(\omega)\,e^{j\varphi_{21}(\omega)}$. At the operating frequency, the objective is to make $\rho\,(\omega_0) = 1$ while reaching the desired phase $\theta\,(\omega_0) = \varphi_{21}(\omega_0)$.

For a two-port shown in Figure 7.1, the phase $\theta(\omega)$ varies depending on the position of the switches.

As described above, at a specified angular frequency ω_0, State-A refers to the phase $\theta_A = \varphi_{21A}\,(\omega_0)$ when all the switches are closed. In State-B, all the switches are opened and the two-port possesses the phase $\theta_B = \varphi_{21B}\,(\omega_0)$.

All the above explanations suggest the following truth stable to define the phase-switching states of the topology shown in Figure 7.1a.

Table 7.1 Description of switching states of LPI-DPS topology

	D_1	D_2	Phase θ of the two-port
State-A	1	1	$\theta_A = \varphi_{21B}(\omega_0)$
State-B	0	0	$\theta_B = \varphi_{21B}(\omega_0)$
Phase Shift Between States			$\Delta\theta = \theta_A - \theta_B$

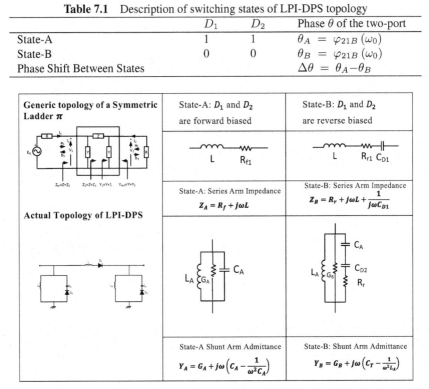

Figure 7.2 Immittance states of the low-pass-based *PI*-Section digital phase shifter topology.

In view of Table 7.1, Figure 7.1 is named as low-pass-based PI-Section digital phase shifter. It possesses two-phase switching states, namely State-A and State-B. The net phase shift between the states is given by

$$\Delta\theta = \theta_A - \theta_B \tag{7.2a}$$

However, it should be noted that

(a) By its nature, in State-B, at the angular frequency ω_0, θ_B is forced to be zero; and

(b) Range of θ_A must be less than $180°$ since $C = \dfrac{1}{\omega_0}\tan\left(\dfrac{|\varphi_{21A}(\omega_0)|}{2}\right)$ goes to infinity when $\varphi_{21A}(\omega_0) = 180°$.

In Figure 7.2, equivalent immittance states of Figure 7.1 are depicted. Thus, we propose the following design algorithm to construct LPI-DPS.

7.3 Algorithm to Design LPI-DPS for the Phase Range $180° < \Delta\theta = \theta_A < 0°$

In this algorithm, we present the design procedure to construct a low-pass π-Section Digital Phase Shifter, which can only cover the phase range of $0° - 180°$. Using this phase shifter topology, it is not possible to obtain exact $180°$ phase shifting unit.

Inputs:

f_{1a}: Lower end of the passband of interest

f_{2a}: Upper end of the passband of interest

f_{0a}: Actual operating frequency such that $f_{1a} \leq f_{0a} \leq f_{2a}$

R: Normalization resistance. For most practical situations, R is selected as the termination resistor, which is perhaps 50Ω.

$\Delta\theta = \varphi_{21A}(\omega_0)$: Desired phase shift

G_A: Overall estimated normalized conductive loss of the shunt arms with respect to R in State-A.

G_B: Overall estimated normalized conductive loss of the shunt arm with respect to R in State-B.

Computational Steps:

Step 1: Select a normalized angular frequency ω_0 such that it is confined within

$$f_{1a}/f_{0a} \leq \omega_0 \leq f_{2a}/f_{0a}$$

It is useful to know that for many cases ω_0 is set to unity (i.e. $\omega_0 = 1$).

Step 2: Compute the normalized shunt arm capacitor $C = \frac{1}{\omega_0}\tan\left(\frac{|\varphi_{21A}(\omega_0)|}{2}\right)$ of the ideal symmetric low-pass π-Section of Figure 5.6.

Step 3: Compute the normalized series arm inductor L of the ideal symmetric low-pass PI-Section of Figure 5.6 as

$$L = \frac{2C}{1 + \omega_0^2 C^2} > 0$$

Step 4: Compute the normalized value of the series arm, reverse-biased diode capacitor C_{D1} such that

$$C_{D1} = \frac{1}{\omega_0^2 L}$$

It is noted that, if C_{D1} is too small or too large to be manufactured, then one can select a reasonable diode capacitor value and adjust its value by connecting either a parallel or series capacitor to D_1, respectively.

Step 5a: Select a proper normalized value for C_{D2} and compute C_A as

$$C_A = \left(\frac{C}{2}\right) + \frac{1}{2}\sqrt{C^2 + 4CC_{D2}} > 0$$

At this point, the designer may wish to select $C_{D2} = C_{D1}$ if appropriate.

Step 5b: Generate C_T as in (6.2e)

$$C_T = \frac{C_A C_{D2}}{C_A + C_2}$$

Step 5c: Generate L_A as in (6.2h)

$$L_A = \frac{1}{\omega_0^2 C_T}$$

Step 6: Computation of S_{21A} of State-A

Step 6a: Obtain the forward bias resistance R_{f1} and reverse bias resistance R_{r1} of D_1, which may be derived from C_{D1} either by computations based on geometric size of the diode or by measurement; or it may be requested from the manufacturer. Then, normalize it with respect to R.

For example, if one employs a CMOS diode, then $R_r \cong 0$ and for $180\,\mu m$ Si process R_{f1} can be computed using (6.1) as

$$R_{f1}\,(\Omega) \times C_{D1}\,(Farad) = 672 \times 10^{-15}$$

Normalized forward-biased resistance is defined as $r_{f1} = R_{f1}/R$.

Step 6b: Generate the normalized series arm impedance $z_a\,(j\omega) = r_a + j\omega L$, where ω is the normalized angular frequency. It may run from zero to perhaps up to $\omega = 5$. Resistance r_a designates the overall estimated loss of the series arms, which includes the forward bias resistance of the diode D_1, loss of the inductor L as well as the resistive loss of the connections. For many practical cases, $r_a = r_{f1}$ is selected.

Step 6c: Generate the normalized shunt arm admittances as $y_a = G_A + j\omega\left(C_A - \frac{1}{\omega^2 L_A}\right)$ of State-A, where G_A is the overall estimated shunt conductive loss of the parallel combination inductor L_A and capacitor C_A. It is noted that, usually, G_A is negligible as compared to the loss effect of r_{f1}. In other words, if noticeable, r_{f1} is more dominant loss factor over G_A. Therefore, G_A may be set to zero in the course of performance computations.

Step 6d: Compute $S_{21A}(j\omega) = \rho_{21A}(\omega)e^{j\varphi_{21A}(\omega)}$ and $S_{11A}(j\omega) = \rho_{11A}(\omega)e^{j\varphi_{11A}(\omega)}$ as in (5.24) and (5.25), respectively

$$S_{11} = \frac{z\left(1 - y^2\right) - 2y}{zy^2 + 2zy + 2y + z + 2}; \quad \left\{ \begin{array}{l} z = z_a \\ y = y_a \end{array} \right\}$$

$$S_{21} = \frac{2}{zy^2 + 2zy + 2y + z + 2}; \quad \left\{ \begin{array}{l} z = z_a \\ y = y_a \end{array} \right\}$$

Step 7: Computation of S_{21B} and S_{11B} of State-B

Step 7a: Generate the normalized series arm impedance $z_b(j\omega) = r_b + j\omega L + \frac{1}{j\omega C_{D1}}$, where ω is the normalized angular frequency. It may run from zero to perhaps up to $\omega = 5$. The resistance r_b is associated overall loss of the series arms. For many practical cases, r_b is selected as the reverse bias resistance of the diodes D_1.

Step 7b: Generate the normalized shunt arm admittance $y_b = G_B + j\omega\left(C_T - \frac{1}{\omega^2 L_A}\right)$, where G_B is the estimated overall conductive loss of the shunt arm in State-B.

Step 7c: Compute $S_{21B}(j\omega) = \rho_{21B}(\omega)e^{j\varphi_{21B}(\omega)}$ and $S_{11B}(j\omega) = \rho_{11B}(\omega)e^{j\varphi_{11B}(\omega)}$ as in (5.9a) and (5.9b), respectively, as

$$S_{11} = \frac{z\left(1 - y^2\right) - 2y}{zy^2 + 2zy + 2y + z + 2}; \quad \left\{ \begin{array}{l} z = z_b \\ y = y_b \end{array} \right\}$$

$$S_{21} = \frac{2}{zy^2 + 2zy + 2y + z + 2}; \quad \left\{ \begin{array}{l} z = z_b \\ y = y_b \end{array} \right\}$$

Step 7d: Plot the results

Now, let us run an example to implement the above design algorithm.

Example 7.1: Let $\omega_0 = 1$, $R = 1$, $r_a = r_b = 0$, $G_A = G_B = 0$.

(a) Design a narrow bandwidth, $0° - 180°$ Phase Range LPπ-DPG for the net-phase shift $\Delta\theta = 45°$.
(b) Investigate the effect of the 10% phase deviation on the frequency band.
(c) Investigate the gain characteristic of the LPI-DPS.

Solution

Let us follow the computational steps of the design algorithm

Step 1: Normalized angular frequency is already provided, which is $\omega_0 = 1$.

Step 2: For $\varphi_{21A}(\omega_0) = 45°$, normalized shunt arm capacitor C is given by

$$C = \frac{1}{\omega_0}\tan\left(\frac{|\varphi_{21A}(\omega_0)|}{2}\right) = 0.4142$$

Step 3: Normalized ideal series arm inductor L is

$$L = \frac{2C}{1 + \omega_0^2 L^2} = 0.7071$$

Step 4: Normalized series arm reverse bias diode capacitance is given by

$$C_{D1} = \frac{1}{\omega_0^2 L} = 1.4142$$

Step 5a: By selecting $C_{D2} = C_{D1} = 2.4142$, normalized shunt arm capacitor C_A is found as

$$C_A = \left(\frac{C}{2}\right) + \frac{1}{2}\sqrt{C^2 + 4CC_{D2}} = 1$$

Step 5b: Compute C_T

$$C_T = \frac{C_A C_{D2}}{C_A + C_2} = 0.5858$$

Step 5c: Compute L_A

$$L_A = \frac{1}{\omega_0^2 C_T} = 1$$

Steps 6 and 7 is implemented under a MatLab program called

$$\text{``}Main\_Lowpass\_PI\_Section\_DPS.m\text{''}$$

This program computes the scattering parameters of the $180°$ phase range LPπ-DPS topology both for State-A and State-B and prints the performance of the digital phase sifter under consideration.

In Figure 7.3a, the phase variations of S_{21A} and S_{21B} are depicted.

A close examination of Figure 6.4a reveals that at $\omega_0 = 1$, $\Delta\theta = 45°$ as it should be. Furthermore, $\varphi_{21B}(1) = 0$ as expected.

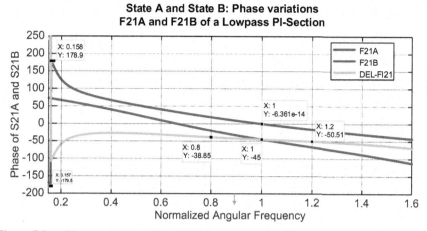

Figure 7.3a   Phase variations of LPπ-DPS for State-A, State-B and $\Delta\theta = \varphi_{21A} - \varphi_{21B}$ of Example 7.1.

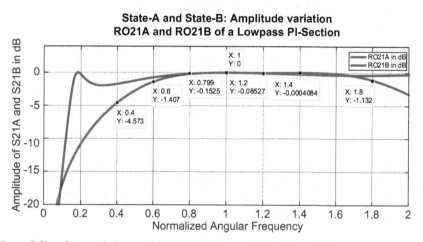

Figure 7.3b   Gain variations of LPπ-DPS for State-A, State-B and $\Delta\theta = \varphi_{21A} - \varphi_{21B}$ of Example 7.1.

In Figure 7.3b, gain performances of State-A and State-B are shown.

A close examination of Figure (7.3) reveals that bandwidth of the LPπ-DPS is about $\mp 20\%$ about the center frequency $\omega_0 = 1$. This is the ideal solution. Certainly, bandwidth is reduced with addition component losses.

Component losses can be included within MatLab programs as in Section 6.5 of Chapter 6 to assess the practical performance of the LPπ-DPS.

Let us propose the following algorithm to generate the phase shifting performance of the LPπ-DPS with component losses introduced in Section 6.4 of the previous chapter.

7.4 Algorithm to Compute Component Lossless of LPI-DPS

This algorithm generates the component losses of LPI-DPS, which in turn yields series and shunt arm immittances both in State-A and State-B. Here, we presume that LPπ-DPS is realized as an MMIC on a silicon substrate using the 0.18u process technology of TMSC [6].

Inputs:

f_{0a}:   Actual center frequency

ω_0:   Normalized angular frequency

R: Normalization resistor

δ_{IND}: Multiplying factor to determine the series loss resistance r_L of an inductor L

It is noted that, quality factor Q_L of an inductor is specified by
$$Q_L = \frac{L\omega_0}{r_L} = \frac{1}{\delta_{IND}}.$$

δ_{CAP}: Multiplying factor to determine the shunt loss conductance G_C of a capacitor C.

(Note: Quality factor Q_C of a capacitor is specified by
$$Q_C = \frac{C\omega_0}{G_C} = \frac{1}{\delta_{CAP}}.$$

δ_D: Multiplying factor to determine the series loss resistance r_r of a reverse bias diode D

Computational Steps:

Step 1: Compute
$$C = \frac{1}{\omega_0}\tan\left(\frac{|\varphi_{21A}(\omega_0)|}{2}\right)$$

Step 2: Compute
$$L = \frac{2C}{1 + \omega_0^2 C^2}$$

Step 3: Compute C_{D1}

$$C_{D1} = \frac{1}{\omega_0^2 L}$$

Step 4: Select C_{D2} and compute

$$C_A = \left(\frac{C}{2}\right) + \frac{1}{2}\sqrt{C^2 + 4CC_{D2}}$$

Note: One may wish to set $C_{D2} = C_{D1}$
Step 5: Compute

$$C_T = \frac{C_A C_{D2}}{C_A + C_2}$$

Step 6: Compute

$$L_A = \frac{1}{\omega_0^2 C_T}$$

Step 7: Generate actual values for C_{D1} and C_{D2}

$$C_{D1a} = \frac{C_{D1}}{(2\pi f_{0a})\,R}$$

$$C_{D2a} = \frac{C_{D2}}{(2\pi f_{0a})\,R}$$

Step 8: Generate normalized values for r_{f1} and r_{f2}

$$R_{f1}\,(\Omega) = 672 \times 10^{-15}/C_{D1a}\,(Farad)$$
$$R_{f2}\,(\Omega) = 672 \times 10^{-15}/C_{D2a}\,(Farad)$$
$$r_{f1} = \frac{R_{f1}}{R}$$
$$r_{f2} = \frac{R_{f2}}{R}$$

Step 9: Generate r_{r1} and r_{r2}

$$r_{r1} = (\omega_0 C_{D1})\,\delta_D$$

$$r_{r2} = (\omega_0 C_{D2})\,\delta_D$$

Step 10: Generate r_{LA} and G_{CA}

$$r_{LA} = (\omega_0 L_A)\,\delta_L$$

$$G_{CA} = (\omega_0 C_A) \delta_C$$

Step 11: Generate r_L, r_a, and G_A of State-A

$$r_L = (\omega_0 L) \delta_L$$

$$r_a = r_L + r_{f1}$$

$$G_A = G_{CA} + \frac{r_{LA}}{r_{LA}^2 + \omega_0^2 L_A^2} + \frac{\omega_0^2 r_{f2} C_A^2}{1 + \omega_0^2 r_{f2}^2 C_A^2}$$

Step 12: Generate r_b and G_B of State-B

$$r_b = r_L + r_{r1}$$

$$G_B = G_{CA} + \frac{r_{LA}}{r_{LA}^2 + \omega_0^2 L_A^2} + \frac{\omega_0^2 r_{r2} C_T^2}{1 + \omega_0^2 r_{r2}^2 C_T^2}$$

Step 13: Generate z_a and y_a of State-A

$$z_a (j\omega) = r_a + j\omega L$$

$$y_a = G_A + \frac{1}{j\omega L_A} + j\omega C_A$$

Step 14: Generate z_b and y_b of State-B

$$z_b (j\omega) = r_b + j\omega L + \frac{1}{j\omega C_{D1}}$$

$$y_b = G_B + \frac{1}{j\omega L_A} + j\omega C_T$$

Step 15: Generate phase and gain performances of LPT-DPS for State-A and State-B

$$S_{21(A,B)} = \frac{2}{zy^2 + 2zy + 2y + z + 2}; \quad \left\{ \begin{array}{l} \text{In State A}: \ z = z_a, \ y = y_a \\ \text{In State B}: \ z = z_b, \ y = y_b \end{array} \right\}$$

Step 16: Plot the phase and gain performance

Now, let us run an example to evaluate the performance of an LPT-DPS constructed with lossy elements.

Example 7.2: Let $\delta_{IND} = 10\%$ and $\delta_{CAP} = \delta_D = 1\%$. Let $R = 50 \ \Omega$.

At $f_{0a} = 5 \ GHz$, repeat Example 7.1 with computed loss of the components as follows.

(a) Compute the ideal element values of the LPπ-DPS.
(b) Compute the resistive and conductive losses associated with the ideal elements
(c) Evaluate the phase shifting performance of the phase shifter constructed with lossy elements.
(d) Compute the actual component values

Solution

(a) Ideal element values of the LPπ-DPS under consideration are given as in Example 7.1.

$$L = 0.7071, C = 0.4142, CD1 = 1.4142, CD1 = CD2, CA = 1,$$
$$CT = 0.5858, LA = 1.7071,$$

(b) Losses associated with component values:
Let us follow steps of the algorithm.

Step 8: Generate normalized values for r_{f1} and r_{f2}

$$R_{f1}\,(\Omega) \times C_{D1a}\,(Farad) = 672 \times 10^{-15}; \quad R_{f1} = 0.7464\,\Omega$$

$$R_{f2}\,(\Omega) \times C_{D2a}\,(Farad) = 672 \times 10^{-15}; \quad R_{f2} = 0.7464\,\Omega$$

$$r_{f1} = \frac{R_{f1}}{R} = 0.0149$$

$$r_{f2} = \frac{R_{f2}}{R} = 0.0149$$

Step 9: Generate r_{r1} and r_{r2}

$$r_{r1} = (\omega_0 C_{D1})\,\delta_D = 0.0141$$

$$r_{r2} = (\omega_0 C_{D2})\,\delta_D = 0.0141$$

Step 10: Generate r_{LA} and G_{CA}

$$r_{LA} = (\omega_0 L_A)\,\delta_L = 0.1707$$

$$G_{CA} = (\omega_0 C_A)\,\delta_C = 0.0100$$

Step 11: Generate r_L, r_a, and G_A of State-A

$$r_L = (\omega_0 L)\,\delta_L = 0.0707$$

$$r_a = r_L + r_{f1} = 0.0927$$

$$G_A = G_{CA} + \frac{r_{LA}}{r_{LA}^2 + \omega_0^2 L_A^2} + \frac{\omega_0^2 r_{f2} C_A^2}{1 + \omega_0^2 r_{f2}^2 C_A^2} = 0.0829$$

Step 12: Generate r_b and G_B of State-B

$$r_b = r_L + r_{r1} = 0.0849$$

$$G_B = G_{CA} + \frac{r_{LA}}{r_{LA}^2 + \omega_0^2 L_A^2} + \frac{\omega_0^2 r_{r2} C_T^2}{1 + \omega_0^2 r_{r2}^2 C_T^2} = 0.0731$$

Steps 13–15: All the above and rest of the steps are programmed under the MatLab program called

$$\text{``}Main\_Example\_7\_2.m\text{''}$$

Execution of "$Main\_Example\_7\_2.m$ delivers the phase performance of the LPπ-DPS under consideration, and it is depicted in Figure 7.4a.

A close examination of Figure 7.4a reveals that, at $\omega_0 = 1$, the phase shift $\Delta\theta$ between the switching states is $-44.92°$ instead of exact $45°$. So, introduction of lossy elements perturbed the phase performance of the phase shifter about 0.18%, which is tolerable. The lossy components did not affect the bandwidth of the phase shifter. It is still $\mp20\%$ about the center frequency.

Gain performance of the phase shifter topology is shown in Figure 7.4b. Over the $\mp20\%$ bandwidth gain is better than -1.1 dB, which is acceptable.

MatLab codes of this Example 7.2 are given under Program Lists.

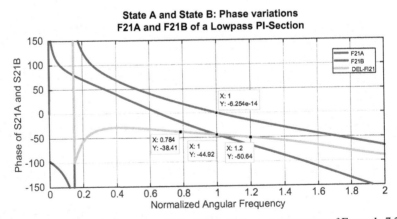

Figure 7.4a   Phase performance of LPπ-DPS with lossy components of Example 7.2.

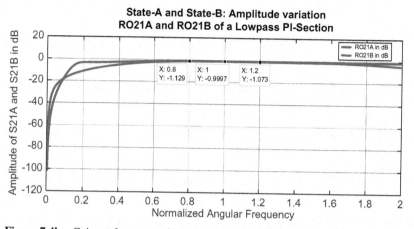

Figure 7.4b Gain performance of LPπ-DPS with lossy components of Example 7.2.

(d) Computation of actual element values

Actual element values for the capacitors are given as

$$C_{actual} = \frac{C_{normalized}}{2\pi f_{0a} R}$$

Similarly, actual element values for the inductors are computed as

$$L_{actual} = \frac{R L_{normalized}}{2\pi f_{0a}}$$

Hence, based on the above equations, actual diode capacitances are given
as

$$C_{D1a} = C_{D2a} = 9.0032e - 13 = 0.9 \ pF$$

Actual series arm inductor L_a is

$$L_a = 1.1254e - 09 = 1.125 \ nH$$

In the shunt arm, shunt inductor L_{Aa} is given by

$$L_{Aa} = 2.7169e - 09 = 2.77 \ nH$$

And finally, the shunt arm capacitor is

$$C_{Aa} = 6.3662e - 13 = 0.63 \ pF$$

All the above values are realizable.

Example 7.3: Let $\delta_{IND} = 10\%$ and $\delta_{CAP} = \delta_D = 1\%$. Let $R = 50 \ \Omega$.

At $f_{0a} = 5 \ GHz$, repeat Example 7.2 for $\Delta\theta (1) = 90°$ with computed loss of the components as follows.

(a) Compute the ideal element values of the LPπ-DPS.
(b) Compute the resistive and conductive losses associated with the ideal elements
(c) Evaluate the phase shifting performance of the phase shifter constructed with lossy elements.
(d) Compute the actual component values

Solution

(a) Ideal element values of the LPπ-DPS under consideration are given as

$$L = 1, C = 1, CD1 = 1, CD1 = CD2, CA = 1.6180,$$
$$CT = 0.6180, LA = 1.6180,$$

(b) Losses associated with component values:
Let us follow steps of the algorithm to generate the component losses.

Step 8: Generate normalized values for r_{f1} and r_{f2}

$$R_{f1} (\Omega) \times C_{D1a} (Farad) = 672 \times 10^{-15}; \quad R_{f1} = 1.0556 \ \Omega$$

$$R_{f2} (\Omega) \times C_{D2a} (Farad) = 672 \times 10^{-15}; \quad R_{f2} = 1.0556 \ \Omega$$

$$r_{f1} = \frac{R_{f1}}{R} = 0.0211$$

$$r_{f2} = \frac{R_{f2}}{R} = 0.0211$$

Step 9: Generate r_{r1} and r_{r2}

$$r_{r1} = (\omega_0 C_{D1}) \delta_D = 0.01618$$

$$r_{r2} = (\omega_0 C_{D2}) \delta_D = 0.01618$$

Step 10: Generate r_{LA} and G_{CA}

$$r_{LA} = (\omega_0 L_A) \delta_L = 0.1707$$

$$G_{CA} = (\omega_0 C_A) \delta_C = 0.0100$$

Step 11: Generate r_L, r_a, and G_A of State-A

$$r_L = (\omega_0 L)\, \delta_L = 0.1618$$

$$r_a = r_L + r_{f1} = 0.1311$$

$$G_A = G_{CA} + \frac{r_L A}{r_{LA}^2 + \omega_0^2 L_A^2} + \frac{\omega_0^2 r_{f2} C_A^2}{1 + \omega_0^2 r_{f2}^2 C_A^2} = 0.1326$$

Step 12: Generate r_b and G_B of State-B

$$r_b = r_L + r_{r1} = 0.1200$$

$$G_B = G_{CA} + \frac{r_L A}{r_{LA}^2 + \omega_0^2 L_A^2} + \frac{\omega_0^2 r_{r2} C_T^2}{1 + \omega_0^2 r_{r2}^2 C_T^2} = 0.0854$$

Steps 13–15: Execution of $Main\_Example\_7\_2.m$ results in the phase performance of the LPπ-DPS of 90°, and it is depicted in Figure 7.5a.

A close examination of Figure 7.5a reveals that, at $\omega_0 = 1$, the phase shift $\Delta\theta$ between the switching states is $-89.10°$ instead of exact 90°. So, introduction of lossy elements perturbed the phase performance of the phase shifter about 0.9%, which is tolerable. In this example, lossy components shrink the absolute bandwidth $\Delta\omega = 1.2 - 0.8 = 0.4$ down to $\Delta\omega = 1.115 - 0.885 = 0.23$, which corresponds to 42.5% reduction. Therefore, we say that $LP\pi - DPS$ provides phase shifts within 10–20% bandwidth in the vicinity of a center frequency up to 90°.

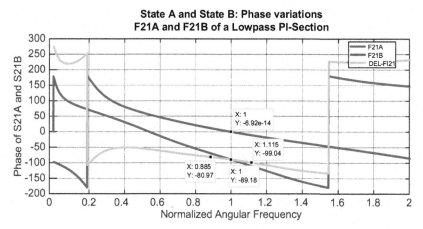

Figure 7.5a Phase performance of LPπ-DPS with lossy components of Example 7.2.

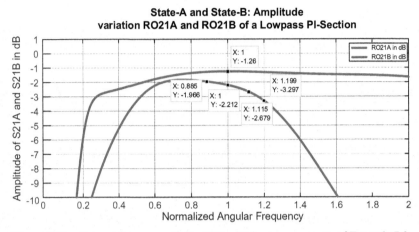

Figure 7.5b   Gain performance of LPπ-DPS with lossy components of Example 7.2.

Gain performance of the phase shifter topology is shown in Figure 7.5b. Over the $\mp 20\%$ bandwidth, gain is better than -1.1 dB, which is acceptable.

(d) Computation of actual element values

Actual element values for the capacitors are given as

$$C_{actual} = \frac{C_{normalized}}{2\pi f_{0a} R}$$

Similarly, actual element values for the inductors are computed as

$$L_{actual} = \frac{R L_{normalized}}{2\pi f_{0a}}$$

Hence, based on the above equations, actual diode capacitances are given as

$$C_{D1a} = C_{D2a} = 6.3662e - 13 = 0.6366\ nF$$

Actual series arm inductor L_a is

$$L_a = 1.5915e - 09 \cong 1.59\ nH$$

In the shunt arm, shunt inductor L_{Aa} is given by

$$L_{Aa} = 2.5752e - 09 \cong 2.57\ nH$$

and finally, the shunt arm capacitor is

$$C_{Aa} = 1.0301e - 12 \cong 1\ pF$$

All the above values are realizable.

It should be noted that when the MatLab program Example_7_2.m is run for $\Delta\theta = 120°$, at $\omega_0 = 1$, the State-B gain of $LP\pi - DPS$ becomes less than -3 dB, which is not acceptable. Therefore, we say that, $LP\pi - DPS$ yields acceptable solutions up to 90°. Beyond 90°, we can use the phase shifter topologies presented in Chapters 8–10.

7.5 General Comments and Conclusion

The symmetric 3-element low-pass-based π-Section Digital Phase Shifter Topology (LPI-DPS) is easy to implement as an MMIC circuit. It is compact and safely provides 10% bandwidth with reasonable loss. On the other hand, LPI-DPS is not suitable to provide 180° phase shift. Nevertheless, two units of 90° LPI-DPS can be cascaded to design a 180° phase shifting cell.

Examples presented in this chapter verify the below statements.

1. LPI-DPS is good for narrow phase range and small bandwidth requirements. It offers reliable phase shift up to 90°.
2. Bandwidth of the LPI-DPS decreases while phase shift ranges increases.
3. Component losses do not much affect the phase shifting performance of the $LP\pi - DPS$ topology up to 90°.
4. LPT-DPS topology is suitable for MMIC implementation.

Appendix 7: MatLab Programs for Chapter 7

Program List 7.1: Main Program: Main_Lowpass_PI_Section_DPS.m

```
% Main Program: Main_Lowpass_PI_Section_DPS.m
% February 24, 2019
% Developed by BS Yarman, Vanikoy, Istanbul
% Enter Positive value
% ------------------------------------------------------------------
clc; close all
% Inputs:
teta=input('Alternative Formulas to Design Lowpass T Section. Enter
positive value for Teta in Degree=')
% Inputs w0=1;
ra=0;rb=0; GA=0;GB=0;
% ------------------------------------------------------------------
% Alternative way to generate inductor L
C=tand(teta/2)/w0;
L=2*C/(1+w0*w0*C*C);
% ------------------------------------------------------------------
% Component Values
CD1=1/(w0^2*L);
```

```
CD2=CD1;
Delta=sqrt(C*C+4*C*CD2);
CA=C/2+Delta/2;
CT=CA*CD2/(CA+CD2);
LA=1/w0/w0/CT;
j=sqrt(-1);
w=0;N=2000;w1=0;w2=2;DW=(w2-w1)/N;
FRI(1:(N+1))=zeros;
for i=1:N+1
    WA(i)=w;
% ----------------
% State A:
% ----------------
za=ra+j*w*L;
ya=GA+1/j/w/LA+j*w*CA;
 [ S11a,S21a,RO11a,F11a,RO21a,F21a ] = S_Par_LP_PI_DPS (za,ya);
    F21A(i)=F21a;
    RO21A(i)=20*log10(RO21a);
    F11A(i)=F11a;
    RO11A(i)=20*log10(RO11a);
% ------------------
% State-B
% ------------------
  zb=rb+j*w*L+1/j/w0/CD1;
  yb=GB+1/j/w/LA+j*w*CT;
  [ S11b,S21b,RO11b,F11b,RO21b,F21b ] = S_Par_LP_PI_DPS (zb,yb);
    F21B(i)=F21b;
    RO21B(i)=20*log10(RO21b);
    F11B(i)=F11b;
    RO11B(i)=20*log10(RO11b);
% ----------------------------------------------------------------
DEL_FI21(i)=F21A(i)-F21B(i);
    w=w+DW;
end
% ----------------------------------------------------------------
Plot_State_AB_LP_PI_DPS(WA,F21A,RO21A, F11A,RO11A,F21B
RO21B, F11B,RO11B,DEL_FI21)
```

Program List 7.2: function S_Par_LP_PI_DPS

```
function [ S11,S21,RO11,F11,RO21,F21 ] = S_Par_LP_PI_DPS (z,y)
% This function generates the S-Parameters of a Lowpass T Section
% Phase Shifter from the series arm impedance Z(jw) and
% the shunt arm admittance Y(jw)
% computed series arm inductor L and the shunt capacitor C
% Developed by BS Yarman: Feb 20, 2019, Vanikoy, Istanbul
% See Equations (5.9)
% ----------------------------------------------------------------
% D=zy^2+2zy+2y+z+2 of S11 = N11/D and S21=2/D
% N11=z(1-y^2)-2y of S11 = N11/D
D=z*y*y+2*z*y+2*y+z+2;
S11=((1-y*y)*z-2*y)/D;
```

```
S21=2/D;
R11=real(S11); X11=imag(S11);F11=atan2d(X11,R11);
R21=real(S21); X21=imag(S21);F21=atan2d(X21,R21);
RO11=abs(S11);
RO21=abs(S21);

end
```

Program List 7.3: Main Program: Main_Example_7_2.m

```
% Main Program: Main_Example_7_2.m
% February 24, 2019
% Developed by BS Yarman, Vanikoy, Istanbul
% This program evaluates the lossy performance of a LPT-DPS for a
  specified
% center actual frequency f0 in Hz.
% --------------------------------------------------------------------
clc; close all
% Inputs:
teta=input('Alternative Formulas to Design Lowpass T Section. Enter
positive value for Teta in Degree=')
% Inputs
f0a=input('Enter the actual center frequency in Hz f0a =')
w0=input('At f0a, enter the normalized angular frequency w0 =')
R=input('Enter the normaliziation Resistor R =')
% --------------------------------------------------------------------
% Compute the normalized element values of LPT-DPS
% Alternative way to generate inductor L
C=tand(abs(teta)/2)/w0;
L=2*C/(1+w0*w0*C*C);
% --------------------------------------------------------------------
% Component Values and their related resistive losses:
CD1=1/(w0^2*L);  % See equation (6.2d) of Chapter 6
% ASSUMPTION 1:
% It is assumed that the series loss Ron of a forward biased diode is
% equal to the on channel resistor of an CMOS Switch
% [see Equation (6.1) of Chapter 6].
[ RF1,rf1 ] =Channel_Resistance_of_a_CMOS(CD1,R,f0a);
% ASSUMPTION 2:
% Reverse biased resistive loss of a diode is the "CrDEL_D: Cronicel
  Delta"
% amount of its reverse baised impedance at w0.
CrDel_D=100;
rr1=1/w0/CD1/CrDel_D;
CD2=CD1;
[ RF2,rf2 ] =Channel_Resistance_of_a_CMOS(CD2,R,f0a);
rr2=1/w0/CD2/CrDel_D;
%
DEN=2;
Delta=sqrt(C*C+4*C*CD2);
CA=(C)/DEN+Delta/DEN;
CT=CA*CD2/(CA+CD2);
LA=1/w0/w0/CT;
```

```
% -----------------------------------------------------------------------
% Loss Computations for both State-A and State-B:
% Assumption 3: Loss of an inductor is CrDEL_L amount of its impedance
% value at w0.
% Assumption 4: Connectivity loss of an inductor is "CrDEL_S" amount of
% its impedance value at w0.
% Assumption 5: Conductive loss of a Capacitor is "CrDEL_C" amount of
% its admittance value at w0.
CrDel_S=100;
CrDEL_L=10;
CrDel_C=100;
% -----------------------------------------------------------------------
% Resistive loss of the series arms in State-A:
rL=w0*L/CrDEL_L;
rs=w0*L/CrDel_S;
ra=rL+rf1+rs;
% -----------------------------------------------------------------------
% Conductive loss of the Shunt arm in State-A:
rLA=w0*LA/CrDEL_L;
GCA=w0*CA/CrDel_C;
GA=GCA+rLA/(rLA*rLA+w0*w0*LA*LA)+(w0*w0*rf2*CA*CA)/(1+w0*w0*rf2*rf2
*CA*CA);
% -----------------------------------------------------------------------
% Resistive loss of the series arms in State-B: rb=rL+rr1+rs;
% -----------------------------------------------------------------------
% Conductive loss of the Shunt arm in State-:
% -----------------------------------------------------------------------
GB=GCA+rLA/(rLA*rLA+w0*w0*LA*LA)+(w0*w0*rf2*CT*CT)/(1+w0*w0*rr2*rr2*CT*
    CT);
% -----------------------------------------------------------------------
j=sqrt(-1);
w=0;N=2000;w1=0;w2=2;DW=(w2-w1)/N;
FRI(1:(N+1))=zeros;
for i=1:N+1
    WA(i)=w;
% ------------------
% State A:
% ------------------
za=ra+j*w*L;
ya=GA+1/j/w/LA+j*w*CA;
[ S11a,S21a,RO11a,F11a,RO21a,F21a ] = S_Par_LP_PI_DPS (za,ya);
    F21A(i)=F21a; RO21A(i)=20*log10(RO21a);
    F11A(i)=F11a;
    RO11A(i)=20*log10(RO11a);
    % ------------------
    % State-B
    % ------------------
    zb=rb+j*w*L+1/j/w0/CD1;
    yb=GB+1/j/w/LA+j*w*CT;
    [ S11b,S21b,RO11b,F11b,RO21b,F21b ] = S_Par_LP_PI_DPS (zb,yb);
    F21B(i)=F21b;
    RO21B(i)=20*log10(RO21b);
    F11B(i)=F11b;
```

```
   RO11B(i)=20*log10(RO11b);
% -----------------------------------------------------------------
DEL_FI21(i)=F21A(i)-F21B(i);
      w=w+DW;
end
% -----------------------------------------------------------------
Plot_State_AB_LP_PI_DPS(WA,F21A,RO21A, F11A,RO11A,F21B,RO21B, F11B,
      RO11B,DEL_FI21)
[CD1a] = ActualValues_of_a_Capacitor(CD1,R,f0a)
[CAa] = ActualValues_of_a_Capacitor(CA,R,f0a)
[LAa] = ActualValues_of_an_Inductor(LA,R,f0a)
[La] = ActualValues_of_an_Inductor(L,R,f0a)
% -----------------------------------------------------------------
ra_actual=ra*R
rb_actual=rb*R
% -----------------------------------------------------------------
GA_actual=GA/R; RA_actual=1/GA_actual
GB_actual=GB/R; RB_actual=1/GB_actual
```

Program List 7.4: function Plot_State_AB_LP_PI_DPS

```
function Plot_State_AB_LP_PI_DPS(WA,F21A,RO21A, F11A,RO11A,F21B,RO21B,
F11B,RO11B,DEL_FI21)
figure
plot(WA,F21A,WA,F21B,WA,DEL_FI21)
title('State A and State B: Phase variations F21A and F21B of a
Lowpass PI-Section') legend('F21A','F21B','DEL-FI21')
xlabel('Normalized Angular Frequency')
ylabel('Phase of S21A and S21B')
% Amplitude of S21
figure
plot(WA,RO21A,WA,RO21B)
title('State-A and State-B: Amplitude variation RO21A and RO21B of a
Lowpass PI-Section')
legend('RO21A in dB','RO21B in dB')
xlabel('Normalized Angular Frequency')
ylabel('Amplitude of S21A and S21B in dB')
% -----------------------------------------------------------
figure
plot(WA,F11A, WA, F11B)
title('State-A and State-B: Phase variation F11A and F11B of a Lowpass
PI-Section') legend('F11A')
xlabel('Normalized Angular Frequency')
ylabel('Phase of S11A and S11B')
% Amplitude of S11
figure
plot(WA,RO11A, WA, RO11B)
title('State-A and State-B: Amplitude variation RO11A and RO11B of a
Lowpass PI-Section')
legend('RO11A in dB','RO11B in dB')
xlabel('Normalized Angular Frequency')
ylabel('Amplitude of S11A and S11B in dB')

end
```

References

[1] B. S. Yarman, "Low pass π-section digital phase shifter apparatus," U.S. Patent: 4614921, Washington DC, USA, September 30, 1986.

[2] B. Yarman, A. Rosen and P. Stabile, "Lowloss EHF Digital Phase Shifters Suitable for Monolithic Implementation," in IEEE ISCAS, Montreal, 1984.

[3] B. S. Yarman, "Design of Digital Phase Shifters Suitable for Monolithic Implementations," Bull. of Technical University of Istanbul, pp. 185–205, 1985.

[4] B. S. Yarman, "Novel circuit configurations to design loss balanced $0°$– $360°$ digital phase shifters," AEU, vol. 45, no. 2, pp. 96–104, 1991.

[5] B. S. Yarman, "New Circuit Configurations for Designing 0–180 degree Digital Phase Shifters," IEE Proceeding H, Vol.134, pp. 253–260, 1987.

[6] TSMC, "Taiwanese Semiconductor Manufacturing Company," (https://www.tsmc.com/english/aboutTSMC/company_profile.htm), Taiwan, 2019.

8

180° High-pass-based T-Section Digital Phase Shifter Topology (HPT-DPS)

Summary

In this chapter, a symmetric 3-element LC high-pass-based T-Section Digital Phase Shifter configuration is presented. The proposed phase shifter topology is compact and it is suitable to manufacture with discrete components, or as microwave monolithic integrated circuits (MMIC).

The content of this chapter is based on our research work published in the open literature [1–5].

8.1 High-pass-based Symmetric T-Section Digital Phase Shifter

In Figure 8.1, a symmetric high-pass-based, T-Section, Digital Phase Shifter HPT-DPS is shown.

This configuration includes three switching diodes, namely in the left and right series arms, we have two identical diodes D_1, and in the shunt arm D_2 respectively.

In one state, say, State-A, when the diodes D_1s are reverse-biased, ideally, they act like capacitors $C = C_{D1}$ as described in Chapter 5 of Figure 5.6. In this representation, C_{D1} is the reverse-biased diode capacitance. At the operating frequency f_0 (or equivalently $\omega_0 = 2\pi f_0$), in the shunt arm, D_2 is also reverse-biased, yielding an equivalent shunt admittance $Y(j\omega_0) = 1/(j\omega_0 L)$.

In this mode of operation, ideally reverse-biased resistor of the diode is zero (i.e., $r_r = 0$). At a selected operation angular frequency ω_0 and phase

295

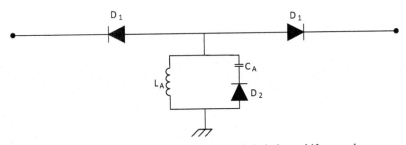

Figure 8.1a   A high-pass-based T-section digital phase shifter topology.

$\varphi_A(\omega_0)$, element values of Figure 8.1 are specified as in (5.23) and Table 5.2 as follows.

$$C = \frac{1}{\omega_0}tan\left(90-\frac{|\theta|}{2}\right) > 0; \qquad 0° < \theta < 180° \qquad (8.1a)$$

and

$$L = \left(\frac{1}{\omega_0^2}\right)\left[\frac{1+\omega_0^2C^2}{2C}\right] > 0 \qquad (8.1b)$$

In this state, shunt arm admittance is given by

$$Y(j\omega_0) = \frac{1}{j\omega_0 L_A} + j\omega_0 C_T = \frac{1}{j\omega_0 L}$$

or

$$L_A = \frac{L}{1 + \omega_0^2 L C_T} \qquad (8.1c)$$

In State-A, at the given operating frequency ω_0, C_{D1} is specified by

$$C_{D1} = C \qquad (8.1d)$$

In this mode of operation, in the shunt arm, the equivalent capacitor C_T is given by

$$C_T = \frac{C_A C_{D2}}{C_A + C_{D2}} \qquad (8.1e)$$

In State-B, all the diodes are forward-biased. Therefore, in the series arms, ideally, we will see shorted diodes. However, in practice, they present a small forward bias resistance r_{f1}. In this state, the shunt arm inductor

L_A resonates with the capacitor C_A. Thus,

$$L_A = \frac{1}{\omega_0^2 C_A} = \frac{L}{1 + \omega_0^2 L C_T} \tag{8.1f}$$

Using (8.1f) in (8.1c), we have

$$\frac{1}{\omega_0^2 C_A} = \frac{L}{1 + \omega_0^2 L \dfrac{C_{D2} C_A}{C_{D2} + C_A}}$$

Once C_{D2} is selected, the capacitor C_A can be determined by solving the following second-order equation

$$\left(\omega_0^2 L\right) C_A^2 - C_A - (C_{D2}) = 0$$

or

$$C_{A(1,2)} = \left(\frac{1}{2\omega_0^2 L}\right) \mp \frac{1}{2\omega_0^2 L} \sqrt{1 + 4\omega_0^2 L C_{D2}} > 0$$

The above equation must yield a positive value for the capacitor C_A. Therefore, we should select the positive "+" sign. Hence,

$$C_A = \left(\frac{1}{2\omega_0^2 L}\right) + \frac{1}{2\omega_0^2 L} \sqrt{1 + 4\omega_0^2 L C_{D2}} > 0 \tag{8.1g}$$

Once C_A is found, L_A is derived from the resonance condition of State-B such that

$$L_A = \frac{1}{\omega_0^2 C_A} \tag{8.1h}$$

8.2 Concept of Digital Phase Shift and Design Algorithm

The phase shifter topology introduced in Figure 8.1 describes two distinct operational states, namely State-A and State-B. The phase of each state is determined based on the position of the switching diodes. For each diode, forward-biased situation may be named as "switch is ON" or "switch is closed", and this position may be designated by the logical-state "1". When a diode is reversed-biased, it may be referred to as "switch is OFF" or "switch is open". This situation may be designated by the logic state "0".

"The phase $\theta(\omega)$ of a lossless two-port" is defined as the phase $\varphi_{21}(\omega)$ of the transfer scattering parameter $S_{21}(j\omega) = \rho(\omega)e^{j\varphi_{21}(\omega)}$. At the operating frequency, the objective is to make $\rho_{21}(\omega_0) = 1$ while reaching the desired phase $\theta(\omega_0) = \varphi_{21}(\omega_0)$.

For a two-port shown in Figure 8.1, the phase $\theta(\omega)$ varies depending on the position of the switches.

As described above, at a specified angular frequency ω_0, State-A refers the phase $\theta_A = \varphi_{21A}(\omega_0)$ when all the switches are opened. In State-B, all the switches are closed and the two-port possess the phase $\theta_B = \varphi_{21B}(\omega_0)$.

All the above explanations suggests the following truth stable to define the phase-switching states of the phase shifter topology shown in Figure 8.1.

In view of Table 8.1 and Figure 8.1, the phase shifter topology is called the "High-pass-based T-Section Digital Phase Shifter (HPT-DPS)". HPT-DPS possesses two phase-switching states, namely State-A and State-B. The net phase shift between the states is given by

$$\Delta\theta = \theta_A - \theta_B \qquad (8.2)$$

Let us make the following comments:

(a) By its nature, in State-B, at the angular frequency ω_0, θ_B is forced to be zero.

(b) Range of θ_A must be less than 180° since $C = \frac{1}{\omega_0}\tan\left(90° - \frac{|\varphi_{21A}(\omega_0)|}{2}\right)$ goes to infinity when $\varphi_{21A}(\omega_0) = 180°$.

In Figure 8.1b, equivalent immittance states of Figure 8.1a is depicted.

Thus, based on the above discussions, we propose the following design algorithm to construct HPT-DPS.

Table 8.1   Description of switching states of HPT-DPS topology

	D1	D2	Phase θ of the Two-port
State-A	0	0	$\theta_A = \varphi_{21B}(\omega_0) > 0$
State-B	1	1	$\theta_B = \varphi_{21B}(\omega_0) = 0$
Net Phase Shift Between States			$\Delta\theta = \theta_A - \theta_B$

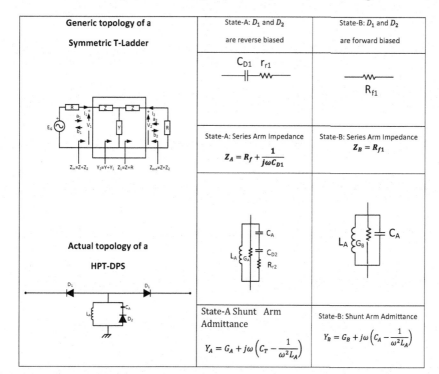

Figure 8.1b Immittance states of the high-pass-based T-section digital phase shifter topology.

8.3 Algorithm to Design HPT-DPS for the Phase Range
$180° < \Delta\theta = \theta_A < 0°$

In this algorithm, we present the design procedure to construct a High-pass T-Section Digital Phase Shifter, which can only cover the phase range of $0° - 180°$. It should be noted that, using this phase shifter topology, it is not possible to obtain exact $180°$ phase shifting unit.

Inputs:

f_{1a}: Lower end of the passband of interest

f_{2a}: Upper end of the passband of interest

f_{0a}: Actual operating frequency. It selected in such a way that
$f_{1a} \leq f_{0a} \leq f_{2a}$.

R: Normalization resistance. For most practical situations, it is selected as the termination resistors, which is perhaps 50Ω.

$\Delta\theta = \varphi_{21A}(\omega_0) - \varphi_{21B}(\omega_0)$: Desired phase shift with $\varphi_{21B}(\omega_0) = 0$.

G_A: In State-A, it is the overall estimated normalized conductive loss of the shunt arm with respect to R.

G_B: In State-B, it is the overall estimated normalized conductive loss of the shunt arm with respect to R.

Computational Steps:

Step 1: Select a normalized angular frequency ω_0 such that it is confined within

$$f_{1a}/f_{0a} \leq \omega_0 \leq f_{2a}/f_{0a}$$

It is useful to know that for many cases, ω_0 is set to unity (i.e., $\omega_0 = 1$)

Step 2: Compute the normalized series arm capacitor $C = \frac{1}{\omega_0}\tan\left(90° - \frac{|\varphi_{21A}(\omega_0)|}{2}\right)$ for the ideal symmetric high-pass T-Section of Figure 5.6.

Step 3: Compute the normalized shunt arm inductor L of the ideal symmetric high-pass T-Section of Figure 5.6 as

$$L = \frac{1 + \omega_0^2 C^2}{2C} > 0$$

Step 4: Compute the normalized value of the series arm, reverse-biased diode capacitor C_{D1} such that

$$C_{D1} = C$$

It is noted that, if C_{D1} is too small or too large to be manufactured, then one can select a reasonable diode capacitor value and adjust its value by connecting either a parallel or a series capacitor to D_1, respectively.

Step 5a: Select a proper normalized value for C_{D2} and compute C_A as in (8.1g)

$$C_A = \left(\frac{1}{2\omega_0^2 L}\right) + \frac{1}{2\omega_0^2 L}\sqrt{1 + 4\omega_0^2 L C_{D2}} > 0$$

At this point, the designer may wish to select $C_{D2} = C_{D1}$ if appropriate.

Step 5b: Generate C_T as in (8.1e)

$$C_T = \frac{C_A C_{D2}}{C_A + C_2}$$

Step 5c: Generate L_A as in (8.1h)

$$L_A = \frac{1}{\omega_0^2 C_A}$$

Step 6: Computation of S_{21A} of State-A

Step 6a: Obtain the forward-bias resistance R_{f1} and the reverse-bias resistance R_{r1} of D_1, which may be derived from C_{D1} either by computations based on geometric size of the diodes, by measurements or obtain it from the manufacturer and normalize it with respect to R. For example, if one employs an MOS diode, then $R_r \cong 0$ and for $180\mu m$ Si process, R_{f1} can be computed using (6.1) as

$$R_{f1}\left(\Omega\right) \times C_{D1}\left(Farad\right) = 672 \times 10^{-15}$$

Normalized forward-biased resistance is defined as

$$r_{f1} = \frac{R_{f1}}{R} = \frac{672 \times 10^{-15}}{C_{D1}R}$$

Step 6b: Generate the normalized series arm impedance $z_a\left(j\omega\right) = r_a + \frac{1}{j\omega C_{D1}}$, where ω is the normalized angular frequency. It may run from zero and perhaps up to $\omega = 5$.

Resistance r_a designates the overall estimated loss of the series arms, which includes the reverse-bias resistance of diodes D_1 and the resistive loss of the connections. For many practical cases, $r_a = r_{r1}$ is selected.

Step 6c: Generate the normalized shunt arm admittance $y_a = G_A + j\omega\left(C_T - \frac{1}{\omega^2 L_A}\right)$ of State-A, where G_A is the overall estimated shunt conductive loss of the parallel combination inductor L_A and capacitor C_A.

Step 6d: Compute $S_{21A}\left(j\omega\right) = \rho_{21A}\left(\omega\right)e^{j\varphi_{21A}\left(\omega\right)}$ and $S_{11A}\left(j\omega\right) = \rho_{11A}\left(\omega\right)e^{j\varphi_{11A}\left(\omega\right)}$ as in (5.9a) and (5.9b)

$$S_{21A} = \frac{2}{z^2 y + 2zy + 2z + y + 2}; \quad \left\{ \begin{array}{c} z = z_a \\ y = y_a \end{array} \right\}$$

$$S_{11A} = \frac{\left(1 - z^2\right)y - 2z}{z^2 y + 2zy + 2z + y + 2}; \quad \left\{ \begin{array}{c} z = z_a \\ y = y_a \end{array} \right\}$$

Step 7: Computation of S_{21B} and S_{11B} of State-B

Step 7a: Generate the normalized series arm impedance $z_b\left(j\omega\right) = r_b$, where ω is the normalized angular frequency. It may run from zero, to perhaps up to $\omega = 5$. The resistance r_b is associated with overall loss of the series arms. For many practical cases, r_b is selected as the forward-bias resistance of the diodes D_1.

Step 7b: Generate the normalized shunt arm admittance $y_b = G_B + j\omega \left(C_A - \frac{1}{\omega^2 L_A} \right)$, where G_B is the estimated overall conductive loss of the shunt arm in State-B.

Step 7c: Compute $S_{21B}(j\omega) = \rho_{21B}(\omega) e^{j\varphi_{21B}(\omega)}$ and $S_{11B}(j\omega) = \rho_{11B}(\omega) e^{j\varphi_{11B}(\omega)}$ as in (5.9a) and (5.9b), respectively, as

$$S_{21B} = \frac{2}{z^2 y + 2zy + 2z + y + 2}; \quad \left. \begin{cases} z = z_b \\ y = y_b \end{cases} \right\}$$

$$S_{11B} = \frac{\left(1 - z^2\right) y - 2z}{z^2 y + 2zy + 2z + y + 2}; \quad \left. \begin{cases} z = z_b \\ y = y_b \end{cases} \right\}$$

Step 7d: Plot the results

Now, let us run an example to implement the above design algorithm.

Example 8.1: Let $\omega_0 = 1$, $R = 1$, $r_a = r_b = 0$, $G_A = G_B = 0$.

(a) Design a narrow bandwidth, $0° - 180°$ Phase Range HPT-DPS for the net-phase shift $\Delta\theta = 45°$.
(b) Investigate the effect of the 10% phase deviation on the frequency band.
(c) Investigate the gain characteristic of the HPT-DPS.

Solution

Let us follow the computational steps of the design algorithm

Step 1: Normalized angular frequency is already provided, which is $\omega_0 = 1$.

Step 2: For $\varphi_{21A}(\omega_0) = 45°$ normalized series arm inductor L is given by

$$C = \frac{1}{\omega_0} \tan \left(90° - \frac{|\varphi_{21A}(\omega_0)|}{2} \right) = 2.4142$$

Step 3: Normalized ideal shunt arm capacitor C is

$$L = \frac{1 + \omega_0^2 C^2}{2C} = 1.4142$$

Step 4: Normalized series arm reverse-bias diode capacitance is given by

$$C_{D1} = C = 2.4142$$

Step 5a: By selecting $C_{D2} = C_{D1} = 2.4142$, normalized shunt arm capacitor C_A is found as

$$C_A = \left(\frac{1}{2\omega_0^2 L}\right) + \frac{1}{2\omega_0^2 L}\sqrt{1 + 4\omega_0^2 L C_{D2}} = 1.7071$$

Step 5b: Compute C_T

$$C_T = \frac{C_A C_{D2}}{C_A + C_2} = 1$$

Step 5c: Compute L_A

$$L_A = \frac{1}{\omega_0^2 C_A} = 0.5858$$

Step 6 and Step 7 is implemented under a MatLab program called

"Main_Highpass_TSection_DPS.m"

This program computes the scattering parameters of the $180°$ phase range HPT-DPS topology both for State-A and State-B and prints the performance of the digital phase sifter under consideration.

In Figure 8.2a, the phase variations of S_{21A} and S_{21B} are depicted.

A close examination of Figure 8.2a reveals that at $\omega_0 = 1$, $\Delta\theta = 45°$ as it should be. Furthermore, $\varphi_{21B}(1) = 0$ as expected.

In Figure 8.2b, gain performance of State-A and State-B is shown.

(b) A close examination of Figure 8.2a indicates that 10% phase fluctuation about $\omega_0 = 1$ results in the frequency band over $\omega_1 = 0.845$ and $\omega_2 = 1.574$. Actually, the bandwidth is approximately computed as $\Delta\omega/\omega_0 \cong 0.73\%$. Thus, high-pass T-Section Digital Phase Shifter yields much wider bandwidth than that of Lowpass based-T section Digital Phase shifter topology for $\Delta\theta = 45°$.

(c) Referring to Figure 8.2b, for 10% fluctuations of the phase shift performance, gain performance is better than that of −2.5 dB. Therefore, we can confidently state that HPT-DPS provides better than 70% bandwidth for 10% phase fluctuation when $\Delta\theta(1) = 45°$.

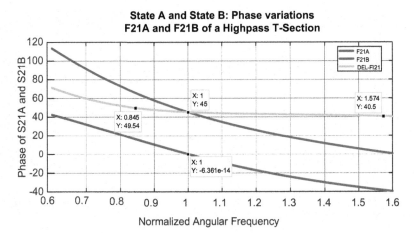

Figure 8.2a   Phase variations of HPT-DPS for State-A, State-B, and $\Delta\theta = \varphi_{21A} - \varphi_{21B}$ of Example 8.1.

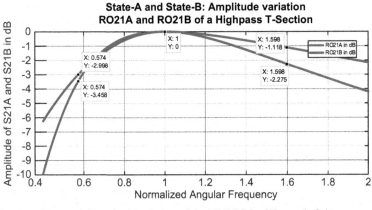

Figure 8.2b   Gain variation of the HPT-DPS of Example 8.1.

Example 8.2: Repeat Example 8.1 for $\Delta\theta = 90°$.

Solution

In Figure 8.3a, phase characteristics of HPT-DPS are shown.

For the topology under consideration, the following component values are found.

Inputs:

$$\omega_0 = 1 \, (Given)$$

$$r_{f1} = r_{r2} = 0 \, (Given)$$

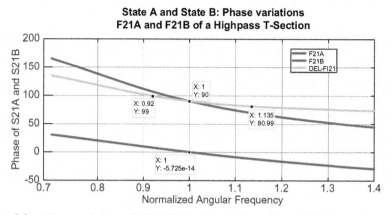

Figure 8.3a Phase variations of LPT-DPS for State-A, State-B, and $\Delta\theta = \varphi_{21A} - \varphi_{21B}$ of Example 6.2.

$$G_A = G_B = 0 \ (Given)$$

$$C = \frac{1}{\omega_0}\tan\left(90° - \frac{|\varphi_{21A}(\omega_0)|}{2}\right) = 1$$

Computational steps:

$\omega_0 = 1$ $(It\ is\ given\ as\ normalized)$

$$L = \frac{1 + \omega_0^2 L^2}{2C} = 1$$

$$C_{D1} = C = 1$$

$$C_{D2} = C_{D1} \ (our\ selection)$$

$$C_A = \left(\frac{1}{2\omega_0^2 L}\right) + \frac{1}{2\omega_0^2 L}\sqrt{1 + 4\omega_0^2 LC_{D2}} = 1.6180$$

$$C_T = \frac{C_A C_{D2}}{C_A + C_2} = 0.6180$$

$$L_A = \frac{1}{\omega_0^2 C_T} = 0.6180$$

Phase characteristic of the 90° HPT-DPS is depicted in Figure 8.3a.

A close examination of Figure 8.3a reveals that 10% phase shift fluctuations about $\omega_0 = 1$ shrinks the bandwidth approximately 21.5%. As compared to Example 8.1, we can confidently state that "larger the phase shift is smaller the bandwidth".

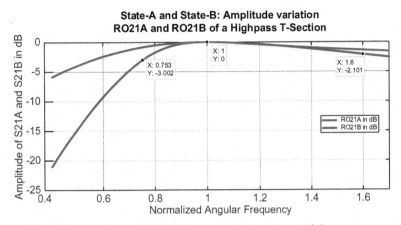

Figure 8.3b   Gain variations of LPT-DPS for State-A, State-B, and $\Delta\theta = \varphi_{21A} - \varphi_{21B}$ of Example 8.2.

In Figure 8.3b, gain variation of the HPT-DPS is depicted. Over the 21.5% phase bandwidth, gain is above $-2\ dB$. Therefore, we say that "gain characteristics is more comfortable than that of phase the characteristics of LPT-DPS".

8.4 Effect of Circuit Component Losses on the Electric Performance of the LPT-DPS

In this section, we investigate the phase and the gain performance of HPT-DPS with respect to diode and other component lossless. For this purpose, we presume that in State-A, the series arm impedance is approximated as

$$z_a\,(j\omega) = r_a + \frac{1}{C_{D1}} \tag{8.3a}$$

where r_a includes forward-bias diode resistance and other related component and connection lossless in the series arm.

Similarly, the shunt arm admittance of State-A is approximated by

$$y_a = G_A + \frac{1}{j\omega L_A} + j\omega C_T \tag{8.3b}$$

where G_A is the overall estimated conductive losses of the circuit components and connections.

In State-B, the symmetric series arm impedance may be approximated as

$$z_b(j\omega) = r_b \tag{8.3c}$$

and the shunt arm impedance is estimated as

$$y_b = G_B + \frac{1}{j\omega L_A} + j\omega C_A \tag{8.3d}$$

In a similar manner to State-A, r_b and G_B are associated with the resistive and conductive overall losses of the series and shunt arms, respectively.

The phase shift and the gain performance of HPT-DPS can be generated using (5.9) as follows

$$S_{11(A,B)} = \frac{(1-z^2)y - 2z}{z^2y + 2zy + 2z + y + 2}; \quad \left\{ \begin{array}{l} \text{In State A}: z = z_a, \ y = y_a \\ \text{In State B}: z = z_b, \ y = y_b \end{array} \right\} \tag{8.3e}$$

and

$$S_{21(A,B)} = \frac{2}{z^2y + 2zy + 2z + y + 2}; \quad \left\{ \begin{array}{l} \text{In State A}: z = z_a, \ y = y_a \\ \text{In State B}: z = z_b, \ y = y_b \end{array} \right\} \tag{8.3f}$$

Now let us estimate r_a, r_b, G_A, and G_B as follows.

In practice, a single inductor L includes a series resistive loss r_L. At the center frequency ω_0, the value of the loss resistance is proportional with inductance impedance $\omega_0 L$.

Let δ_{IND} be the multiplying constant of $\omega_0 L$ such that

$$r_L = (\omega_0 L)\,\delta_{IND} \tag{8.3g}$$

The size of δ_{IND} depends on the manufacturing technology of the inductor. For example, if one employs discrete component manufacturing technology, δ_{IND} could be less than 10%. If MMIC technology is used, then it goes beyond 12.5%. Note that quality factor Q_L of the inductor is given by

$$\delta_{IND} = \frac{1}{Q_L} \tag{8.3h}$$

In a similar manner, a capacitor C includes a a shunt conductive loss G_C. At the center frequency ω_0, G_C is proportional to the admittance $\omega_0 C$.

In this case, one can find a proper multiplying constant δ_{CAP}, which estimates the conductive loss of a capacitor such that

$$G_C = (\omega_0 C)\, \delta_{CAP} \tag{8.3i}$$

For discrete capacitors, δ_{CAP} is much less than 10%. However, for MMIC implementation, the δ_{CAP} value may even go beyond 15%. If Q_C is the quality factor of a capacitor, then

$$\delta_{CAP} = \frac{1}{Q_C} \tag{8.3j}$$

In State-A, when the diode D_1 is forward-biased, the series arm impedance becomes

$$z_a = (r_{on1} + r_s) + \frac{1}{j\omega C_{D1}} = r_a + \frac{1}{j\omega C_{D1}}$$

In this case, the equivalent series arm resistive loss is estimated as

$$r_a = r_{on1} + r_s \tag{8.4a}$$

where r_s is the estimated resistive loss of all the connections. In (8.4a), the dominant term may be r_{on1}.

Therefore, r_s can be neglected as compared to r_{on1} and r_s.

In the shunt arm, the State-A inductor impedance is given by $r_{LA} + j\omega L_A$. Capacitor C_A is connected to reverse-biased diode capacitance C_{D2}. In this case, let the conductive loss of C_A be specified as G_{CA}, then the resulting admittance of the shunt capacitive arm is expressed as

$$Y_{CA} = G_{CA} + \frac{j\omega C_T}{1 + j\omega C_T r_{r2}} \cong G_{CA} + \frac{\omega_0^2 C_T^2 r_{r2}}{1 + \omega^2 C_T^2 r_{r2}^2} + j\,\frac{\omega C_T}{1 + \omega^2 C_T^2 r_{r2}^2}$$

In this case, equivalent admittance y_a is given by

$$
\begin{aligned}
y_a &= \frac{1}{r_{LA} + j\omega L_A} + G_{CA} + \frac{\omega_0^2 C_T^2 r_{r2}}{1 + \omega^2 C_T^2 r_{r2}^2} + j\,\frac{\omega C_T}{1 + \omega^2 C_T^2 r_{r2}^2} \\
&= \left(G_{CA} + \frac{\omega_0^2 C_T^2 r_{r2}}{1 + \omega^2 C_T^2 r_{r2}^2} + \frac{r_{LA}}{r_{LA}^2 + \omega^2 L_A^2} \right) \\
&\quad + j\omega \left(\frac{C_T}{1 + \omega^2 C_T^2 r_{r2}^2} - \frac{L_A}{r_{LA}^2 + \omega^2 L_A^2} \right)
\end{aligned}
$$

Here, we assumed that r_{LA} also includes the series connection resistance r_s of the shunt arm.

In the light of above comments, y_a may be expressed in the neighborhood of ω_0 as

$$
\begin{aligned}
y_a &= \left(G_{CA} + \frac{\omega^2 C_T^2 r_{r2}}{1 + \omega^2 C_T^2 r_{r2}^2} + \frac{r_{LA}}{r_{LA}^2 + \omega^2 L_A^2} \right) \\
&\quad + jw \left(\frac{C_T}{1 + \omega^2 C_T^2 r_{r2}^2} - \frac{L_A}{r_{LA}^2 + \omega^2 L_A^2} \right) \\
&= G_A + jw \left(\frac{C_T}{1 + \omega^2 C_T^2 r_{r2}^2} - \frac{L_A}{r_{LA}^2 + \omega^2 L_A^2} \right)
\end{aligned}
$$

For most practical cases, the reverse-bias resistance r_{r2} is small compared to other resistive quantities. Therefore, it may be appropriate to set to zero. Hence, in the neighborhood of ω_0, equivalent shunt conductive loss is approximated as

$$
\begin{aligned}
G_A &= G_{CA} + \frac{\omega_0^2 C_T^2 r_{r2}}{1 + \omega^2 C_T^2 r_{r2}^2} + \frac{r_{LA}}{r_{LA}^2 + \omega^2 L_A^2} \\
&\cong G_{CA} + \frac{r_{LA}}{r_{LA}^2 + \omega_0^2 L_A^2}
\end{aligned} \tag{8.4b}
$$

Similarly, in State-B, the series arm loss r_b is estimated as

$$
r_b \cong r_{f1} \tag{8.4c}
$$

and y_B and G_B are approximately as

$$
y_b = G_B + jw \left(\frac{C_A}{1 + \omega^2 C_A^2 r_{f2}^2} - \frac{L_A}{r_{LA}^2 + \omega^2 L_A^2} \right) \tag{8.4d}
$$

$$
G_B = G_{CA} + \frac{\omega_0^2 C_A^2 r_{f2}}{1 + \omega^2 C_A^2 r_{f2}^2} + \frac{r_{LA}}{r_{LA}^2 + \omega_0^2 L_A^2} \tag{8.4e}
$$

In the above equations, reverse-bias diode loss resistance r_r may be estimated as

$$
r_{ri} = (\omega_0 C_{Di})\,\delta_D; \quad \{i = 1, 2\} \tag{8.4f}
$$

Let us propose the following algorithm to develop component losses introduced above.

8.5 Algorithm: Design of a Lossy HPT-DPS

This algorithm generates the component losses of HPT-DPS, which in turn yields series arm and shunt arm immittances both in State-A and State-B. Here, we presume that HPT-DPS is realized as MMIC on a silicon substrate using the 0.18u technology of TSMC[1].

Inputs:

f_{0a}: Actual center frequency

ω_0: Normalized angular frequency

R: Normalization resistor

δ_{IND}: Multiplying factor to determine the series loss resistance r_L of an inductor L

δ_{CAP}: Multiplying factor to determine the shunt loss conductance G_C of a capacitor C

δ_D: Multiplying factor to determine the series loss resistance r_r of a reverse-bias diode D

Computational Steps:

Step 1: Compute

$$C = \frac{1}{\omega_0} \tan\left(90° - \frac{|\varphi_{21A}(\omega_0)|}{2} \right)$$

Step 2: Compute

$$L = \frac{1 + \omega_0^2 L^2}{2C}$$

Step 3: Compute C_{D1}

$$C_{D1} = C$$

Step 4: Select C_{D2} and compute

$$C_A = \left(\frac{1}{2\omega_0^2 L} \right) + \frac{1}{2\omega_0^2 L} \sqrt{1 + 4\omega_0^2 L C_{D2}}$$

Note: One may wish to set $C_{D2} = C_{D1}$

Step 5: Compute

$$C_T = \frac{C_A C_{D2}}{C_A + C_2}$$

---

[1]Taiwanese Semiconductor Manufacturing Company (TSMC); (https://www.tsmc.com/english/aboutTSMC/company_profile.htm)

Step 6: Compute

$$L_A = \frac{1}{\omega_0^2 C_A}$$

Step 7: Generate actual values for C_{D1} and C_{D2}

$$C_{D1a} = \frac{C_{D1}}{(2\pi f_{0a}) R}$$

$$C_{D2a} = \frac{C_{D2}}{(2\pi f_{0a}) R}$$

Step 8: Generate normalized values for r_{f1} and r_{f2}

$$R_{f1} (\Omega) = \frac{672 \times 10^{-15}}{C_{D1a} (Farad)}$$

$$R_{f2} (\Omega) = \frac{672 \times 10^{-15}}{C_{D2a} (Farad)}$$

$$r_{f1} = \frac{R_{f1}}{R}$$

$$r_{f2} = \frac{R_{f2}}{R}$$

Step 9: Generate r_{r1} and r_{r2}

$$r_{r1} = (\omega_0 C_{D1}) \delta_D$$

$$r_{r2} = (\omega_0 C_{D2}) \delta_D$$

Step 10: Generate r_{LA} and G_{CA}

$$r_{LA} = (\omega_0 L_A) \delta_L$$

$$G_{CA} = (\omega_0 C_A) \delta_C$$

Step 11: Generate r_L, r_a, and G_A of State-A

$$r_{LA} = (\omega_0 L_A) \delta_L$$

$$r_a = r_{r1}$$

$$G_A = G_{CA} + \frac{\omega_0^2 C_T^2 r_{r2}}{1 + \omega^2 C_T^2 r_{r2}^2} + \frac{r_{LA}}{r_{LA}^2 + \omega^2 L_A^2}$$

Step 12: Generate r_b and G_B of State-B

$$r_b = r_{f1}$$

$$G_B = G_{CA} + \frac{\omega_0^2 C_A^2 r_{f2}}{1 + \omega^2 C_A^2 r_{f2}^2} + \frac{r_{LA}}{r_{LA}^2 + \omega_0^2 L_A^2}$$

Step 13: Generate z_a and y_a of State-A

$$z_a\left(j\omega\right) = r_a + j\omega L$$

$$y_a = G_A + j\omega\left(\frac{C_T}{1 + \omega^2 C_T^2 r_{r2}^2} - \frac{L_A}{r_{LA}^2 + \omega^2 L_A^2}\right)$$

Step 14: Generate z_b and y_b of State-B

$$z_b\left(j\omega\right) = r_b$$

$$y_b = G_B + j\omega\left(\frac{C_A}{1 + \omega^2 C_A^2 r_{f2}^2} - \frac{L_A}{r_{LA}^2 + \omega^2 L_A^2}\right)$$

Step 15: Generate phase and gain performance of LPT-DPS for State-A and State-B

$$S_{21(A,B)} = \frac{2}{z^2 y + 2zy + 2z + y + 2}; \quad \left\{\begin{array}{l} \text{In State A}: z = z_a, \ y = y_a \\ \text{In State B}: z = z_b, \ y = y_b \end{array}\right\}$$

Step 16: Plot the phase and gain performance

Now, let us run an example to evaluate the performance of an LPT-DPS constructed with lossy elements.

Example 8.3: Referring to Example 8.1, let the switching diodes are made of 0.180 um CMOS process of TSMC[2]. Let the actual design frequency be $f_{0a} = 5\ GHz = 5 \times 10^9$ Hz.

Let $\delta_{IND} = 10\%$, $\delta_{CAP} = 1\%$, $\delta_D = 100$,

(a) Compute the ideal element values of the HPT-DPS.
(b) Compute the resistive and conductive losses associated with the ideal elements

---

[2]Taiwanese Semiconductor Manufacturing Company. (https://www.tsmc.com/english/aboutTSMC/company_profile.htm)

(c) Evaluate the phase shifting performance of the phase shifter constructed with lossy elements.

(d) Compute the actual component values

Solution

(a) Ideal element values of the HPT-DPS under consideration are given as in Example 8.1.

$L = 1.4142$, $C = 2.4142$, $CD1 = 2.4142$, $CD2 = CD1$, $CA = 1.7071$, $CT = 1$, $LA = 0.5858$,

(b) Losses associated with component values:
Let us follow steps of the algorithm.

Step 8: Generate normalized values for r_{f1} and r_{f2}

$$R_{f1}\,(\Omega) \times C_{D1a}\,(Farad) = 672 \times 10^{-15}; \quad R_{f1} = 0.4372\,\Omega$$

$$R_{f2}\,(\Omega) \times C_{D2a}\,(Farad) = 672 \times 10^{-15}; \quad R_{f2} = 0.4372\,\Omega$$

$$r_{f1} = \frac{R_{f1}}{R} = 0.0087$$

$$r_{f2} = \frac{R_{f2}}{R} = 0.0087$$

Step 9: Generate r_{r1} and r_{r2}

$$r_{r1} = (\omega_0 C_{D1})\,\delta_D = 0.0241$$

$$r_{r2} = (\omega_0 C_{D2})\,\delta_D = 0.0241$$

Step 10: Generate r_{LA} and G_{CA}

$$r_{LA} = (\omega_0 L_A)\,\delta_L = 0.0586$$

$$G_{CA} = (\omega_0 C_A)\,\delta_C = 0.0171$$

Step 11: Generate r_L, r_a, and G_A of State-A

$$r_L = (\omega_0 L)\,\delta_L = 0.1414$$

$$r_a = r_{r1} = 0.0041$$

$$G_A = G_{CA} + \frac{\omega_0^2 C_T^2 r_{r2}}{1 + \omega^2 C_T^2 r_{r2}^2} + \frac{r_{LA}}{r_{LA}^2 + \omega^2 L_A^2} = 0.1902$$

Step 12: Generate r_b and G_B of State-B

$$r_b = r_{f1} =$$

$$G_B = G_{CA} + \frac{\omega_0^2 C_A^2 r_{f2}}{1 + \omega^2 C_A^2 r_{f2}^2} + \frac{r_{LA}}{r_{LA}^2 + \omega_0^2 L_A^2} = 0.2116$$

Step 13–15: All the above and rest of the steps are programmed under the MatLab Program called

"Main_Example_8_3.m"

Execution of *"Main_Example_8_3.m"* reveals the phase performance of the HPT-DPS under consideration as depicted in Figure 8.4a.

A close examination of Figure 8.4a reveals that at $\omega_0 = 1$, the phase shift between the switching states is 44.53°. So, introduction of lossy elements perturbed the phase performance of the phase shifter about 1.04%, which is tolerable. The lossy components did not affect the bandwidth of the phase shifter. It is still $\mp 23.3\%$ about the center frequency.

Gain performance of the phase shifter topology is shown in Figure 8.4b. Over the $\mp 20\%$ bandwidth, gain is better than –2.8 dB, which is acceptable. As far as gain curves are concerned, over the angular frequency interval of $0.812 < \omega < 1.512$, the gain is above –3 dB.

MatLab codes of Example 8.3 are given under Program Lists.

(d) Computation of actual element values

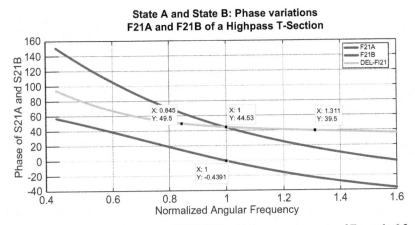

Figure 8.4a　Phase performance of HPT-DPS with lossy components of Example 6.3.

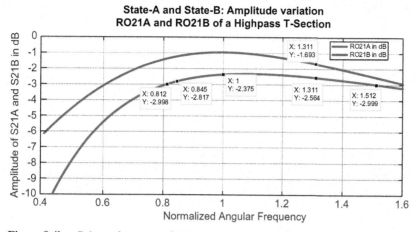

Figure 8.4b Gain performance of HPT-DPS with lossy components of Example 6.3.

Actual element values for the capacitors are given as

$$C_{actual} = \frac{C_{normalized}}{2\pi f_{0a} R}$$

Similarly, actual element values for the inductors are computed as

$$L_{actual} = \frac{R L_{normalized}}{2\pi f_{0a}}$$

Hence, based on the above equations, actual diode capacitances are given as

$$C_{D1a} = C_{D2a} = 1.5369e - 12 = 1.54 \, pF$$

Idealized actual shunt arm inductor L_a is

$$L_a = 2.2508e - 09 = 2.25 \, nH$$

In the shunt arm, shunt inductor L_{Aa} is given by

$$L_{Aa} = 9.3231e - 10 = 0.9323 \, nH$$

And finally, the shunt arm capacitor is

$$C_{Aa} = 1.0868e - 12 = 1.086 \, pF$$

All the above values are realizable.

In the next example, we will investigate the effect of large phase shift on the phase and gain characteristics.

Example 8.4: Using MatLab Program "*Main_Example_8_3.m*", Repeat Example 8.3 for $\theta = 120°$ and report the results.

Solution

Execution of "*Main_Example_8_3.m*" results in the following actual element values

$CD1a = 0.3676\ pF;\ CAa = 0.835\ pF;\ LAa = 1.26\ nH;\ La = 1.8378\ nH$

Actual values of r_a and r_b are given by

$$r_{a-actual} = 8.1792\ \Omega$$

$$r_{b-actual} = 1.8283\ \Omega$$

And actual values of G_A and G_B are computed as

$$G_{A-actual} = 0.0028\ siemens\ (or\ shunt\ resistor\ R_{A-actual}$$

$$= \frac{1}{G_{A-actual}} = 356.3617\ \Omega)$$

$$G_{B-actual} = 0.0039\ siemens\ (or\ shunt\ resistor\ R_{B-actual}$$

$$= \frac{1}{G_{B-actual}} = 255.4701\ \Omega)$$

The phase performance of the LPT-DPS is depicted in Figure 8.5a.

A close examination of Figure 8.5a reveals that the 10% phase variation is confined over the bandwidth of $0.916 \leq \omega \leq 1.07$. In this interval, the phase changes from 108° to 132°. Target phase of 120° shift occurs at $\omega_0 = 0.987$. So, there is a small drift of $\Delta\omega = -0.0130$ from the center frequency $\omega_0 = 1$. This shift is tolerable.

Gain curves of both State-A and State-B are given in Figure 8.5b. For $\Delta\theta = 120°$, component losses heavily penalize the gain below half power or –3 dB. At the center frequency, gain is about –3.37 dB; which is not acceptable. At $\omega = 0.916$, gain is –4.157. At $\omega = 1.07$, gain becomes –3.221 dB.

8.6 General Comments and Conclusion

The Symmetric 3-Element HCL type – Highpass based T-Section Digital Phase Shifter Topology (HPT-DPS) is easy to implement as an MMIC circuit.

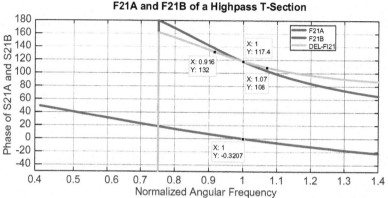

Figure 8.5a Phase performance of HPT-DPS with lossy components of Example 8.4.

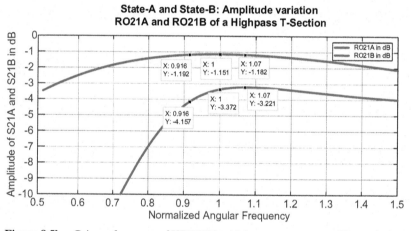

Figure 8.5b Gain performance of HPT-DPS with lossy components of Example 8.4.

It is compact and safely provides 10% bandwidth with reasonable loss. On the other hand, HPT-DPS is not suitable to provide 180° phase shift. Nevertheless, two units of 90° HPT-DPS can be cascaded to design a 180° phase shifting cell.

Examples presented in this chapter verify the below statements.

1. HPT-DPS is good for narrow phase range and small bandwidth requirements. It offers reliable phase shift up 90°
2. Bandwidth of the HPT-DPS decreases while phase shift range increases.

3. Component losses do not much effect the phase shifting performance of the HPT-DPS topology.
4. HPT-DPS topology is suitable for MMIC implementation.

Appendix 8: MatLab Programs for Chapter 8

Program List 8.1. Main_Highpass_TSection_DPS.m

```
% Main Program: Main_Highpass_TSection_DPS.m
% February 20, 2019
% Developed by BS Yarman, Vanikoy, Istanbul
% Enter Positive value
%
clc; close all
% Inputs:
teta=input('Alternative Formulas to Design Lowpass T Section. Enter
                        positive value for Teta in Degree=')
% Inputs
w0=1;
ra=0;rb=0;  GA=0;GB=0;
%
% Computation of the ideal component values of an Highpass T-Section
C=tand(90-teta/2)/w0;
L=(1+w0*w0*C*C)/2/C;
%
% Component Values
CD1=C;
CD2=CD1;
% Delta=1+4*w0*w0*L*CD2
% CA=(1+sqrt(Delta))/2/w0/w0/L
Delta=1+4*w0*w0*L*CD2;
CA=(1+sqrt(Delta))/2/w0/w0/L;
% HPT State-A: D1 is revered biased; D2 is reversed biased
CT=CA*CD2/(CA+CD2);
LA=1/w0/w0/CA;
j=sqrt(-1);
w=0;N=2000;w1=0;w2=2;DW=(w2-w1)/N;
FRI(1:(N+1))=zeros;
for i=1:N+1
    WA(i)=w;
%  -------------------
% State A:
%  -------------------
za=ra+1/j/w/CD1;
ya=GA+1/j/w/LA+j*w*CT;
[ S11a,S21a,RO11a,F11a,RO21a,F21a ] = S_Par_HPT_DPS ( za,ya );
    F21A(i)=F21a;
    RO21A(i)=20*log10(RO21a);
    F11A(i)=F11a;
    RO11A(i)=20*log10(RO11a);
%  -------------------
```

```
% State-B
%  ----------------------------
  zb=rb;
  yb=GB+1/j/w/LA+j*w*CA;
  [ S11b,S21b,RO11b,F11b,RO21b,F21b ] = S_Par_HPT_DPS ( zb,yb );
    F21B(i)=F21b;
    RO21B(i)=20*log10(RO21b);
    F11B(i)=F11b;
    RO11B(i)=20*log10(RO11b);
  %  ----------------------------
DEL_FI21(i)=F21A(i)-F21B(i);
    w=w+DW;
end
%  ----------------------------
Plot_State_AB_HPT_DPS(WA,F21A,RO21A, F11A,RO11A,F21B,RO21B, F11B,RO11B
  ,DEL_FI21)
```

Program List 8.2. S_Par_HPT_DPS

```
function [ S11,S21,RO11,F11,RO21,F21 ] = S_Par_HPT_DPS ( z,y )

% This function generates the S-Parameters of a Lowpass T Section
% Phase Shifter from the series arm impedance Z(jw) and
% the shunt arm admittance Y(jw)
% computed series arm inductor L and the shunt capacitor C
% Developed by BS Yarman: Feb 20, 2019,Vanikoy, Istanbul
% See Equations (5.9)
%  ----------------------------
D=z*z*y+2*z*y+2*z+y+2;
S11=((1-z*z)*y-2*z)/D;
S21=2/D;
R11=real(S11); X11=imag(S11);F11=atan2d(X11,R11);
R21=real(S21); X21=imag(S21);F21=atan2d(X21,R21);
RO11=abs(S11);
RO21=abs(S21);

  end
```

Program List 8.3. function Plot_State_AB_HPT_DPS

```
function Plot_State_AB_HPT_DPS(WA,F21A,RO21A, F11A,RO11A,F21B,RO21B,
    F11B,RO11B,DEL_FI21)
 figure
 plot(WA,F21A,WA,F21B,WA,DEL_FI21)
 title('State A and State B: Phase variations F21A and F21B
                    of a Highpass T-Section')
 legend('F21A','F21B','DEL-FI21')
 xlabel('Normalized Angular Frequency')
 ylabel('Phase of S21A and S21B')
 % Amplitude of S21
 figure
```

```
plot(WA,RO21A,WA,RO21B)
title('State-A and State-B: Amplitude variation RO21A and RO21B
                     of a Highpass T-Section')
legend('RO21A in dB','RO21B in dB')
xlabel('Normalized Angular Frequency')
ylabel('Amplitude of S21A and S21B in dB')
%-----------------------------------------------------------

figure
plot(WA,F11A, WA, F11B)
title('State-A and State-B: Phase variation F11A and F11B
                     of a Highpass T-Section')
legend('F11A')
xlabel('Normalized Angular Frequency')
ylabel('Phase of S11A and S11B')
% Amplitude of S11
figure
plot(WA,RO11A, WA, RO11B)
title('State-A and State-B: Amplitude variation RO11A and RO11B
                     of a Highpass T-Section')
legend('RO11A in dB','RO11B in dB')
xlabel('Normalized Angular Frequency')
ylabel('Amplitude of S11A and S11B in dB')

end
```

Program List 8.4. Main Program:Main_Example_8_3.m

```
% Main Program: Main_Example_8_3.m
% March 3, 2019
% Developed by BS Yarman, Vanikoy, Istanbul
% This program evaluates the lossy performance of a Highpass
                         T-Section DPS
% for a specified actual center frequency f0 in Hz.
% -----------------------------------------------------------
clc; close all
% Inputs:
teta=input('Design of a Lossy Highpass T Section DPS. Enter positive
                          value for Teta in Degree=')
% Inputs
f0a=input('Enter the actual center frequency in Hz f0a =')
w0=input('At f0a, enter the normalized angular frequency w0 =')
R=input('Enter the normaliziation Resistor R=')
%-----------------------------------------------------------
% Compute the normalized element values of LPT-DPS
% Compute the ideal component values C & L of a higpass T-Section
C=tand(90-teta/2)/w0;
L=(1+w0*w0*C*C)/2/C;
%-----------------------------------------------------------
% Compute the unknown Component Values
CD1=C;
CD2=CD1;
Delta=1+4*w0*w0*L*CD2;
```

```
CA=(1+sqrt(Delta))/2/w0/w0/L;
%
% HPT State-A: D1 is revered biased; D2 is reversed biased
CT=CA*CD2/(CA+CD2);
LA=1/w0/w0/CA;
%--------------------------------------------------------------
% Component Values and their related resistive losses:
CD1=C; % See equation (6.2d) of Chapter 6
% ASSUMPTION 1:
% It is assumed that the series loss Ron of a forward biased diode is
% equal to the on channel resistor of an CMOS Switch
% [see Equation (6.1) of Chapter 6].
[ RF1,rf1 ] =Channel_Resistance_of_a_CMOS( CD1,R,f0a );
% ASSUMPTION 2:
% Reverse biased resistive loss of a diode is the "Percent_RVS"
%                     amount of its reverse baised impedance at w0
CrDel_D=100;
rr1=1/w0/CD1/CrDel_D;
CD2=CD1;
[ RF2,rf2 ] =Channel_Resistance_of_a_CMOS( CD2,R,f0a );
rr2=1/w0/CD2/CrDel_D;
%
%--------------------------------------------------------------
% Loss Computations for both State-A and State-B:
% Assumption 3: Loss of an inductor is Percent_L amount of
its impedance
% value at w0.
% Assumption 4: Connectivity loss of an inductor is "Percent_S"
%                     amount of its impedance value at w0.
% Assumption 5: Conductive loss of a Capacitor is "Percent_C" amount
%                     of its admittance value at w0.
CrDel_S=100;
CrDEL_L=10;
CrDel_C=100;
%--------------------------------------------------------------
% Resistive loss of the series arms in State-A:
rL=w0*L/CrDEL_L;
rs=w0*L/CrDel_S;
ra=rr1;
%--------------------------------------------------------------
% Conductive loss of the Shunt arm in State-A:
rLA=w0*LA/CrDEL_L;
GCA=w0*CA/CrDel_C;
GA=GCA+rLA/(rLA*rLA+w0*w0*LA*LA)+(w0*w0*rr2*CT*CT)/(1+w0*w0*rr2*rr2*
CT*CT);
%--------------------------------------------------------------
% Resistive loss of the series arms in State-B:
rb=rf1;
%--------------------------------------------------------------
% Conductive loss of the Shunt arm in State-:
%--------------------------------------------------------------
GB=GCA+rLA/(rLA*rLA+w0*w0*LA*LA)+(w0*w0*rf2*CA*CA)/(1+w0*w0*rf2*rf2
*CA*CA);
```

```
%
j=sqrt(-1);
w=0;N=2000;w1=0;w2=2;DW=(w2-w1)/N;
FRI(1:(N+1))=zeros;
for i=1:N+1
    WA(i)=w;
%  ---------------------
% State A:
%
za=ra+1/j/w/CD1;

ya=GA+j*w*(CT/(1+w*w*rr2*rr2*CT*CT)-LA/(rLA*rLA+w*w*LA*LA));
[ S11a,S21a,RO11a,F11a,RO21a,F21a ] = S_Par_HPT_DPS ( za,ya );
    F21A(i)=F21a;
    RO21A(i)=20*log10(RO21a);
    F11A(i)=F11a;
    RO11A(i)=20*log10(RO11a);
%  ---------------------
% State-B
%  ---------------------
 zb=rb;
 yb=GB++j*w*(CA/(1+w*w*rr2*rr2*CA*CA)-LA/(rLA*rLA+w*w*LA*LA));
 [ S11b,S21b,RO11b,F11b,RO21b,F21b ] = S_Par_HPT_DPS ( zb,yb );
    F21B(i)=F21b;
    RO21B(i)=20*log10(RO21b);
    F11B(i)=F11b;
    RO11B(i)=20*log10(RO11b);
%
DEL_FI21(i)=F21A(i)-F21B(i);
    w=w+DW;
end
%
Plot_State_AB_HPT_DPS(WA,F21A,RO21A, F11A,RO11A,F21B,RO21B,F11B,RO11B,
    DEL_FI21)
[ CD1a ] = ActualValues_of_a_Capacitor( CD1,R,f0a )
[ CAa ] = ActualValues_of_a_Capacitor( CA,R,f0a )
[ LAa ] =ActualValues_of_an_Inductor( LA,R,f0a )
[ La ] =ActualValues_of_an_Inductor( L,R,f0a )
%
ra_actual=ra*R
rb_actual=rb*R
%
GA_actual=GA/R; RA_actual=1/GA_actual
GB_actual=GB/R; RB_actual=1/GB_actual
```

Program List 8.5. Channel_Resistance_of_a_CMOS

```
function [ Ron,ron ] = Channel_Resistance_of_a_CMOS( Coff,R,f0 )
% This function generates the on forward biased channel
     resistance of a
% 0.180um CMOS switch manufactured by TSMC
% February 22, 2019, Vanikoy, Istanbul, Turkey
% Developed by BS Yarman
```

```
% Inputs:
% Coff: Normalized value of the Inductor L
% f0=Actual operating frequency (Normalization frequency)
% R: Actual terminatons resistors (Normalization Resistor)
% Output:
% Ron: Actual value of the channel resistor
% ron: Normalized Value of the channel resistro
%-----------------------------------------------------------
[ Coff_Actual ] =ActualValues_of_a_Capacitor( Coff,R,f0 );
% Ron (?)×C_D1 (Farad)=672×10^(-15)
Ron=672e-15/Coff_Actual;
ron=Ron/R;

end
```

Program List 8.6. ActualValues_of_a_Capacitor

```
function [ C_actual ] = ActualValues_of_a_Capacitor( C,R,f0 )
% February 22, 2019, Vanikoy, Istanbul,Turkey
% Developed by BS Yarman
% Inputs:
% C: Normalized value of the capacitor C
% f0=Actual operating frequency(Normalization frequency)
% R: Actual terminatons resistors(Normalization Resistor)
% Output:
% C_actual: Actual value of the capacitor
%-----------------------------------------------------------
C_actual=C/2/pi/f0/R;
end
```

Program List 8.7. ActualValues_of_an_Inductor

```
function [ L_actual ]=ActualValues_of_an_Inductor( L,R,f0 )
% February 22, 2019, Vanikoy, Istanbul,Turkey
% Developed by BS Yarman
% Inputs:
% L: Normalized value of the Inductor L
% f0=Actual operating frequency (Normalization frequency)
% R: Actual terminatons resistors (Normalization Resistor)
% Output:
% L_actual: Actual value of the Inductor
%-----------------------------------------------------------
L_actual=L*R/2/pi/f0;
end
```

References

[1] B. S. Yarman, "T-section digital phase shifter apparatus," U.S. Patent: 4630010, Washington, DC, USA, December 16, 1986.

[2] B. Yarman, A. Rosen ve P. Stabile, "Lowloss EHF Digital Phase Shifters Suitable for Monolithic Implementation," *IEEE ISCAS,* Montreal, 1984.

[3] B. S. Yarman, "Design of Digital Phase Shifters Suitable for Monolithic Implementations," *Bull. of Technical University of Istanbul,* pp. 185–205, 1985.

[4] B. S. Yarman, "New Circuit Configurations for Designing 0°–180° Digital Phase Shifters," *IEE Proceeding H,* Vol. 134, pp. 253–260, 1987.

[5] B. S. Yarman, "Novel circuit configurations to design loss balanced 0°–360° digital phase shifters," AEU, cilt 45, no. 2, pp. 96–104, 1991.

[6] TSMC, "Taiwanese Semiconductor Manufacturing Company;" (https://www.tsmc.com/english/aboutTSMC/company_profile.htm), Taiwan, 2019.

9

A Symmetric Lattice-based Wide Band – Wide Phase Range Digital Phase Shifter Topology

Summary

In this chapter, a novel digital phase shifter topology, which achieves wide band and wide phase range with high linearity, is proposed. Wide frequency band operation is accomplished employing symmetrical all-pass lattice structures. Compact phase shifter size is obtained utilizing the miniaturized MMIC design implementation technology. Therefore, resulting phase shifter units are suitable for various communication systems such as radar and cellular communication smart antenna arrays.

This chapter provides the complete design equations for the selected phase shift and the center frequency. Moreover, necessary performance analysis tools are provided as they are developed on MatLab environment.

It is shown that employing the commercially available 0.18 μm silicon CMOS MMIC technology, $0°-360°$ phase shift range is possible starting from C-Band to X-Band, and even beyond.

It should be mentioned that the content of this chapter is based on the Ph.D. thesis of Dr. Ercan Atasoy of Istanbul University, 2015, and Dr. Celal Avci of Technical University of Istanbul, 2019, under the guidance of Prof. Dr. Binboga Siddik Yarman of Istanbul University and Technical University of Istanbul and Prof. Dr. Ece Olcay Gunes Technical University of Istanbul [1–6].

9.1 Introduction

Digital phase shifters are the major building blocks of electronically steered smart antenna array systems. Many thousands of them are placed on a single plate. Therefore, their power consumptions are highly crucial.

In many applications, such as software-defined radios, radars, electronic warfare systems, point to point or directed communication systems, etc.; use of "low loss, low power consumptions, broadband, and wide phase range" passive digital phase shifting cells or units are inevitable.

Thus, in this chapter, a novel passive, broadband, wide phase range, compact digital phase shifter topology is introduced.

In the classical digital phase shifter literature, P-I-N diodes are used as switching elements to switch from low pass- to high pass-based, filter like, $LC - T$ *or* π network topologies [1–8]. For narrow-band and narrow phase-shift applications, one may wish to employ digital phase shifters constructed with loaded transmission lines [9–14]. These types of phase shifter topologies are tuned at a single frequency f_0, which yields a few percent bandwidth. Reflective type phase shifter topologies are presented in [10–11]. As far as compact-size implementation is concerned, lumped elements may be preferred over transmission-line or branch-line couplers; however, these types of replacements result in narrowed frequency band with higher losses. Active vector modulator-based phase shifter is presented in [12]. This phase shifter topology cannot offer wide frequency band and wide phase range.

An MMIC active phase shifter topology that utilizes tunable resonant circuit is reported in [13]. This circuit can only achieve narrow band with 2 dB insertion loss by consuming more than 90 mW power, which is not suitable for the next-generation wireless communication systems.

In this chapter, a single, novel, compact digital phase shifter topology is proposed. In the new topology, CMOS transistors is used as switching elements. By properly switching, proposed topology resembles the operation of either "symmetrical LC-all pass" with lagging-phase or "symmetrical LC-all pass" with leading-phase, as shown in Figures 9.2 and 9.3, respectively. Proposed compact topology can provide any phase shift between 0 and 360° by properly computing the passive component values. Operational details are presented in the following sections. Our early results are briefly discussed in [14, 15]. In this chapter, however, design details and practical MMIC implementations issues are covered with electromagnetic simulations. Eventually, complete design of 45°, 90°, and 180° bit digital phase shifting cells are presented. It is exhibited that proposed digital phase shifter topology provides wide phase shifting capability over broad frequency band with reasonable losses.

In the following, first, we present the properties of symmetrical LC lattice structures (Section 9.1), In Section 9.2, lattice-based phase shifter topologies are covered. Section 9.3 is devoted to the generic form of the proposed

"Simple and Single Digital Phase Shifter" topology. In short, this architecture is referred as "SSS-DPS or equivalently 3S-DPS". MMIC implementation of 3S-DPS is given in Section 9.4. In Section 9.5, we introduce the design algorithm of 3S-DPS. In Section 9.6, design examples for 3S-DPS units, namely 45°, 90°, and 180° phase shifters are presented. Finally, in Section 9.7, we compare the electric performance of our proposed phase shifters with those of the similar ones given in the literature.

The simulation results reveal that the proposed "novel 3S-DPS" topology covers the complete phase plane (i.e., over 0°–360°) from C to X-Band (more specifically, over 3–13 GHz) with acceptable loss.

9.2 Properties of Lossless Symmetric Lattice Structures

In this section, properties of a symmetrical lattice structure are re-visited.

In Figure 9.1, a typical symmetrical lattice section, which can be utilized as the building blocks of broadband digital phase shifters, is shown. The actual impedance Z_A is called the series arm impedance. Similarly, the actual impedance Z_B is referred as the cross-arm impedance. Ideally, phase shifter circuits are lossless two-ports. Therefore, impedances Z_A and Z_B are Foster impedances.

In Figure 9.1, the lattice is derived by an independent voltage source E_G with an internal resistor R. At the output port, it is terminated in the same resistor R.

In practice, at RF and microwave frequencies, a two-port, such as a lossless lattice, is described in terms of its real normalized scattering parameters. For the case under consideration, port normalization numbers is selected as

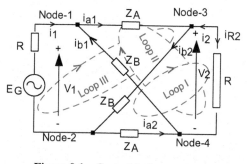

Figure 9.1   General lattice structure.

the terminating resistors R. In terms of the actual impedance values, it can be shown that scattering parameters of a symmetrical lattice is given by

$$S_{11}(p) = S_{22}(p) = \frac{Z_A Z_B - R^2}{Z_A Z_B + (Z_A + Z_B)R + R^2} \tag{9.1a}$$

$$S_{21}(p) = S_{12}(p) = \frac{(Z_B - Z_A)R}{Z_A Z_B + (Z_A + Z_B)R + R^2} \tag{9.1b}$$

Over the entire actual frequencies ($p = j\omega_a$), it is interesting to note that, from input to output, perfect signal transmission requires

$$S_{11}(j\omega_a) = S_{22}(j\omega_a) = 0 \tag{9.2}$$

where $\omega_a = 2\pi f_a$ is the actual angular frequency with f_a being the actual frequency itself. The above condition yields that

$$Z_A(p)\, Z_B(p) = R^2 \tag{9.3}$$

Let the normalized impedances be defined as

$$z_a(p) = \frac{Z_A(p)}{R} = \frac{N_a(p)}{D_a(p)} = O_a(p) \tag{9.4a}$$

$$z_b(p) = \frac{Z_B(p)}{R} = \frac{N_b(p)}{D_b(p)} = O_b(p) \tag{9.4b}$$

where the letters N and D designate the numerator and denominator polynomials in complex variable $p = \sigma + j\omega$ of $z_a(p)$ and $z_b(p)$, respectively. Notice that $O_a(p)$ and $O_b(p)$ must be odd functions since $z_a(p)$ and $z_b(p)$ are Foster. In this regard, it is immediately concluded that

$$z_a^\dagger = z_a(-p) = O_a(-p) = -O_a(p) = -z_a(p) \tag{9.5a}$$

$$z_b^\dagger = z_b(-p) = O_b(-p) = -O_b(p) = -z_b(p) \tag{9.5b}$$

where the sign "$\dagger$" designates the para-conjugate of the complex variable p. Then, perfect transmission condition of (3) yields that

$$\left[\frac{N_a(p)}{D_a(p)}\right]\left[\frac{N_b(p)}{D_b(p)}\right] = 1 \tag{9.6a}$$

or normalized algebraic form of $z_a(p)$ and $y_b(p) = \frac{1}{z_b(p)}$ must satisfy the following equation

$$z_a(p)\, z_b(p) = 1 \tag{9.7a}$$

$$z_a(p) = \frac{1}{z_b(p)} = y_b(p) \tag{9.7b}$$

Referring to Figure 9.1, let us recapitulate the properties of a lossless symmetrical lattice two-port [N].

- Ideally, a lossless symmetrical lattice two-port [N] must possess Foster impedances $Z_A(p)$ and $Z_B(p)$ at the series and cross arms, respectively.

- Two-port [N] must possess a real normalized, bounded-real para unitary scattering matrix $S(p) = \begin{bmatrix} S_{11}(p) & S_{12}(p) \\ S_{21}(p) & S_{22}(p) \end{bmatrix} = \begin{bmatrix} S_{11}(p) & S_{21}(p) \\ S_{21}(p) & S_{11}(p) \end{bmatrix}$ as described by (9.1) such that

$$S(p) S^T(-p) = S(p) S^\dagger(p)$$

$$= \begin{bmatrix} S_{11}(p) & S_{12}(p) \\ S_{21}(p) & S_{22}(p) \end{bmatrix} \begin{bmatrix} S_{11}(-p) & S_{21}(-p) \\ S_{12}(-p) & S_{22}(-p) \end{bmatrix} = I = \begin{bmatrix} 1 & 0 \\ 0 & 1 \end{bmatrix}$$

$$(9.8)$$

In (9.8), the sign "$\dagger$" designates the para conjugate-transpose of a matrix; the letter "I" refers to identity matrix. For the case under consideration, termination resistors R is selected as the port normalization numbers for $S(p)$.

Open form of (9.8) yields that

$$S_{11}(p) S_{11}(-p) + S_{21}(p) S_{21}(-p) = 1 \qquad (9.9)$$

- An ideal symmetrical lattice is designed as a phase shifting unit under perfect transmission over the entire frequency axis such that $S_{11}(p) = 0$, which yields $z_a(p) z_b(p) = 1$ as in (9.7). Furthermore, (9.9) results in

$$S_{21}(p) S_{21}(-p) = 1 \qquad (9.10)$$

Employing the normalized impedances in (1), scattering parameters of $[N]$ is re-written as,

$$S_{11}(p) = S_{22}(p) = \frac{z_a z_b - 1}{z_a z_b + (z_a + z_b) + 1} \qquad (9.11a)$$

$$S_{21}(p) = S_{12}(p) = \frac{(z_b - z_a)}{z_a z_b + (z_a + z_b) + 1} \qquad (9.11b)$$

- Under perfect transmission (i.e., for Foster z_a and z_b satisfying $z_b = \frac{1}{z_a}$ condition), (9.11) becomes

$$S_{11}(p) = S_{22}(p) = 0 \qquad (9.12a)$$

$$S_{21}(p) = S_{12}(p) = \frac{\left(1 - z_a^2\right)}{\left(1 + z_a^2\right) + 2z_a} = \frac{\left(1 - z_a^2\right)}{\left(1 + z_a\right)^2}$$

$$= \frac{\left(1 - z_a\right)\left(1 + z_a\right)}{\left(1 + z_a\right)\left(1 + z_a\right)} = \frac{1 - z_a(p)}{1 + z_a(p)} \tag{9.12b}$$

At this point, (9.10) must be verified for $z_b = \frac{1}{z_a}$ case. This proof is straightforward.

Normalized impedances are odd functions of p as described by

$$z_a(p) = O_a(p)$$

and

$$z_a(-p) = -O_a(p).$$

Then, using (9.5) and (9.11), we construct $S_{21}(p) S_{21}(-p)$ as follows

$$S_{21}(p) S_{21}(-p) = \left[\frac{1 - O_a}{1 + O_a}\right]\left[\frac{1 + O_a}{1 - O_a}\right] = 1$$

Hence, verification is completed.

In the following, we will investigate the phase shifting properties of lossless symmetrical lattice structures with simple impedance forms.

9.3 A Lossless Symmetric Lattice Utilized as a Phase Shifter

As discussed in the above section, perfect transmission condition or equivalently "all pass condition" imposes two major restrictions on the series and the cross arm impedances of a lossless symmetrical lattice section.

1. Symmetrical arm impedances Z_A and Z_B or equivalently normalized impedances $z_a(p)$ and $z_b(p)$ must be Foster impedances. Therefore, they must be odd functions in complex variable p such that $z_a(-p) = -z_a(p)$ and $z_b(-p) = -z_b(p)$.
2. Algebraic form of the normalized Foster impedance $z_a(p)$ must be equal to the inverse of $z_b(p)$. In other words, $z_a(p) = 1/z_b(p)$.

Thus, one can construct a variety of lossless symmetrical lattice all-pass sections as long as z_a and z_b are Foster functions and they satisfy $z_a(p) = 1/z_b(p)$ condition.

Regarding the design of a phase shifter unit, let $x_1, x_2, \ldots, x_m$ be the unknown element values of the normalized Foster impedance $z_a(j\omega, x_1, x_2, \ldots, x_m) = O_a(j\omega) = jx_a(\omega)$. Then, by (9.12)

$$S_{21}(p) = \frac{1 - jx_a}{1 + jx_a} = 1.e^{j\varphi_{21}(\omega)} = e^{-2jtan^{-1}[x_a(\omega)]} \tag{9.13}$$

or

$$\varphi_{21}(\omega) = -2tan^{-1}[x_a(\omega)] \tag{9.14}$$

Hence, at a specified frequency f_0 or equivalently angular frequency $\omega_0 = 2\pi f_0$, desired phase shift θ is obtained by setting

$$\theta = -2tan^{-1}[x_a(\omega_0, x_1, x_2, \ldots, x_m)] \tag{9.15}$$

(9.15) leads to the solution of unknown components values $x_1, x_2, \ldots, x_m$.

In this chapter, we concern with two simple lossless (L) symmetrical (S) lattice (L) sections (S) or in short ($LSLS$). The first one utilizes a simple-single (SS) normalized inductor $x_1 = L_1$ in the series arms, and a simple single (SS) capacitor $C_1 = L_1$ in the cross arms. This is a lagging phase shifter, which provides almost linear negative phase shift in the neighborhood of an operating frequency f_0. The second one is called the leading $LSLS$ with capacitors C_2 in the series arms and inductors $L_2 = C_2$ in the cross arms. This structure yields almost linear positive phase shift in the neighborhood f_o.

In the following sections, operation of lagging and leading $LSLS$ is detailed.

9.4 Lagging LSLS

The lagging $LSLS$ is depicted in Figure 9.2. This configuration is also referred as "$Type - I$" phase shifting unit or equivalently phase shifting cell.

For the above unit, let the actual impedances be $Z_A(j\omega_a) = j\omega_a L_A$ and $Z_B = \frac{1}{j\omega_a C_B}$.

Then, by perfect transmission, we have,

$$\frac{(j\omega_a L_A)/R}{(j\omega_a C_B)R} = 1 \tag{9.16}$$

In the above formulation, $\omega_a = 2\pi f_a$ refers to actual angular frequency for which f_a is the actual frequency itself. At a given specific actual frequency

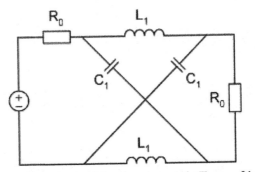

Figure 9.2   LC lattice topology with lagging phase shift (*Type – I* lattice topology).

f_0 or equivalently angular frequency $\omega_0 = 2\pi f_0$, normalized inductor L_1 and capacitor C_1 are defined as

$$L_1 = \frac{\omega_0 L_A}{R} = \frac{2\pi f_0 L_A}{R}$$
$$C_1 = \omega_0 R C_A = 2\pi f_0 R C_A \qquad (9.17)$$

Let the normalized frequency f be

$$f = f_a / f_0 \qquad (9.18)$$

Then, the normalized angular frequency ω is given by

$$\omega = 2\pi f = \frac{\omega_a}{\omega_0} = \frac{f_a}{f_0} \qquad (9.19)$$

Based on the above nomenclature, (9.17) yields that

$$L_1 = C_1 \qquad (9.20)$$

Thus, in (9.13) by setting $x_{an}(\omega) = \omega L_1$, we have the following transfer scattering parameter S_{21n}

$$S_{21n}(j\omega) = \frac{1 - j\omega L_1}{1 + j\omega L_1} = 1.e^{j\varphi_{21n}(\omega)} \qquad (9.21)$$

where

$$\varphi_{21n}(\omega) = -2tan^{-1}(\omega L_1) \qquad (9.22)$$

In the above formulation, subscript "n" refers to negative phase shifting property of *Type – I* structure.

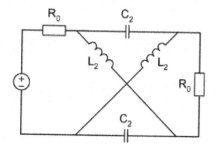

Figure 9.3 LC lattice topology with leading phase shift (Type-II lattice).

9.5 Leading LSLS

As opposed to lagging phase shifter ($Type - I$), leading phase shifter topology has capacitors on the series paths and inductors on the cross paths, as shown in Figure 9.3. This configuration is called $Type - II$.

In (9.13), by setting $x_{ap}(\omega) = -\frac{1}{\omega C_2}$ and $L_2 = C_2$, $Type - II$ phase shifter configuration must possess the following transfer scattering parameter $S_{21p}(j\omega) = 1.e^{j\varphi_{21p}(\omega)}$.

$$S_{21p}(j\omega) = S_{12p}(j\omega) = \frac{1 - 1/j\omega C_2}{1 + 1/j\omega C_2} = \frac{1 + j\frac{1}{\omega C_2}}{1 - j\frac{1}{\omega C_2}} = 1.e^{j\varphi_{21p}} \quad (9.23)$$

where

$$\varphi_{21p}(\omega) = +2tan^{-1}\left[\frac{1}{\omega C_2}\right] \quad (9.24)$$

In the above formulation, subscript "p" refers to positive phase shifting property of $Type - II$ configuration.

9.6 Switching between the Lattice Topologies

Referring to Figure 9.4, one can obtain a wide phase shift, switching between $Type - I$ and $Type - II$ sections. In doing so, one is able to cover complete phase plane (i.e., $0° - 360°$). The phase state of $Type - I$ section is called "State-A". Similarly, the phase state of $Type - II$ section is called "State-B". Then, the phase difference between States B and A is given by

$$\triangle\theta(\omega) = \theta_B(\omega) - \theta_A(\omega) = 2[arctan\left(\frac{1}{\omega C_2}\right) + arctan(\omega L_1)] \quad (9.25)$$

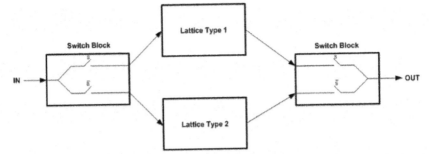

Figure 9.4   Switching between lattice structures.

where θ_A and θ_B refer to phase states of $Type - I$ and $Type - II$ sections, respectively.

Thus, (9.25) indicates that by switching from State A to State B, one is able to cover complete 360° phase range.

At this point, let us discuss the practical implementation of the above digital phase shifter configuration. At the first glance, our target is to design multi-purpose compact digital phase shifter units to be utilized over the entire C and X bands (perhaps, from 3 GHz to 13 GHz) to be utilized in variety of different commercial and military applications such as smart antenna smart antenna array systems. Considering the overall cost of the physical implementation, choice of 0.18 μm. Silicon-based Microwave Monolithic Integrated Circuit (MMIC) technology may be attractive. In this regard, straightforward implementation of Figure 9.4 requires two-bulky "single pole-double throw" switches (one is for the input and the other is for the output). Furthermore, layout of $Type - I$ and $Type - II$ lattice topologies must utilize four printed inductors, which may occupy large chip area together with several parasitic elements that highly complicates the realization. Therefore, this approach is not feasible by cost and technology wise. Hence, one needs to come up with a new idea to reduce the utilization of number of inductors and bulky switches in the circuit design. This way of thinking leads us, perhaps, to reduce the number of lattice sections from two to one, which saves substantial chip area, reduces the number of parasitic elements, which in turn reduces the overall loss of the circuit, and thus, improves the bandwidth. In this chapter, we propose the following novel circuit topology together with a switching scheme to implement the operation of Figure 9.4. The new approach offers wide phase shift range (WPSR) over broad frequency band (BFB) with low phase error and insertion loss. The crux of the idea is to combine the operation

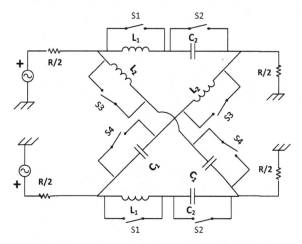

Figure 9.5 Proposed compact phase shifter with ideal elements.

of $Type - I$ and $Type - II$ sections under one-compact symmetrical circuit topology. This circuit is called the "Simple and Single-Symmetric Digital Phase Shifter (or in short SSS-DPS)". In Figure 9.5, SSS-DPS topology is shown.

Series arms (top and bottom) of SSS-DPS include an inductor L_1 with a parallel switch S1 and a series capacitor C_2 with a parallel switch S2. Similarly, cross arms consist of an inductor L_2 with a parallel switch S3 and a series capacitor C_1 with a parallel switch S4. When S1 & S4 are OFF and S2 & S3 are ON (Switching State-A), the circuit resembles the operation of $Type - I$ symmetrical lattice structure as shown in Figure 9.6.

In a similar manner, when S1 & S4 are ON and S2 & S3 are OFF (Switching State-B), the SSS-DPS resembles the operation of $Type - II$ symmetrical lattice section as shown in Figure 9.7.

At the desired normalized angular frequency $\omega_0 = 1$, let $\triangle\theta_0 = \triangle\theta(\omega_0) = \varphi_{12B}(\omega_0) - \varphi_{12a}(\omega_0)$ be the target phase shift between the states B and A, respectively. Let us further assume that, for each state, $\triangle\theta_0$ is evenly distributed. In other words, for $State - A$ $\varphi_{12A}(\omega_0) = -\frac{\triangle\theta(\omega_0)}{2}$ and for $State - B$, $\varphi_{12B}(\omega_0) = \frac{\triangle\theta(\omega_0)}{2}$.

Regarding $State - A$ operation, at $\omega = \omega_0$, desired phase shift $\varphi_{12A} = -\frac{\triangle\theta(\omega_0)}{2}$ is determined using (9.22) as

$$\varphi_{12A} = -2arctan(\omega_0 L_1) \qquad (9.26)$$

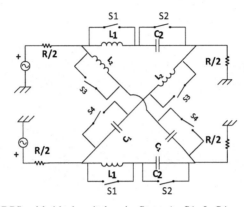

Figure 9.6a SSS-DPS with ideal switches in State A: S1 & S4 are OFF and S2 & S3 are ON.

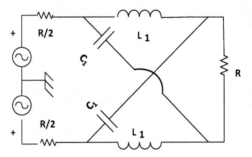

Figure 9.6b Equivalent circuit of SSS-DPS with ideal switches in State A: S1 & S4 are OFF and S2 & S3 are ON.

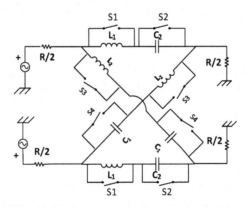

Figure 9.7a SSS-DPS with ideal switches in State B: S1 & S4 are ON and S2 & S3 are OFF.

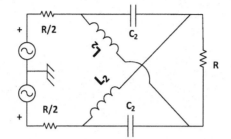

Figure 9.7b Equivalent circuit of SSS-DPS with ideal switches in State B: S1 & S4 are ON and S2 & S3 are OFF.

or normalized value of the series arm inductors L_1 is given by

$$L_1 = \left(\frac{1}{\omega_0}\right) tan\left(\frac{\triangle\theta_0}{4}\right) = tan\left(\frac{\triangle\theta_0}{4}\right) \qquad (9.27)$$

and the normalized value of the cross arm capacitors C_1 is determined as in (9.20). Hence,

$$C_1 = L_1 \qquad (9.28)$$

Similarly, in $State - B$, at the given normalized angular frequency $\omega = \omega_0 = 1$, series arm capacitors C_2 is computed using (9.24) as

$$C_2 = \frac{1}{\omega_0 tan\left(\frac{\triangle\theta_0}{4}\right)} = \frac{1}{tan\left(\frac{\triangle\theta_0}{4}\right)} = \frac{1}{L_1} \qquad (9.29)$$

and the normalized value of the cross-arm inductors L_2 is

$$L_2 = C_2 = \frac{1}{L_1} \qquad (9.30)$$

Thus, the net phase shift $\triangle\theta = \theta_B - \theta_A$ between the states A and B is given by

$$\triangle\theta(\omega) = \varphi_{12B}(\omega) - \varphi_{12A}(\omega) = 2\left[tan^{-1}\left(\frac{1}{\omega C_2}\right) + tan^{-1}(\omega L_1)\right] \qquad (9.31)$$

Clearly, the above equation yields the desired phase shift $\triangle\theta_0 = \triangle\theta(\omega)$ at $\omega = \omega_0 = 1$. However, as we move away from ω_0, a deviation or perturbation δ_θ (*in percentage of* $\triangle\theta_0$) is observed. Literally, phase perturbation is defined as the percentage of the target phase shift $\triangle\theta_0$ such that

$$\delta_\theta = \frac{\triangle\theta(\omega) - \triangle\theta_0}{\triangle\theta_0} \times 100 \qquad (9.32)$$

It is expected that δ_θ is tolerable over at least one octave bandwidth due symmetrical lattice operation. This fact can easily be observed by means of a simple example.

Example 9.1: At $\omega_0 = 1$, let the target phase shift be $\triangle\theta = 45°$.

(a) Find the normalized component values of an SSS-DPS of Figure 9.5.
(b) Plot the phase-shift $\triangle\theta(\omega)$ and observe its perturbation over a large bandwidth.

Solution

(a) Solution of (9.27–9.30) yields the normalized component values of SS-DPS as listed in Table 9.1.

Employing (9.31), the phase shift $\triangle\theta(\omega) = \varphi_{12B}(\omega) - \varphi_{12A}(\omega)$ is depicted in Figure 9.8 and separate plots of $\theta_B(\omega)$ and $\theta_A(\omega)$ are shown in Figure 9.9. A close examination of Figure 9.8 reveals that, at the normalized angular frequency $\omega = 1$, the net phase shift between the states is $\triangle\theta(\omega) = 45°$ as desired. Over an octave bandwidth (i.e., $0.6 \le \omega \le 1.2$), maximum phase deviation δ_θ is about $\delta_\theta = \frac{50.29-45}{45} \times 100 = 11.76\%$. Perturbation of the phase curves away from the center frequency is due to nonlinear behavior of arctangent curves of (9.31). Nonlinearity is dominant at the lower frequencies since

Table 9.1   Normalized component values of Figure 9.7 for Example 9.1

$L_1 = C_1 = \left[\frac{1}{\omega_0}\right]\left[tan\left(\frac{45}{4}\right)\right] = 0.1989$	$C_2 = L_2 = \left[\frac{1}{\omega_0}\right]\frac{1}{tan\left(\frac{5}{\_}\right)} = 5.0273$

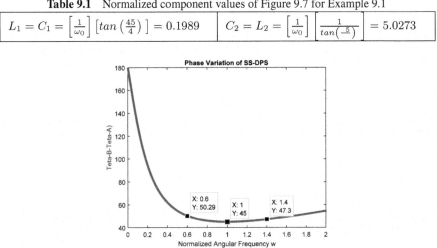

Figure 9.8   Plot of $\triangle\theta(\omega) = \theta_B - \theta_A$ for Example 9.1.

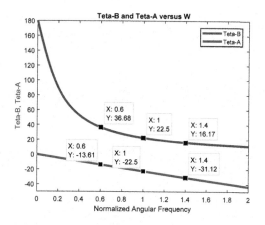

Figure 9.9 Separate plots of θ_B and θ_A for Example 9.1.

Table 9.2 Normalized component values of Figure 9.7 for Example 1

	$L_{p1} = C_{p2} = 1/C_{p1}$	$C_{p1} = L_{p2} = 1/L_{p1}$
$\triangle\theta = 45°$	0.1989	5.0273
$\triangle\theta = 90°$	0.4142	2.4142
$\triangle\theta = 135°$	0.6682	1.4966
$\triangle\theta = 180°$	1	1

$arctan\left(\frac{1}{\omega C_2}\right)$ jumps to $180°$ at $\omega = 0$. The bandwidth improves as the net phase shift between the states becomes larger. It is ideal when $\triangle\theta(\omega) = 180°$.

All the above computations were completed under MatLab program called "$Main\_SSS - DPS\_Example1.m$". This program is listed in Appendix 9 as attached at the end of this chapter.

Let us investigate this situation in the following example.

Example 9.2: Repeat Example 1 for the target phase shifts $\triangle\theta(\omega_0) = 45°$, $90°$, $135°$, and $180°$. Plot the results.

Solution

Solving (9.27–9.30), normalized component values for $\triangle\theta = 45°$, $90°$, $135°$, and $180°$ are found as in Table 9.2.

Corresponding net phase shifts $\triangle\theta$ is depicted in Figure 9.10. As it is seen from this figure, $\triangle\theta_0 = 180°$ case or $\frac{\triangle\theta_0}{4} = 45°$ case yields perfect phase

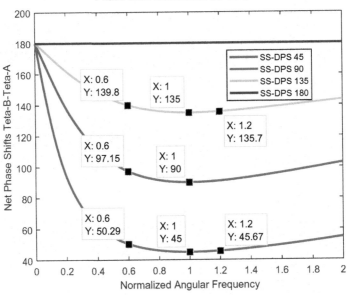

Figure 9.10   Plots of $\triangle\theta\,(\omega) = 45°, 90°, 135°, 180°$ for Example 2.

spread over the entire frequencies due to the nature of (9.31). As the target phase shift $\triangle\theta_0 = \triangle\theta(\omega_0)$ increases from $45°$ to $180°$, phase perturbation decreases down to zero. In Figure 9.11, maximum phase perturbation δ_θ is depicted. In this figure, ordinate is assigned to maximum deviation in percentage over an octave bandwidth. The abscissa is linked with the target phase shifts $45°$, $90°$ $135°$, and $180°$ of the phase curves. In this regard, let $\omega_0 = 1$ be the center frequency of the "one octave bandwidth" of $0.6 \leq \omega \leq 1.2$. Let the maximum phase shift deviation in percentage be δ_θ. It is noted that maximum deviation or perturbation occurs at the lower end of the band. In this case, at $\omega_1 = 0.6$, the phase shift is given by

$$\triangle\theta\,(\omega_1) = \triangle\theta_1 = \triangle\theta_0 + \delta_\theta\triangle\theta_0 \tag{9.33}$$

Maximum phase deviation of (9.33) can be computed for each phase curve as tabulated in Table 9.3.

In this case, perturbation δ_θ may be expressed in terms of target phase shifts $\triangle\theta_0$ and its maximum value $\triangle\theta_1$ over the selected frequency band of operation such that $\delta_\theta = \delta_\theta(\triangle\theta_0, \triangle\theta_1)$. Hence, Table 9.3 is depicted in Figure 9.11 as a three-dimensional plot.

Maximum Phas-Shift Perturbation over an octave bandwidth

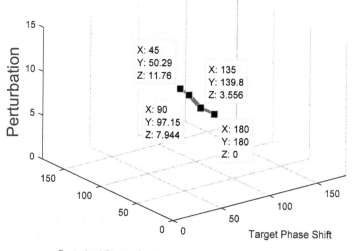

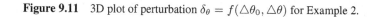

Figure 9.11   3D plot of perturbation $\delta_\theta = f(\triangle\theta_0, \triangle\theta)$ for Example 2.

Table 9.3   Maximum phase perturbation δ_θ on the phase curves

	$\triangle\theta_1$ $(at$ $\omega_1 = 0.6)$	$\delta_\theta = \frac{\triangle\theta_1 - \triangle\theta_0}{\triangle\theta_0}$ $(\%)$
$\triangle\theta_0 = 45°$	50.29	11.76
$\triangle\theta_0 = 90°$	97.15	7.944
$\triangle\theta_0 = 135°$	139.8	3.5556
$\triangle\theta_0 = 180°$	180°	0°

All the above computations were completed under a Matlab program called "Main_SS_DPS_Example 9.2.m". This program is included in the appendix.

It would be handy to summarize the above results in a "basic design algorithm".

9.7 Basic Algorithm to Design Ideal 3S-DPS Section at $\omega_0 = 1$

At the normalized center frequency $\omega_0 = 1$, for a given target phase shift $\triangle\theta_0$, let us define the "major design parameter" μ as

$$\mu = tan\left(\frac{\triangle\theta_0}{4}\right) \tag{9.34}$$

Then, the basic algorithm is given to design 3S-DPS as in Figure 5 with ideal switches and ideal normalized elements values.

Basic Algorithm to design 3S-DPS with Ideal Elements

Inputs: Desired phase shift $\triangle\theta_0$

Computational Steps:

Step-1. Compute the major design parameter μ at ω_0	$\mu = tan\left(\frac{\triangle\theta_0}{4}\right)$
Step-2. Compute the normalized series arm inductor L_1	$L_1 = \frac{1}{\omega_0}\mu$
Step-3. Compute the normalized series arm capacitor C_2	$C_2 = \left(\frac{1}{\omega_0}\right)\left(\frac{1}{\mu}\right)$
Step-4. Compute the normalized cross arm capacitor C_1	$C_1 = L_1$
Step-5. Compute the normalized cross arm inductor L_2	$L_2 = C_2$

An important design remark

It is important to note that a symmetrical lattice network must be derived by means of a symmetrical voltage source so that the ground is isolated from the input and the output port terminals. If the internal impedance of each voltage source is $R_G = 50\ \Omega$, then the equivalent internal impedance of the driving *Thevénin* source will be $R = 2R_G = 100\ \Omega$. In a similar manner, at the output port, in order to isolate the ground from the output terminals, each terminal is connected to a $R_L = 50\ \Omega$ so that equivalent output port termination resistor is $R = 2R_L = 100\ \Omega$. These practical considerations are clearly shown in Figures 9.5, 9.6, and 9.7. In this case, for the sake of simplicity, the scattering parameters of SSS-DPS or symmetrical lattice sections is defined with respect to $R = 100\ \Omega$ and elements values of the actual SSS-DPS is calculated using (9.17) and (9.18) with $R = 100\ \Omega$. Let us run an example to clarify the remark.

Example 9.3: Referring to Example 1, calculate the actual element values SSS-DPS at $10\ GHz$. Here, we assume that SSS-DPS is derived using two identical voltage source each having $R_G = 50\ \Omega$ internal impedance.

Solution

Since $R_G = 50\ \Omega$, normalization number for S parameters must be $R = 2R_G = 100\ \Omega$.

By Example 1, normalized element values are given by

$$L_1 = C_1 = 0.1989$$

$$L_2 = C_2 = \frac{1}{L_1} = 5.0273$$

Then, the actual elements values are computed as

$$L_{1a} = \frac{L_1}{\omega_0} R = \frac{0.1989}{2 \times \pi \times 10 \times 10^9} \times 100 = 0.31656\ nH = 316\ pH$$

$$C_{1a} = \frac{C_1}{\omega_0 R} = \frac{0.1989}{2 \times \pi \times 10 \times 10^9 \times 100} = 0.3165\ fF = 31.65 fF$$

$$L_{2a} = \frac{L_2}{\omega_0} R = \frac{5.0273}{2 \times \pi \times 10 \times 10^9} \times 100 = 8.0012\ nH$$

$$C_{2a} = \frac{C_2}{\omega_0 R} = \frac{5.0273}{2 \times \pi \times 10 \times 10^9 \times 100} = 0.80012\ pF = 800\ fF$$

The above computations were completed under our MatLab program called "*Main_SS_DPS_Example3.m*". This main program calls the function "*Basic_SLLS(Teta, f0a, R)*", which yields both normalized and actual element values of the Basic Symmetrical Lagging and Leading Section (Basic SLLS).

So far, we have dealt with ideal switches and lossless circuit components. Unfortunately, this is not the case in daily practice. Therefore, in the following section, we concern with actual implementation of the proposed SSS-DPS constructed with NMOS Transistors.

The metal oxide semiconductor field effect transistors (MOSFET) are widely used in RF and microwave integrated circuit switch applications. Transistor frequency performance is associated with its transit frequency f_T. f_T is the frequency upon which the current-gain is unity. f_T values for different CMOS processes such as 0.25 µm, 0.18 µm, 0.13 µm, and 90 nm

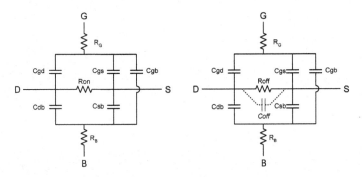

Figure 9.12 MOS switch equivalent circuits: (a) ON state, (b) OFF state.

are 30, 50, 75, and 110 GHz, respectively. In this work, we have the freedom to use TSMC's 0.18 μm NMOS transistors as switching elements. ON and OFF modes of an NMOS 0.18 μm transistor can be modeled as shown in Figure 9.12 [7].

At the "ON" state ($V_{GS} = 1.8\ V$), most effective component of the model is the channel resistor R_{on}. For the process selected in this work, R_{on} varies from 7.5 Ω ohm to 60 Ω. At the "OFF" state ($V_{GS} = 0\ V$), transistor simply exhibits a capacitor $C_{DS} = C_{off}$. For the selected gate length $L = 180$ nm and gate width $W = 200$ μm), the product of $R_{on} \times C_{off}$ is constant and it is given by

$$R_{on}\ (\Omega) \times C_{off}\ (Farad) = 672 \times 10^{-15} \tag{9.35}$$

In this case, for the selected-ON state channel resistor R_{on}, OFF state capacitor C_{off}, actual components of the SSS-DPS Cell is determined. Considering the simplified models of the CMOS switches, let us investigate State-B and State-A modes of an SSS-DPS (3S-DPS) cell. In the following section, details of 3S-DPS topology are presented.

9.8 Operation of 3S-DPS Topology

Referring to Figure 9.13, in State-B, the series arms switch S1 is ON. Therefore, NMOS transistor loads the inductor L_{p1} with the channel resistor R_{on1}. In this mode of operation, S2 is OFF. Then, the capacitor C_{p1} is loaded with the switch capacitor C_{off2}.

In this case, the series arm impedance is given by

$$z_{aB}\ (j\omega) = (R_{on1} \parallel j\omega L_{p1}) + \frac{1}{j\omega\ (C_{p1} + C_{off2})} \tag{9.36a}$$

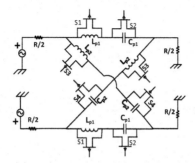

Figure 9.13a Proposed SSS-DPS unit cell topology.

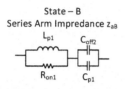

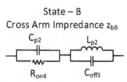

Figure 9.13b State-B: Series and cross arm impedances.

or

$$z_{aB}\left(j\omega\right) = \frac{\omega^2 R_{on1} L_{p1}^2}{R_{on1}^2 + \omega^2 L_{p1}^2} + j\frac{\omega R_{on1}^2 L_{p1}}{R_{on1} + j\omega L_{p1}} + \frac{1}{j\omega\left(C_{p1} + C_{off2}\right)} \quad (9.36b)$$

Using straightforward algebraic derivations, open form of $z_{aB}\left(j\omega\right)$ is found as

$$z_{aB}\left(j\omega\right) = \left[\frac{\omega^2 R_{on1} L_{p1}^2}{R_{on1}^2 + \omega^2 L_{p1}^2}\right] + j\left[\frac{\omega R_{on1}^2 L_{p1}}{R_{on1}^2 + \omega^2 L_{p1}^2} - \frac{1}{\omega\left(C_{p1} + C_{off2}\right)}\right]$$

$$(9.36c)$$

In (9.36), it is expected that R_{on1} is small enough so that it is approximated as

$$z_{aB}\left(j\omega\right) \cong \frac{1}{j\omega\left(C_{p1} + C_{off2}\right)} \quad (9.37)$$

In this mode of operation, at $\omega_0 = 1$, ideally, it is desired that $z_{aB} = \frac{1}{j\omega C_2}$. Thus, the unknown capacitor of the series arms is estimated as

$$C_{p1} = C_2 - C_{off2} \geq 0 \tag{9.38}$$

It is crucial to note that (9.38) imposes a serious restriction on switch S2 such that

$$C_2 \geq C_{off2} \tag{9.39}$$

or in the worst case $C_{p2} = 0$ requires that maximum value of C_{off2} must not exceed C_2 such that

$$(C_{off2})_{max} = C_2 \tag{9.40a}$$

At this point, we may control the value of C_{off2} by defining a positive real number k_2 such that

$$C_{off2} = k_2 (C_{off2})_{max} \tag{9.40b}$$

where k_2 is called the capacitor control coefficient of S2.

Equation (9.40b) guides the design of the second switch S2. Clearly, OFF mode capacitance of NMOS cannot exceed the capacitor C_2. We should also consider an optimum value for the OFF mode NMOS switch capacitors. Let the actual and normalized values of the optimum switch capacitor be C_{a-opt} and C_{n-opt}, respectively. In this case, if one wishes to make the digital phase shifter design based on the optimum value of the switch capacitor $C_{n-opt} = C_{off2}$, then the value of k_2 is determined as

$$k_2 = \frac{C_{off2}}{(C_{off2})_{max}} = \frac{C_{n-opt}}{C_2} \leq 1 \tag{9.40c}$$

If the above condition is not satisfied, then the best choice for k_2 is unity which makes $C_{p1} = 0$.

Let us further continue with the computations in State-B mode when the switches S3 is OFF and S4 is ON. In this mode of operation (State-B), the cross-arm impedances z_{bB} are evaluated as

$$z_{bB}(j\omega) = (R_{on4} \parallel j\omega C_{p2}) + (j\omega L_{p2}) \parallel \left(\frac{1}{j\omega C_{off3}} \right) \tag{9.41}$$

or

$$z_{bB}(j\omega) = \frac{R_{on4}}{1 + j\omega R_{on4} C_{p2}} + j \frac{\omega L_{p2}}{1 - \omega^2 L_{p2} C_{off3}} \tag{9.42a}$$

or

$$z_{bB}(j\omega) = \frac{R_{on4}}{1+(\omega R_{on4}C_{p2})^2} + j\omega \left[\frac{L_{p2}}{1-\omega^2 L_{p2}C_{off3}} - \frac{R_{on4}^2 C_{p2}}{1+(\omega R_{on4}C_{p2})^2}\right]$$

(9.42b)

In a similar manner to (9.36), in (9.42), R_{on4} must be small enough to be neglected as compared to the term $\frac{L_{p2}}{1-\omega^2 L_{p2}C_{off3}}$. Furthermore, in this mode of operation, for the target phase shift $\triangle\theta_0$, which is specified at the normalized angular center frequency ω_0, the term $\frac{L_{p2}}{1-\omega^2 L_{p2}C_{off3}}$ must be equal to ideal cross arm inductors L_2. Hence, the unknown cross arm inductor L_{p2} is estimated as

$$L_{p2} = \frac{L_2}{1+\omega_0^2 L_2 C_{off3}}$$

(9.43)

It is nice to report that (9.43) does not impose any restriction on L_{p2}.

In summary, in State-B operation, for a given phase shift $\triangle\theta_0$, we were able to roughly estimate series arm capacitors as $C_{p1} = C_2 - C_{off2}$ and cross arm inductors as $L_{p2} = L_2/\left(1+\omega_0^2 L_2 C_{off3}\right)$.

Referring to Figure 9.14, let us estimate the unknown component values L_{p1} and C_{p2} in State-A operation.

In State-A, when switches S1 is OFF and S2 is ON, the series arm impedance z_{aA} is given by

$$z_{aA}(j\omega) = (R_{on2} \| j\omega C_{p1}) + (j\omega L_{p1}) \| \left(\frac{1}{j\omega C_{off1}}\right)$$

(9.44a)

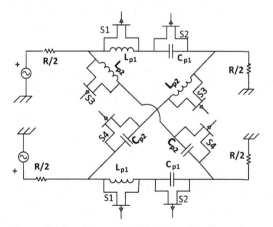

Figure 9.14a Proposed SSS-DPS unit cell topology.

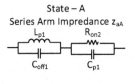

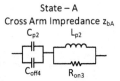

Figure 9.14b   State A: Series and cross arm impedances.

or

$$z_{aA}(j\omega) = \frac{R_{on2}}{1 + j\omega R_{on1} C_{p1}} + j\frac{\omega L_{p1}}{1 - \omega^2 L_{p1} C_{off1}} \qquad (9.44b)$$

By simple algebraic manipulation,

$$z_{aA}(j\omega) = \frac{R_{on2}}{1 + (\omega R_{on2} C_{p1})^2} + j\omega \left[\frac{L_{p1}}{1 - \omega^2 L_{p1} C_{off1}} - \frac{R_{on2}^2 C_{p1}}{1 + (\omega R_{on2} C_{p1})^2} \right]$$

$$(9.44c)$$

In a similar manner to (9.36), the term $(R_{on2} \parallel 1/j\omega C_{p1})$ can be neglected due to small size of R_{on2}. At $\omega_0 = 1$, the second term $(j\omega L_{p1}) \parallel \left(\frac{1}{j\omega C_{off1}} \right) = \frac{L_{p1}}{1 - \omega^2 L_{p1} C_{off1}}$ must yield the ideal series arm inductor L_1. Thus, L_{p1} is estimated without any restriction as

$$L_{p1} = \frac{L_1}{1 + \omega_0^2 L_1 C_{off1}} = \frac{L_1}{1 + L_1 C_{off1}} \qquad (9.45)$$

In cross arms, when switches S3 is ON and S4 is OFF, the equivalent impedance z_{bA} is found as

$$z_{bA}(j\omega) = (R_{on3} \parallel j\omega L_{p2}) + \frac{1}{j\omega (C_{p2} + C_{off4})} \qquad (9.46a)$$

or

$$z_{bA}(j\omega) = \left[\frac{\omega^2 R_{on3} L_{p2}^2}{R_{on3}^2 + \omega^2 L_{p2}^2} \right] + j \left[\frac{\omega R_{on3}^2 L_{p2}}{R_{on3}^2 + \omega^2 L_{p2}^2} - \frac{1}{\omega (C_{p2} + C_{off4})} \right]$$

$$(9.46b)$$

Setting $R_{on3} \cong 0$, and $C_1 = (C_{p2} + C_{off4})$, the unknown cross arm capacitor C_{p2} is estimated as

$$C_{p2} = C_1 - C_{off4} \geq 0 \tag{9.47}$$

As it can be seen from (9.47), maximum value of C_{off4} must be equal to C_1. More explicitly,

$$(C_{off4})_{max} = C_1 \tag{9.48a}$$

This is the restriction imposed on the fourth switch. In a similar manner to (9.40), the value of C_{off4} may be controlled by means of a capacitor control coefficient k_4 such that

$$C_{p2} = C_1 - k_4 (C_{off4})_{max} = C_1 - k_4 C_1 \geq 0 \tag{9.48b}$$

where

$$k_4 = \frac{C_{off4-opt}}{C_1} \leq 1 \tag{9.48c}$$

As in (9.40), the best value of k_4 is unity which makes $C_{p2} = 0$.

It should be mentioned that from the practical point of view, NMOS switches must be designed in such a way that the optimum value of the OFF-mode capacitor should satisfy the following condition:

$$C_{off-opt} \leq min(C_1, C_2) \tag{9.49}$$

All the above derivations suggest the following practical design algorithm to estimate the element values of a practical SSS-DPS Cell.

9.9 Practical Design Algorithm: Estimation of the Normalized Element Values of an $3S-DPS$ CELL

This algorithm leads the designer to estimate the normalized element values of the proposed digital phase shifter topology $3S - DPS$.

Practical design algorithm consists of four parts.

In the first part, component values $\{L_1, C_1, L_2, C_2\}$ of the basic designs of symmetrical lagging and leading sections are computed.

In the second part, NMOS switches S1, S2, S3, and S4 are designed considering the selected manufacturing technology. Switch or MMIC production technology dictates the optimum value of the OFF-Mode capacitances of the transistors, which in turn yields the ON State resistors of the switches.

Therefore, in this part of the algorithm, the designer makes sure that optimum values of the OFF-state capacitors C_{off2} and C_{off4} must be less than or equal to *minimum value of* (C_1, C_2) and capacitor control coefficients k_2 and k_4 must be selected accordingly as in (9.40) and (9.47), respectively. If each switch transistor is biased with different V_{GS}, then $C_{off2} \leq C_2$, $C_{off4} \leq C_1$. Furthermore, to reduce the loss of S1 and S3 C_{off1} and C_{off3} must be selected as large as possible, which in turn reduces the value of the channel resistors R_{on1} and R_{on3}.

In the third part of the algorithm, normalized component values (L_{p1}, C_{p1}) and (L_{p2}, C_{p2}) of 3S-DPS are computed.

Finally, actual element values are determined in the fourth part.

Algorithm 1. Computations of the Elements Values of a 3S-DPS for a given Phase-Shift $\triangle\theta$, which is Evenly Distributed Between States

In the following algorithm, actual component values are indicated by subscript "*a*"

Inputs:

Desired phase shift $\triangle\theta_0$ specified at the normalized centered frequency $\omega_0 = 2\pi f_0 = 1$. It should be noted that $\triangle\theta_0$ is evenly distributed between the state A and B.

Actual center frequency f_{0a} $(in\ Hz)$,

Actual normalizing Resistor R_a $(It\ may\ be\ selected\ as\ R_a = 100\ \Omega)$

Optimum switch capacitor values for S1 and S3, which is designated by $C_{off-opt}$ (for the $TSMC$ 0.18 μm process, it may be appropriate to choose it about 90 fF). Note that $C_{off2} \geq C_2$ and $C_{off4} \geq C_1$.

Computational Steps:

Part I: Design of Basic Lattice Sections: Type-I (Lagging) and Type-II (Leading) with normalized element values	$\mu = \left[tan\left(\frac{\triangle\theta_0}{4}\right)\right]$
	$L_1 = \left[\frac{1}{\omega_0}\right]\mu$
	$C_1 = L_1$
	$C_2 = \left[\frac{1}{\omega_0}\right]\left[\frac{1}{\mu}\right]$
	$L_2 = C_2$

Part II: NMOS Switch Designs	
S1: Select the optimum value for off-mode switch capacitor C_{off1a} and compute its normalized value C_{off1}	$C_{off1a} = C_{off-opt}$ $C_{off1} = (2\pi f_{0a})(R_a C_{off1a})$
Compute the ON State switch cannel resistor R_{on1a}	$R_{on1a} = \dfrac{672\ (fF-\Omega)}{C_{off1a}\ (fF)}$
S2: Select the normalized value of switch capacitor of S2; ; $(C_{off2})_{max} \leq C_2$	$(C_{off2})_{max} = C_2$ $(C_{off2a})_{max} = \dfrac{(C_{off2})_{max}}{(R_a)(2\pi f_{0a})}$
Compute the actual value of $(C_{off2a})_{max}$	
Compute the ON State Resistor R_{on2}	$R_{on2a} = \dfrac{672\ (fF-\Omega)}{C_{off2a}(fF)}$ $C_{off3a} = C_{off1a} = C_{off-opt}$
S3: Select the optimum value for C_{off3a}	$C_{off3} = (2\pi f_{0a})(R_a C_{off3a})$ $R_{on3a} = \dfrac{672\ (fF-\Omega)}{C_{off3a}\ (fF)}$
Compute the normalized values C_{off3}	
Compute the ON State Resistor R_{on3a}	$(C_{off4})_{max} = C_1$
S4: Select the normalized value of $(C_{off4})_{max} \leq C_1$	$(C_{off4a})_{max} = \dfrac{(C_{off4a})_{max}}{(2\pi f_{0a})R_a}$
Compute the actual value of $(C_{off4a})_{max}$	
Part III: Computation of series & cross arm impedances	
Compute the series arm component values: L_{p1} & C_{p1}	$L_{p1} = \dfrac{L_1}{1+\omega_0^2 C_{off1n}}$ $C_{p1n} = C_2 - k_2 C_{off2-max} \geq 0$

Note that C_{off2} is selected as $k_2 \times (C_{off2})_{max}$ where $k_2 \leq 1$ is a control number $(k_2 = 1 \; makes \; C_{p1} = 0)$ Compute the cross-arm component values: L_{p2} & C_{p2} Note that C_{off4} is selected as $k_4 \times (C_{off4})_{max}$ where $k_4 \leq 1$ is a control number $(k_4 = 1 \; makes \; C_{p2} = 0)$	$L_{p2} = \frac{L_2}{1 + \omega_0^2 C_{off3}}$ $C_{p2} = C_1 - k_4 C_{off4-max} \geq 0$
Part IV: Compute the actual Element Values of SSS-DPS $L_{p(1,2)a}$ $C_{p(1,2)a}$	$L_{p(1,2)a} = \frac{L_{p(1,2)}}{2\pi f_{0a}} R_a$ $C_{p(1,2)a} = \frac{C_{p(1,2)}}{2\pi f_{0a} R_a}$

Using the above algorithm, let us design a practical $3S - DPS$ cell for $\triangle\theta = 45°$ in the following example.

Example 9.4: Let $\triangle\theta_0 = 45°$ at $f_{0a} = 10\,GHz$. Select $C_{off-opt} = 90\,fF$. Compute the normalized and actual elements of the SSS-DPS unit. Assume that $\triangle\theta$ is evenly distributed between the states.

Solution

The above algorithm is programmed under a *Matlab* main program called "*Main_Example4.m*"

As in the algorithm, program consists of four major parts. In part I, normalized element values are computed as in Example 1. This part of the program calls the MatLab function

$$[\,L1, C1, L1a, C1a, L2, C2, L2a, C2a\,] = Basic\_SLLS(Teta, f0a, R_a)$$

and returns with the normalized and actual element values as follows:

$$L1 = C1 = 0.1989, \; C2 = L2 = 5.0273$$

$$L1a = 3.1658e - 10\,H, C1a = 3.1658e - 14\,F, C2a = 8.0013e - 13\,F,$$

$$L2a = 8.0013e - 09\,H$$

In the second part of the computations, NMOS switches are designed using the MatLab function called "$NMOS\_Switch\_Design$".

S1 is designed based on $Coff1a = 90\ fF$ and $Ron1a = \frac{672}{C_{off1}} = 7.46\ \Omega$ or $Ron1 = 0.0746$.

S2 is designed for $Coff2a-max = C_{2a} = 8.0013e-13\ F$, $k_2 = 1$, $Coff2 = Coff2-max$ which yields $Ron2a = \frac{672}{C_{off2a}} = 0.8399\ \Omega$ or normalized $Ron2 = 0.0084$.

S3 is identical to S1 or equivalently, $Coff3a = 90\ fF$ and $Ron3a = 7.46\ \Omega$ or $Ron3 = 0.0746$.

S4 is designed for $Coff1a-max = C_{1a} = 3.1658e-14\ F$, $k_4 = 1$ and $Ron4a = \frac{672}{C_{off4a}} = 21.22\ \Omega$ or $Ron4 = 0.212$

In Part III, normalized element values of 3S-PDS is computed at $\omega_0 = 1$ as follows:

$$Lp1 = \frac{L1}{1 + L1 * Coff1} = 0.1788,\ Cp1 = C2 - Coff2 = 0,$$
$$Lp2 = L2/(1 + L2 * Coff3) = 1.3082,\ Cp2 = C1 - Coff4 = 0.$$

In Part IV, actual elements are computed as follows:

$$[Lp1a] = Actual\_Inductor(Lp1, R, f0a) = 2.8457e-10\ H$$
$$[Lp2a] = Actual\_Inductor(Lp2, R, f0a) = 2.0821e-09\ H$$
$$[Cp1a] = Actual\_Capacitor(Cp1, R, f0a)) = 0\ F$$
$$[Cp2a] = Actual\_Capacitor(Cp2, R, f0a) = 0\ F$$

Note: Series and cross arm capacitors C_{p1a} and C_{p2a} is set to zero to minimize the overall loss of the phase shifting unit.

It is expected that performance analysis of the above practical-design coincides with the performance of MMIC layout, which includes identical passive components and NMOS switches of the same kind. Therefore, next section is devoted to the performance analysis of the practical phase shifting units.

9.10 Analysis of the Phase Shifting Performance of 3S-DPS

Once the actual element values are computed, phase shifting performance of an 3S-DPS cell can be investigated by plotting the transfer scattering parameter $S_{21}(j\omega)$ over a wide frequency band for both State-A and State-B. In this regard, generic form of the scattering parameters of (9.11) must be programmed as a function of the normalized angular frequency ω such that

$$
\begin{aligned}
S_{21(A \text{ or } B)}(j\omega) &= \frac{z_a - z_b}{z_a z_b + (z_a + z_b) + 1} \\
&= R_{21(A \text{ or } B)}(\omega) + jX_{21(A \text{ or } B)}(\omega) \\
&= \rho_{21(A \text{ or } B)}(j\omega) e^{\varphi_{21(A \text{ or } B)}(\omega)}
\end{aligned}
\tag{9.50a}
$$

In the above representation, subscripts A and B refer to S_{21} at State-A and State-B, respectively. Signal attenuation performance of the phase shifting sections is measured by means of the transducer power gain (TPG). TPG is generated by

$$
Gain(\omega) = \rho_{21}^2(\omega) = R_{21}^2(\omega) + X_{21}^2(\omega)
\tag{9.50b}
$$

or in dB,

$$
TPG_{dB}(\omega) = 10 log_{10}\rho_{21}^2 = 20 log_{10}\rho_{21}(\omega)
\tag{9.50c}
$$

The phase response is

$$
\varphi_{21}(\omega) = arctan\left[\frac{X_{21}(\omega)}{R_{21}(\omega)}\right]
\tag{9.50d}
$$

For State-B, practical lossy forms of series and cross arm impedances z_{aB} and z_{bB} are specified by (9.36) and (9.42), respectively. They are depicted in Figure 9.13.

Similarly, for State-A, lossy forms of series and cross arm impedances z_{aA} and z_{bA} are given by (9.44) and (9.46), respectively, and they are shown in Figure 9.14.

For an ideal digital phase shifter, the phase shift $\Delta\theta(\omega) = \varphi_{21B}(\omega) - \varphi_{21A}(\omega)$ must be flat and $TPG(\omega) = \rho_{21}^2(\omega)$ must be unity for both State-A and State-B over the frequency band of interest. Flatness of the $\Delta\theta(\omega)$ may be simply achieved if $\varphi_{21B}(\omega)$ and $\varphi_{21A}(\omega)$ are in the form of two parallel functions, shifted from each other by $\Delta\theta_0$. The simplest form of these functions may be two parallel lines. These situations are depicted in Figure 9.15.

Phase performance analysis can be completed by programming (9.50) as a function of normalized angular frequency ω employing the series and cross arm impedances for State-A and State-B as specified by (9.36), (9.42), (9.44), and (9.46). For this purpose, we have developed the following MatLab functions.

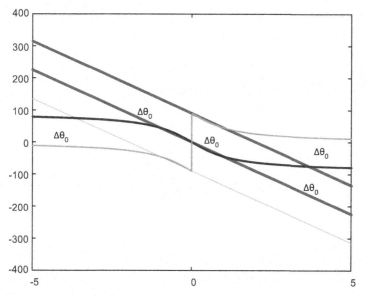

Figure 9.15 Two parallel phase curves shifted from each other by desired amount of constant.

- $SSS\_DPS\_Component\_Values$
- $Symmetrical\_Lattice\_State\_A$
- $Symmetrical\_Lattice\_State\_B$

MatLab function "$SSS\_DPS\_Component\_Values$" generates normalized and actual component values of an 3S-DPS section for a given evenly distributed phase shift $\triangle\theta_0$ between the states, which is specified at the actual center frequency f_{oa}. In this function, for a specified normalized-optimum value of the NMOS OFF state capacitor $C_{off-opt}$, ON mode switch channel resistors $R_{on(i)}$ and OFF mode NMOS switch capacitors $C_{off(i)}$ are automatically determined to minimize the overall loss of the phase shifting section. In this regard, series arm capacitor is determined as

$$C_{p1} = C_2 - C_{off2} \geq 0$$

where

$$C_{off2} = k_2 C_2$$

The reasonable choice for k_2 is given by

$$k_2 = \frac{C_{off-opt}}{C_2} \leq 1$$

At this point, one must be careful since C_2 is calculated from $\triangle\theta_0$ and $C_{off-opt}$ is selected by considering the technology process to manufacture the digital phase shifting cell as an MMIC. As the result of these restrictions, k_2 may turn out to be greater than 1. If it is the case, it is set to unity automatically, which in turns makes the capacitor $C_{p1} = 0$.

Similarly, C_{p2} is determined by

$$C_{p2} = C_1 - C_{off4} \geq 0$$

where

$$C_{off4} = k_4 C_1$$

and

$$k_4 = \frac{C_{off-opt}}{C_1} \leq 1$$

If k_4 becomes greater than one, then it is set to unity for which $C_{p2} = 0$. Computation of

$$L_{p1} = \frac{L_1}{1 + \omega_0^2 L_1 C_{off-opt}}$$

and

$$L_{p2} = \frac{L_2}{1 + \omega_0^2 L_2 C_{off-opt}}$$

is straightforward as derived by (9.45) and (9.43), respectively.

MatLab functions $Symmetrical\_Lattice\_State\_A$ and $Symmetrical\_Lattice\_State\_B$ generate the transfer scattering parameters of State-A and State-B as functions of normalized angular frequency ω and return the phase performance parameters over the frequency band of interest. These parameters are listed as follows:

- $\varphi_{21A}(\omega)$, $\rho_{21A}(\omega)$
- $\varphi_{21B}(\omega)$, $\rho_{21B}(\omega)$
- $\triangle\theta(\omega) = \varphi_{21B}(\omega) - \varphi_{21A}(\omega)$

9.11 Performance Measure of Digital Phase Shifters

Performance of a digital phase shifter is measured by means of two major parameters, namely phase and power transfer or equivalently loss measures.

(a) Phase Performance Measure

In all the above derivations, phase performance of a 3S-DPS may be described by means of an Average Phase Shift θ_{AVP}. Over a prescribed frequency bandwidth $\Delta\omega = \omega_2 - \omega_1$, let the minimum and the maximum values of $\Delta\theta(\omega)$ be

$$\Delta\theta_{min} = \Delta\theta\left(\omega_{min}\right)$$

$$\Delta\theta_{max} = \Delta\theta\left(\omega_{max}\right)$$

Then, the average phase shift θ_{AVP} is described by

$$\theta_{AVP} = \frac{\Delta\theta_{max} + \Delta\theta_{min}}{2}$$

Further, let the average phase fluctuation Δ_θ be defined as

$$\Delta_\theta = \Delta\theta_{max} - \theta_{AVP}$$

or

$$\Delta_\theta = \theta_{AVP} - \Delta\theta_{min}$$

Then, over $\Delta\omega$, phase shift is expressed as

$$\Delta\theta\left(\omega\right) = \theta_{AVP} \mp \Delta_\theta$$

In practice, it is desirable to have Δ_θ less than 20% of the target $\Delta\theta_0 \cong \theta_{AVP}$.

(b) Power Transfer or Loss Performance Measure

Power transfer capability of a passive two-port is described by its transducer power gain (TPG), which is given by

$$TPG\left(\omega\right) = \left|S_{21}\left(j\omega\right)\right|^2 = \rho_{21}^2 \leq 1$$

Ideal transmission is obtained when $TPG\left(\omega\right) = \rho_{21}^2 = 1$ over $\Delta\omega$; meaning that 100% of the available power of the source is transmitted to the output port terminated in the load resistor R_a. In this concept, voltage source E_G is assumed to have an internal impedance $Z_G = R_a$.

In dB, TPG is given by

$$TPG\left(dB\right) = 10log_{10}\left|S_{21}(j\omega)\right|^2 = 20log_{10}\left|S_{21}(j\omega)\right|$$

For ideal transmission, $TPG\left(dB\right) = 0\ dB$.

An equivalent definition is given by means of insertion loss $IL(dB)$, which is defined by

$$IL(dB) = -20log_{10}\left|S_{21}\left(j\omega\right)\right| \geq 0$$

In practice, for a specified bandwidth, it is always desirable to have $TPG \geq 0.5$ ($or\ TPG\ (dB) \leq -3dB$), meaning that more than 50% of the available power is transmitted to the actual termination load R_a. Equivalently, it is preferable to have the insertion loss $IL = TPG\ (dB) \geq 3\ dB$ in the passband.

It is worth mentioning that, usually, phase shifting cells are passive lossless two-ports. In this case, input and out scattering parameters, namely $S_{11}(j\omega) = \rho_{11}(\omega)e^{j\varphi_{11}(\omega)}$ and $S_{22}(j\omega) = \rho_{22}(\omega)e^{j\varphi_{22}(\omega)}$, are given by

$$\rho_{11} = \rho_{22} = 1 - \rho_{21}^2$$

$$\varphi_{21} = 180° - \frac{\varphi_{11} + \varphi_{22}}{2}$$

For an ideal phase shifter, $\rho_{21}^2 = 1$ condition demands that

$$\rho_{11} = \rho_{22} = 1$$

Let us investigate the details of the phase response of a 3S-DPS by means of an example.

Example 9.5:

(a) Let $\triangle\theta_0$ be evenly distributed $\triangle\theta_0 = 45°$ at $f_{0a} = 10\ GHz$. Compute the component values of the 3S-DPS.

(b) Analyze the phase shifting performance of 3S-DPS.

Solution

For this example, we developed a MatLab program called "$Main\_SSS\_DPS\_Example5.m$".

In this program, first, normalized and actual element values of 3S-DPS are computed under the MatLab function "$SSS\_DPS\_Component\_Values$". At this step, OFF mode capacitor of the NMOS switch transistors is selected as $C_{offa-opt} = 90\ fF$. ON mode channel resistor is calculated using the form $R_{on}(\Omega) \times C_{off}(fF) = 672(fF \times \Omega)$. Execution of "$Main\_SSS\_DPS\_Example5.m$" results in the component values as listed in Table 9.4.

It should be noticed that, in the above design, the series arm capacitor C_{p2} is set to zero (i.e., $k_4 = 0$ chosen) and the off-state capacitor of switch 4 is found as $31.66\ fF$. Corresponding on state resistor is $R_{on4} = 21.22\ \Omega$. Obviously, this resistor introduces considerable amount of loss in State-B.

Table 9.4 Actual component values of SSS-DPS for $\triangle\theta_0 = 45°$ at $f_0a = 10GHz$

Components	Actual Values	Channel Resistors	Actual Values in Ω	Off-State Capacitors	Actual Values Values in fF
$L_{p1}(nH)$	0.28457	R_{on1}	7.4667	C_{off1}	90
$C_{p1}(nF)$	0.71013	R_{on2}	7.4667	C_{off2}	90
$L_{p2}(nH)$	2.0821	R_{on3}	7.4667	C_{off3}	90
$C_{p2}(fF)$	0	R_{on4}	21.2269	C_{off4}	31.66

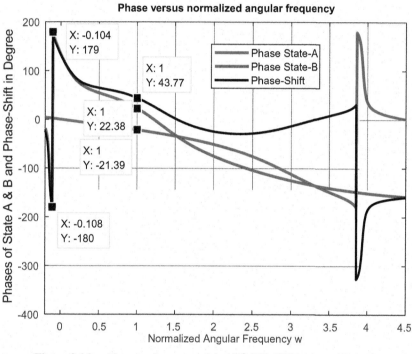

Figure 9.16a Phase performance of the 45° SSS-DPS over a wide band.

Phase performance of the 3S-DPS is depicted in Figure 9.16. In Figure 9.16a, we can analyze the phase variations as a function of normalized angular frequency ω. At $\omega = 0$, in State-B, phase $\varphi_{21B}(\omega)$ jumps from $-180°$ to $+180°$. This sudden jump is due to the nature of the arctangent function of (9.18) and it may be interpreted as a discontinuity.

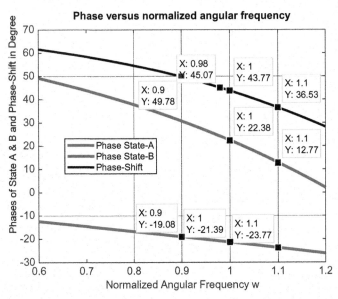

Figure 9.16b   Zoomed-in phase performance of the 45° 3S-DPS over $0.6 \leq \omega \leq 1.2$.

However, $\mp 180°$ are the same points on the phase plane. It is expected that at $\omega = 1$, leading phase of State-B should come down from $+180°$ to $+22.5°$. However, a close examination of Figure 9.16a reveals that, at $\omega = 1$, $\varphi_{21B}(1) = +22.38 \ degree$. The small discrepancy $\varepsilon_{\varphi_{21B}}(1) = 0.12$ is due to the losses of the switches, as expected. When we zoom into the phase curves of Figure 9.16b, the useful bandwidth may be selected from 0.9 to 1.1 or equivalently 9 to 11 GHz). In this interval, it is found that

for State-B:

- maximum of φ_{21B} is $\varphi_{21B-max} = \varphi_{21B}(0.9) = 30.70$
- minimum of φ_{21B} is $\varphi_{21B-min} = \varphi_{21B}(1.1) = 12.77$

for State-A:

- maximum of φ_{21A} is $\varphi_{21A-max} = \varphi_{21A}(0.9) = -19.08$
- minimum of φ_{21A} is $\varphi_{21A-min} = \varphi_{21A}(1.1) = -23.77$

Referring to Figure 9.16b, overall phase performance of 3S-DPS of 45° section is given by

- maximum of $\triangle \theta$ is $\triangle \theta_{max} = \triangle \theta(0.9) = 49.78$; $\omega_1 = 0.9$
- minimum of $\triangle \theta$ is $\triangle \theta_{min} = \triangle \theta(1.1) = 36.53$; $\omega_2 = 1.1$

- Bandwidth is $\triangle\omega_{Even\ Dist} = |0.9 - 1.1| = 0.2\ (2\ GHz)$
- The Average Phase Shift is $\theta_{AVP} = \left(\frac{\triangle\theta_{max}+\triangle\theta_{min}}{2}\right) = 43.15$
- Phase fluctuation is $\triangle_\theta = 6.6$
- $\triangle\theta(\omega) = \theta_{AVP} \mp \triangle_\theta = 43.15 \mp 6.6$

Power transfer performance curves of 3S-DPS-45° is depicted in Figure 9.17.

In State-A, TPG is almost flat and above $-0.6\ dB$ over a wide frequency band ($0 \leq \omega \leq 4$).

In State-B, TPG is above $-3\ dB$ over one octave bandwidth of ($0.6 \leq \omega \leq 1.2$).

Now, let us compare the above results for the case where the switches are perfect. In other words, ON mode channel resistors of NMOS switches are all zero. In fact, we completed this exercise in Example 1 and 2 and plot the results in Figures 9.8 and 9.10. For the ideal case, TPG for both states are $0\ dB$ and the bandwidth is one octave from 0.6 to 1.2 (6 GHz to 12 GHz).

As we observe from the above results, switch losses severely penalized the bandwidth and shrink it from [0.6–1.2] down to [0.9–1.1].

For the ideal case, in State-A,

$$\varphi_{21A}(\omega) = -2tan^{-1}(\omega L_1) \tag{9.51a}$$

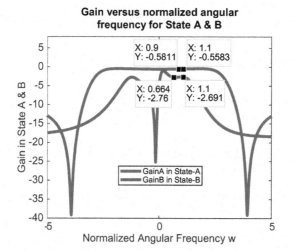

Figure 9.17a Power transfer performance of 45° SSS-DPS over $-5 \leq \omega \leq +5$.

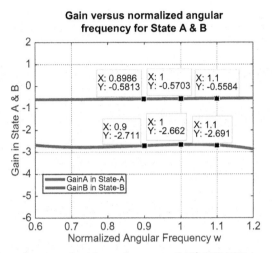

Figure 9.17b   Zoom-In power transfer performance of $45°$ SSS-DPS over $0.6 \leq \omega \leq +1.2$.

is always negative since it belongs to a lagging symmetrical all pass lattice section. Furthermore, the major design parameter μ_A is defined as

$$\mu_A = tan\left[\frac{|\varphi_{21A}(\omega_0)|}{2}\right] = \omega L_1|_{\omega=1} = \omega_0 L_1 = L_1 \qquad (9.51b)$$

or the series arm normalized-inductor L_1 and normalized-cross arm capacitor C_1 is given by

$$L_1 = \frac{1}{\omega_0}tan^{-1}(\mu_A) = C_1 \qquad (9.51c)$$

and the function

$$f(\omega L_1) = -2tan^{-1}(\omega L_1) \qquad (9.51d)$$

may be approximated by a straight line such that

$$y_A(\omega) = [\varphi_{21A}(1)][\omega] \qquad (9.51e)$$

On the other hand, wild behavior of State-B phase response stems from the quasi-hyperbolic behavior of

$$f\left(\frac{1}{\omega C_2}\right) = 2tan^{-1}\left(\frac{1}{\omega C_2}\right) \qquad (9.51f)$$

which may be approximated by the hyperbola

$$f(\omega) = [\varphi_{21B}(1)]\left[\frac{1}{\omega}\right] \qquad (9.51g)$$

where C_2 represents the equivalent normalized series arm capacitor evaluated at $\omega = 1$ and it is determined in terms of the major design parameter

$$\mu_B = tan\left(\frac{\varphi_{21B}(1)}{2}\right) \tag{9.52a}$$

such that

$$C_2 = \frac{1}{\mu_B} = L_2 \tag{9.52b}$$

Similarly, in State-B, the function

$$f\left(\frac{1}{\omega C_2}\right) = 2tan^{-1}\left(\frac{1}{\omega C_2}\right) \tag{9.52c}$$

can be approximated by a shifted line

$$y_B(\omega) = [\triangle\theta_0] - [\varphi_{21A}(1)][\omega] \tag{9.52d}$$

All the above phase functions and their approximations are depicted in Figure 9.18 for $\triangle\theta_0 = 45°$.

A close examination of Figure 9.18 reveals the following practical rules:

(a) In State-B, the function $f(\omega) = 2tan^{-1}\left(\frac{1}{\omega C_2}\right)$ pretty much traces the locus of the hyperbola given by $f(\omega) = [\varphi_{21B}(1)]\left[\frac{1}{\omega}\right]$, and both functions may be approximated by a straight line $y_B(\omega) = [\triangle\theta_0] - [\varphi_{21A}(1)][\omega]$ over a wide band such as $0.2 \leq \omega \leq 1.5$.

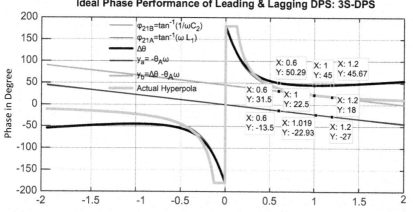

Figure 9.18 Ideal phase characteristic of 3S-DPS for 45° over wide band & wide phase range.

(b) In State-A, function $f(\omega) = -2tan^{-1}(\omega L_1)$ is a well-behaved-smooth function and it is closely approximated by a straight line $y_A(\omega) = [\varphi_{21A}(1)][\omega]$ over broad band $(-2 \leq \omega \leq +2)$.

(c) Due to quasi-hyperbolic behavior of $f(\omega) = 2tan^{-1}\left(\frac{1}{\omega C_2}\right)$, one would avoid utilizing the lower edge of the frequency axis ω in State-B, perhaps up $\omega = 0.1$.

(d) The useful frequency band can be selected in the region where the leading phase hyperbola overlaps with line $y_B(\omega)$, which secures the constant phase shift over line y_A.

Up to this point, we presume that, the phase shift $\triangle\theta_0$ is evenly distributed between the switching states. However, all the above points suggest that, if we work towards the high frequency end $(\omega > 0.5)$ of the hyperbolic function (i.e. in State-B) and distribute $\triangle\theta_0$ unevenly between the states, then we may have a chance to improve the phase range of 3S-DPS. In this regard, we may even consider moving the center frequency from unity to $\omega = 1 + \delta_\omega$, where δ_ω small positive number. Therefore, in the following section, we will investigate the effect of uneven distribution on the phase shift performance of 3S-DPS.

9.12 Investigation of Unequal Phase Distributions between the States

In State-A (ideal lagging symmetric phase section), the phase $\varphi_{21A}(\omega)$ is always a negative quantity. Let $\theta_A(\omega)$ be

$$\theta_A(\omega) = |\varphi_{21A}(\omega)| = \left|-2tan^1(\omega L_1)\right| \geq 0; \; \forall\omega > 0$$

$$\varphi_{21A}(\omega) = -\theta_A(\omega) \leq 0; \; \forall\omega$$

Similarly, in State-B (ideal leading symmetric phase section). $\varphi_{21B}(\omega)$ is always a positive quantity. For the sake of using balanced nomenclature, let $\theta_B(\omega)$ be

$$\theta_B(\omega) = \varphi_{21B}(\omega) = 2tan^1\left(\frac{1}{\omega c_2}\right) \geq 0; \; \forall\omega$$

Then, the phase shift between State-B and State-A is given by

$$\triangle\theta(\omega) = \varphi_{21B}(\omega) - \varphi_{21A}(\omega) = \theta_B(\omega) + \theta_A(\omega)$$

or at $\omega = 1$, for specified $\triangle\theta_0$ and θ_{B0}, θ_{A0} is given by

$$\theta_{A0} = \triangle\theta_0 - \theta_{B0} > 0$$

From the practical implementation point of view, all the switches are selected as NMOS transistor with actual on-state channel resistors $R_{ona(i)}$ and actual off-state capacitor $C_{offa(i)}$. Then, one can design an 3S-DPS using unevenly distributed phase shift as in the following algorithm.

Design algorithm consists of three parts.

In the first part, the basic symmetrical lattice structures are designed using the following algorithm.

Algorithm 2. Design of a Basic Symmetrical Lattice Sections with unevenly distributed phases between the states

Inputs:

$\triangle\theta_0$: Desired phase shift between the states in degree at a specified actual center frequency f_{0a}

Note: At f_{0a} normalized angular frequency is $\omega_0 = \omega = 1$.

θ_{B0}: Desired positive phase-shift of the leading symmetrical lattice structure (State-B)

R: Normalization number for the Scattering Parameters. For our applications, it is set to $R = 100 \, \Omega$.

Computation Steps:

Step 1: Set θ_{A0} as $\theta_{A0} = \triangle\theta_0 - \theta_{B0} \geq 0$.

Step 2: Compute the major design parameters as $\mu_A = tand\left(\frac{\theta_A}{2}\right)$ and $\mu_B = tand\left(\frac{\theta_B}{2}\right)$ for state A & B respectively.

Step 3: Compute Major Component Values for Lagging and Leading Symmetrical LC Lattice as

$$L_1 = \left(\frac{1}{\omega_0}\right)\mu_A$$
$$C_1 = L_1$$
$$C_2 = \left(\frac{1}{\omega_0}\right)\left(\frac{1}{\mu_B}\right) = L_2$$

Step 4: Compute the actual component values as

- $L_{1a} = Actual\_Indutor(L_1, R, f_{0a})$
- $L_{2a} = Actual\_Indutor(L_2, R, f_{0a})$
- $C_{1a} = Actual\_Capacitor(C_1, R, f_{0a})$
- $C_{2a} = Actual\_Capacitor(C_2, R, f_{0a})$

The above steps are gathered under a MatLab function called "$UnevenTeta\_Basic\_SLLS$".

In the second part of the design, normalized component values of a 3S-DPS are computed as in the following algorithm.

Algorithm 3. Design of a 3S-DPS for Unequally Distributed Phases between the States

Inputs:

Component values of the basic lagging and leading sections L_1, C_1, C_2, L_2 & L_{1a}, C_{1a}, C_{2a}, L_{2a},

Normalization Resistor R_a,

Actual Center Frequency f_{0a},

Optimum Off-state capacitors for the NMOS switches $C_{offa-opt(i)}$ for $i = 1, 2, 3, 4$.

Note: One may select all $C_{off-opt(i)} = C_{off-opt}$

Computation Steps:

Step 1: Compute the normalized value of the OFF-State Capacitor $C_{offa-opt}$

$$C_{off-opt} = (2\pi f_{0a})(RC_{offa-opt})$$

Step 2a: Design all NMOS switches using our MatLab function

$[Coff(i),\ Ron(i),\ Ron(i)a\,] = NMOS\_Switch\_Design\ (Coff(i)$ $a, R, f0a)$ for $i = 1, 2, 3, 4$.

Step 2b: Compute the switch control coefficients k_2 and k_4.

$$k_2 = \frac{C_{off-opt(2)}}{C_2}$$

$$k_4 = \frac{C_{off-opt(4)}}{C_1}$$

$$C_{off(2)} = k_2 C_2$$

$$C_{off(1)} = k_4 C_1$$

Note: One may select $C_{off-opt(2)} = C_{off-opt(4)} = C_{off-opt}$

Step 3: Compute the normalized values of series and cross arm components.

Step 3a: Compute the series arm capacitors C_{p1}

$$C_{p1} = C_2 - C_{off-opt(2)} \geq 0$$

If C_{p1} is negative, then set $k_2 = 0$, $C_{p1} = 0$ and $C_{off(2)} = C_2$.
Then, re-compute $R_{on(2)}$ using

$$[Coff(2), \ Ron(2), \ Ron(2)a] \ = \ NMOS\_Switch\_Design \ (C_{2a}, R, f0a)$$

Step 3b: Compute the cross-arm capacitors C_{p2}

$$C_{p2} = C_1 - C_{off(4)} \geq 0$$

If C_{p2} is negative, then set $k_4 = 0$, $C_{p2} = 0$ and $C_{off(4)} = C_1$.
Then, re-compute $R_{on(4)}$ using

$$[Coff(4), \ Ron(4), \ Ron(4)a] \ = \ NMOS\_Switch\_Design \ (C_{1a}, R, f0a)$$

Step 3c: Compute the series arm inductors

$$L_{p1} = \frac{L_1}{1 + \omega_0^2 L_1 C_{off(1)}}$$

Step 3d: Compute the cross-arm inductors

$$L_{p2} = \frac{L_2}{1 + \omega_0^2 L_2 C_{off(3)}}$$

The above algorithm is gathered under the MatLab function called

"UnevenPhase_DPS_Component_Values"

Finally, in the third part of the computations, actual component values are determined by

- $L_{p1a} = Actual\_Indutor(L_{p1}, R, f_{0a})$
- $L_{p2a} = Actual\_Indutor(L_{p2}, R, f_{0a})$
- $C_{p1a} = Actual\_Capacitor(C_{p1}, R, f_{0a})$
- $C_{p2a} = Actual\_Capacitor(C_{p2}, R, f_{0a})$

Once, all the actual component values are found, and the phase-shifting performance of 3S-DPS can be generated as in the previous section using our MatLab functions.

- *Symmetrical_Lattice_State_A*
- *Symmetrical_Lattice_State_B*

Let us investigate the performance of the newly proposed 3S-DPS by means of an example.

Example 9.6:

(a) Referring to Figures 9.13 and 9.14, design an 3S-DPS at $f_{0a} = 10\, GHz$ center frequency, for

$$\triangle \theta_0 = 45° \ and \ \triangle \theta_{B0} = 5°$$

which in turn yields

$$\triangle \theta_{A0} = \triangle \theta_0 - \triangle \theta_{B0} = 40°$$

For NMOS switches, select $C_{offa-opt} = 90\, fF$ as in Example 5.

(b) Analyze the phase-shifting performance of the 3S-DPS under the input parameters specified as above.

Solution

(a) For this example, we developed a MatLab program called

"Main_SSS_DPS_Example6.m"

This program calls the MatLab function *"UnevenPhase_DPS_Component_Values"*, which returns the component values of 3S-DPS as shown in Table 9.5.

(b) Phase performance analysis is completed using the component values presented in Table 9.5. Resulting plots are depicted in Figures 9.19.

A close examination of Figure 9.19a reveals that

for State-B:

- maximum of φ_{21B} is $\varphi_{21B-max} = \varphi_{21B}\,(0.794) = 20.5$
- minimum of φ_{21B} is $\varphi_{21B-min} = \varphi_{21B}\,(1.798) = -34.68$

for State-A:

- maximum of φ_{21A} is $\varphi_{21A-max} = \varphi_{21A}\,(0.794) = -29.3$
- minimum of φ_{21A} is $\varphi_{21A-min} = \varphi_{21A}\,(1.798) = -83.45$

Table 9.5 Actual component values of SSS-DPS for $\triangle\theta_0 = 45°$ and $\theta_B0 = 5°$ at $f_0a = 10\ GHz$

Components	Actual Values	Channel Channel	Actual Values in Ω	Off-State Capacitors	Actual Values in fF
$L_{p1a}(nH)$	0.4804	R_{on1}	7.4667	C_{off1}	90
$C_{p1}a(pF)$	3.5552	R_{on2}	7.4667	C_{off2}	90
$L_{p2a}(nH)$	2.6127	R_{on3}	7.4667	C_{off3}	90
$C_{p2a}(fF)$	0	R_{on4}	11.6007	C_{off4}	57.93

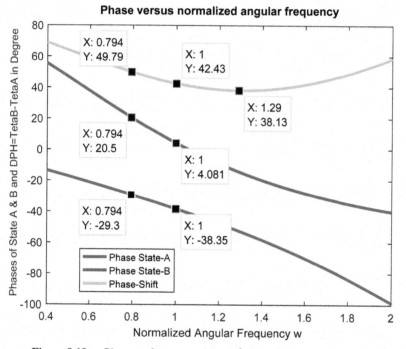

Figure 9.19a Phase performance of the 45° SSS-DPS over wide band.

Overall phase performance of 3S-DPS of 45° section is given by

- maximum of $\triangle\theta$ is $\triangle\theta_{max} = \triangle\theta\,(0.794) = 49.79$; $\omega_1 = 0.794$
- minimum of $\triangle\theta$ is $\triangle\theta_{min} = \triangle\theta\,(1.29) = 38.13$; $\omega_2 = 1.29$

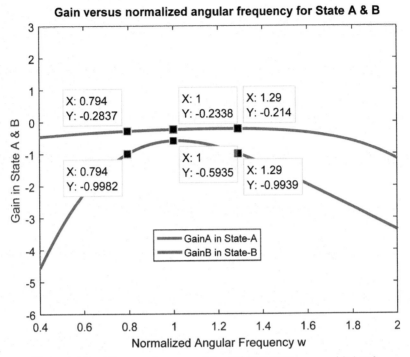

Figure 9.19b   Phase performance of the $45°$ SSS-DPS over a wide band.

- The Average Phase Shift is $\theta_{AVP} = \left(\dfrac{\Delta\theta_{max}+\Delta\theta_{min}}{2}\right) = \dfrac{49.79+36.13}{2} = 42.96$
- The Phase Fluctuation is found as $\Delta_\theta = \Delta\theta_{max} - \theta_{AVP} = 49.79 - 42.96 = 6.83$
- $\Delta\theta(\omega) = \theta_{AVP} \mp \Delta_\theta = 42.96 \mp 6.83$

Bandwidth $\Delta\omega_{45°-Uneven\ Dist.} = |1.29 - 0.794| = 0.496\,(4.96\,GHz)$
Power transfer response of the new design is depicted in Figure 9.19b. In State-A loss performance is smooth and TPG is above –0.6 dB even beyond $\omega = 1.29$. As in Example 5, State-B loss performance is higher than that of State-B (TPG is above –1 dB up to = 1.29).

Comparative performance results are summarized in Table 9.6.

A close examination of Table 9.6 reveals that digital phase shifter designed with unequal phase distribution provides superior frequency bandwidth over the one designed with even phase distribution.

Table 9.6 Performance comparison between digital phase shifters designed with equal and unequal phase distribution

Design Parameters:	3S-DPS Design with Phase Distribution	3S-DPS Design with Equal Phase Distribution
Actual Center Frequency f_{0a}	$10\,GHz$	$10\,GHz$
Target Phase Shift Between States: $\triangle\theta_0 = \theta_{B0} + \theta_{A0}$	$45°$	$45°$
Target Phase Shift θ_{B0} of State-B	$5°$	$22.5°$
Target Phase Shift θ_{A0} of State-A	$40°$	$22.5°$
Average Phase Shift θ_{AVP}	42.96	43.155
Phase Fluctuation $\triangle_\theta$	∓ 4.32	∓ 6.33
Actual Bandwidth $\triangle f$: $\triangle f = (\omega_2 - \omega_1) \times f_{0a}$	$4.96\,GHz$	$2\,GHz$
Minimum TPG in State-A	$-0.293\,dB$	$-0.57\,dB$
Minimum TPG in State-B	$-0.99\,dB$	$-2.79\,dB$

9.13 Practical Lossy Design of a 3D–DPS

So far, element values of an 3S-DPS is estimated using the ideal switches with no loss. In State-B, this approach theoretically yields always a positive leading phase $\varphi_{21B}(\omega) \geq 0$. However, in practice, lossy switches may result in negative $\varphi_{21B}(\omega)$ as the frequency becomes large enough as depicted in Figure 9.19.

In this section, we introduce a method to estimate the component values of a practical 3S-DPS for which all the switches include losses. In this way, we may further increase the bandwidth of an 3S-DPS working in the smooth region of the quasi-hyperbolic phase curve of State-B.

In the following derivations, first, we drive the series arm components L_{P1} and C_{P1}. Then, the cross-arm impedance components L_{P2} and C_{P2} are determined.

Derivation of L_{p1} and C_{p1} for the lossy switch case

Referring to Figures 9.13 and 9.14, in State-B, transfer scattering parameter $S_{21B}(j\omega)$ is given in terms of the series arm impedance Z_{aB} as in (9.12) such that

$$S_{21B}(j\omega) = \frac{1 - z_{aB}}{1 + z_{aB}} = \rho_{21B}(\omega)\, e^{j\varphi_{21B}(\omega)}$$

provided that

$$z_{bB} = \frac{1}{z_{aB}}$$

The series arm impedance z_{aB} can be expressed as in (9.36).

$$z_{aB}(j\omega) = R_{aB}(\omega) + jX_{aB}(\omega) \tag{9.53}$$

where

$$R_{aB}(\omega) = \frac{\omega^2 R_{on1} L_{p1}^2}{R_{on1}^2 + \omega^2 L_{p1}^2} \tag{9.54}$$

and

$$X_{aB}(\omega) = \beta(\omega) - \frac{1}{\omega\,(C_{p1} + C_{off2})} \tag{9.55}$$

and

$$\beta(\omega) = \frac{\omega R_{on1}^2 L_{p1}}{R_{on1}^2 + \omega^2 L_{p1}^2} > 0 \tag{9.56}$$

On the other hand, in Sate-A operation, the series arm inductors L_{p1} is given by (9.45).

$$L_{p1} = \frac{L_1}{1 + \omega_0^2 L_1 C_{off1}}$$

where L_1 is computed for a specified lagging phase $\varphi_{21A}(\omega_0) = -\theta_A < 0$ of State-A or equivalently $\mu_A = \tan(\theta_A) > 0$ as in (9.51) such that

$$L_1 = \frac{1}{\omega_0} \tan^{-1}(\mu_A) = C_1 > 0$$

Going back to Sate-B operation,

$$S_{21B}(j\omega) = \frac{(1 - R_{aB}) - jX_{aB}}{(1 + R_{aB}) + jX_{aB}} = \rho_{21B} e^{j\varphi_{21B}} \tag{9.57a}$$

where

$$\rho_{21B}^2(\omega) = \frac{(1 - R_{aB})^2 + X_{aB}^2}{(1 + R_{aB})^2 + X_{aB}^2} \tag{9.57b}$$

Let

$$\emptyset_{1B} = tan^{-1}\left(\frac{X_{aB}}{1 - R_{aB}}\right) \tag{9.57b}$$

and

$$\emptyset_{2B} = tan^{-1}\left(\frac{X_{aB}}{1 + R_{aB}}\right) \tag{9.57c}$$

Then, the phase φ_{21B} of the transfer scattering parameter S_{21B} is expressed by means of $\emptyset_{1B}$ and $\emptyset_{2B}$ such that

$$\varphi_{21B}(\omega) = \emptyset_{1B}(\omega) - \emptyset_{2B}(\omega) \tag{9.57d}$$

Let $\mu_{1B} = tan(\emptyset_{1B})$. Then,

$$\mu_{1B} = tan(\emptyset_{1B}) = -\frac{X_{aB}}{1 - R_{aB}} = \frac{X_{aB}}{R_{aB} - 1} \tag{9.58a}$$

Let $\mu_{2B} = tan(\emptyset_{2B})$. Then,

$$\mu_{2B} = tan(\emptyset_{2B}) = +\frac{X_{aB}}{1 + R_{aB}} = \frac{X_{aB}}{R_{aB} + 1} \tag{9.58b}$$

Employing the trigonometric identity for $tan[\emptyset_{1B}(\omega) - \emptyset_{2B}(\omega)]$, we can write,

$$
\begin{aligned}
tan[\varphi_{21B}(\omega)] &= tan[\emptyset_{1B}(\omega) - \emptyset_{2B}(\omega)] \\
&= \frac{tan(\emptyset_{1B}) - tan(\emptyset_{2B})}{1 + tan(\emptyset_{1B})tan(\emptyset_{2B})}
\end{aligned}
\tag{9.59a}
$$

or

$$tan[\varphi_{21B}(\omega)] = \frac{\mu_{1B} - \mu_{2B}}{1 + \mu_{1B}\mu_{2B}} \tag{9.59b}$$

Let

$$\gamma_B = tan(\theta_B) \tag{9.59c}$$

where

$$\theta_B = \varphi_{21B}(\omega_0)$$

It is noted that θ_B may take positive or negative values depending on the selected frequency ω_0.

Then, at a specified frequency ω_0,

$$\gamma_B = tan\left(\theta_B\right) = \frac{\frac{X_{aB}}{R_{aB}-1} - \frac{X_{aB}}{R_{aB}+1}}{1 + \frac{X_{aB}}{R_{aB}-1}\frac{X_{aB}}{R_{aB}+1}} = \frac{2X_{aB}}{\left(R_{aB}^2 - 1\right) + X_{aB}^2} \tag{9.60}$$

or

$$\gamma_B X_{aB}^2 - 2X_{aB} + \gamma_B \left(R_{aB}^2 - 1\right) = 0 \tag{9.61}$$

Solution of (9.61) yields two distinct X_{aB}, namely $X_{aB,1}$ and $X_{aB,2}$ such that

$$X_{aB,1} = \frac{1 + \sqrt{1 - \gamma_B^2\left(R_{aB}^2 - 1\right)}}{\gamma_B} \tag{9.62a}$$

and

$$X_{aB,2} = \frac{1 - \sqrt{1 - \gamma_B^2\left(R_{aB}^2 - 1\right)}}{\gamma_B} \tag{9.62b}$$

Notice that the discriminant of the above equations must be always non-negative to end up with real solutions. In other words,

$$\triangle = 1 - \gamma_B^2\left(R_{aB}^2 - 1\right) \geq 0 \ or \ \gamma_B^2\left(R_{aB}^2 - 1\right) \leq 1 \tag{9.62c}$$

Once $X_{aB}\left(\omega_0\right)$ is found, for a selected C_{off2}, C_{p1} is computed as follows.

$$X_{aB} = \beta\left(\omega_0\right) - \frac{1}{\omega_0\left(C_{p1} + C_{off2}\right)}$$

or

$$\frac{1}{\omega_0\left(C_{p1} + C_{off2}\right)} = \beta\left(\omega_0\right) - X_{aB}$$

Thus,

$$C_{p1} = \frac{1}{\omega_0\left[\beta\left(\omega_0\right) - X_{aB}\right]} - C_{off2} \geq 0 \tag{9.63}$$

The following remarks are found helpful for the designers.

Remarks:

(a) The above formulations cover both positive and negative values of θ_B, which is equivalent of having either positive or negative $\gamma = tan\left(\theta_B\right)$.

(b) A close examination of (9.62) reveals that if θ_B is selected as a positive quantity, then we must choose the solution which gives negative value for X_{aB}, which in turn results in positive value for $X_{bB} = -\frac{1}{X_{aB}}$.

On the other hand, if we start the design with negative value of θ_B, then positive value solution for X_{aB} must be selected, which yields negative value for $X_{bB} = -\frac{1}{X_{aB}}$. These choices are mandatory to end up with realizable component values for the 3S-DPS topology.

(c) Solution of the above equations to compute the unknown components $\{L_{p1}, C_{p1}, L_{p2}, C_{p2}\}$ values are not exact. For the sake of simplicity, component values for $\{L_1, C_1\}$ and $\{L_2, C_2\}$ pairs are computed for the ideal switches. However, the imaginary part X_{aB} of $Z_{aB} = R_{aB} + jX_{aB}$ is determined for the switches with on-mode channel resistors. Therefore, at the end of the explicit solutions, one may optimize the performance of the 3S-DPS unit for the target phase shift $\triangle\theta$ while minimizing the overall loss of State-A and B over the frequency band of operation. The optimization can be done manually as described in the following example.

(d) Equation (9.63) must yield a positive C_{p1}. If this is not the case, the designer can set it to zero (i.e. $C_{p1} = 0$). In this regard, switch (S2) must be re-designed in such a way that $C_{off2} = C_2$, which in turn yields a positive $\theta_B = \varphi_{21B}(\omega_0) > 0$. Hence, at the beginning of the computations, θ_B must be selected as a positive quantity, perhaps it is selected as a sufficiently small to have State-B phase curve to operate in the smooth phase region.

(e) If $\triangle$ is positive, then $X_{aB,i}$ of (9.62) could be either positive or negative. If we start designing the 3S-DPS unit from the series arm components using X_{aB}, then (9.63) yields $C_2 = \frac{1}{\omega_0[\beta(\omega_0)-X_{aB}]} = C_{p1} + C_{off2} = L_2 > 0$. In this regard, $\beta(\omega_0)$ of (9.56) is always positive. Therefore, X_{aB} must be a negative quantity to yield positive $C_2 = L_2$. Hence, these computations demand a positive $\theta_B = \varphi_{21B}(\omega_0)$.

(f) In order to obtain a wide phase range over a broad frequency band, phase shift $\triangle\theta$ must be unequally distributed between the states. As indicated above, phase of State-B may be selected small enough to have State-B phase operation in the smooth region of the phase curve. For example, for a $\triangle\theta = 45°$ phase shift, if State-A phase $\varphi_{21A}(\omega_0)$ is selected as $-40°$ (*i.e.* $\theta_A = 40°$), then State-B phase becomes $\varphi_{21B}(\omega_0) = \theta_B = 5°$, which corresponds to 12.5% of θ_A, which may be considered as sufficiently small enough.

Derivation of L_{p2} and C_{p2} for the lossy switch case

At this point, for State-B, we must determine cross arm impedance component values L_{p2} and C_{p2} when the all pass condition of (9.7) is satisfied.

In this case, the cross-arm impedance is given by

$$Z_{bB} = \frac{1}{Z_{aB}} = \frac{1}{R_{aB} + jX_{aB}} = \frac{R_{aB}}{R_{aB}^2 + X_{aB}^2} - j\frac{X_{aB}}{R_{aB}^2 + X_{aB}^2} = R_{bB} + jX_{bB}$$

$$(9.64a)$$

where

$$R_{bB} = \frac{R_{aB}}{R_{aB}^2 + X_{aB}^2} \qquad (9.64b)$$

And if R_{aB} is small enough, then,

$$X_{bB} = -\frac{X_{aB}}{R_{aB}^2 + X_{aB}^2} \cong -\frac{1}{X_{aB}} > 0 \qquad (9.64c)$$

On the other hand, the cross-arm impedance Z_{bB} is specified as

$$Z_{bB} = impedance\ of\ \{L_{p2}//C_{off3}\} + impedance\ of\ \{C_{p2}\ //R_{on4}\}$$

or

$$Z_{bB} = j\left[\frac{\omega L_{P2}}{1 - \omega^2 L_{p2} C_{off3}}\right] + \frac{R_{on4}}{1 + j\omega R_{on4} C_{P2}}$$

or

$$Z_{bB} = \frac{R_{on4}}{1 + (\omega R_{on4} C_{p2})^2} + j\left[\frac{\omega L_{P2}}{1 - \omega^2 L_{P2} C_{off3}} - \frac{\omega R_{on4}^2 C_{p2}}{1 + (\omega R_{on4} C_{p2})^2}\right]$$

$$= R_{bB} + jX_{bB} \qquad (9.65a)$$

where

$$R_{bB}(\omega) = \frac{R_{on4}}{1 + \omega^2 R_{on4}^2 C_{P2}^2};$$

$$X_{bB} = \omega\left[\frac{L_{P2}}{1 - \omega^2 L_{P2} C_{off3}} - \frac{R_{on4}^2 C_{p2}}{1 + (\omega R_{on4} C_{p2})^2}\right] \qquad (9.65b)$$

which is computed by (9.64b). Thus, at $\omega = \omega_0$, (9.65b) results in

$$C_{p2} = \left[\frac{1}{\omega_0 R_{on4}}\right]\left[\sqrt{\frac{R_{on4} - R_{bB}(\omega_0)}{R_{bB}(\omega_0)}}\right] \geq 0 \qquad (9.65c)$$

In (9.65b), if C_{p2} becomes negative, it is appropriate to set it to zero. On the other hand, for State-A, C_{P2} must approximately satisfy (9.48b) such that

$$C_{p2} = C_1 - C_{off4} = \mu_A - C_{off4} \geq 0 \tag{9.65d}$$

Obviously, if (9.65b) and (9.65c) are not coincidentally satisfied, it is difficult to have identical C_{p2} out of (9.65b) and (9.65c). Upon designer's choice, C_{p2} can be set to zero to minimize the overall loss of the phase shifter. If $C_{p2} = 0$ then, Switch 4 (S4) must be re-designed in such a way that off-mode capacitor C_{off4} is equal to C_1 (i.e., $C_{off4} = C_1$). In this state, $\varphi_{21A}(\omega_0)$ is always negative.

Similarly, (9.65a) dictates that

$$X_{bB}(\omega) = \frac{\omega L_{p2}}{1 - \omega^2 L_{p2} C_{off3}} - \omega \alpha(\omega); \quad \alpha(\omega) = \frac{\omega R_{on4}^2 C_{p2}}{1 + (\omega R_{on4} C_{p2})^2} \tag{9.65e}$$

where X_{bB} is already known by (9.64). Thus, for negligible R_{on4} or $\alpha(\omega_0)$ at a specified ω_0, desired cross arm inductor L_{p2} is found as follows

$$\begin{aligned} L_{p2} &= \frac{[X_{bB}(\omega_0) + \omega_0 \alpha(\omega_0)]}{\omega_0 \{1 + \omega_0 [X_{bB}(\omega_0) + \omega_0 \alpha(\omega_0)] C_{off3}\}} \\ &\cong \frac{X_{bB}(\omega_0)}{\omega_0 [1 + \omega_0 X_{bB}(\omega_0) C_{off3}]} \end{aligned} \tag{9.65f}$$

Based on the above derivations, one can design a 3S-DPS for a specified $\theta_B = \varphi_{21B}(\omega_0) > 0$ using the following algorithm.

Algorithm 4. Employing lossy switches, computation of the series arm inductors L_{p1} and capacitors C_{p1} for arbitrary selection of State-A and State-B phases of a 3S-DPS

Inputs: ω_0, $\theta_A = |\varphi_{21}(\omega_0)| > 0$, $\theta_B = \varphi_{21B}(\omega_0) > 0$, C_{off1a}, C_{off2a}, R_a; $0° \leq \theta_B \leq +90°$. It should be noted that off-mode switch capacitors are selected by the designers to ease the n-MOS switch productions on the chip.

Computational Steps:

Step 1: Design Switches 1 & 2.
Compute normalized capacitances C_{off1} and C_{off2}:

$$C_{off1} = (2\pi f_{01}) R_a C_{off1a}$$

$$C_{off2} = (2\pi f_{01}) R_a C_{off2a}$$

Step 2: Compute the actual ON-State channel resistors R_{on1a} and R_{on2a} and their normalized values R_{on1} and R_{on2}, respectively.

$$R_{on1a} = \frac{672(fF \times ohm)}{C_{off1a}(fF)}$$

$$R_{on2a} = \frac{672(fF \times ohm)}{C_{off2a}(fF)}$$

and

$$R_{on1} = R_{on1a}/R_a$$

$$R_{on2} = R_{on2a}/R_a$$

Step 3: Compute μ_A, γ_B L_1, and L_{p1}

$$\mu_A = tand\left(\frac{\theta_A}{2}\right) > 0$$

$$\gamma_B = tan(\theta_B)$$

$$L_1 = \mu_A/\omega_0$$

$$C_1 = L_1$$

$$L_{p1} = \frac{L_1}{1 + \omega_0^2 L_1 C_{off1}}$$

Step 4: Compute $R_{aB}(\omega_0)$ and $\beta(\omega_0)$ as in (9.54) as follows

$$R_{aB}(\omega_0) = \frac{\omega_0^2 R_{on1} L_{p1}^2}{R_{on1}^2 + \omega_0^2 L_{p1}^2} > 0$$

$$\beta(\omega_0) = \frac{\omega_0 R_{on1}^2 L_{p1}}{R_{on1}^2 + \omega_0^2 L_{p1}^2} > 0$$

Step 5: Solve the Equation (9.61) to generate X_{aB}

$$X_{aB,1} = \frac{1 + \sqrt{1 - \gamma_B^2\left(R_{aB}^2 - 1\right)}}{\gamma_B}$$

and

$$X_{aB,2} = \frac{1 - \sqrt{1 - \gamma_B^2 \left(R_{aB}^2 - 1 \right)}}{\gamma_B}$$

At this point, we must check if $X_{aB,1}$ and $X_{aB,2}$ are real. If yes, then we have to check if they are positive or negative. For the case under consideration, negative value of X_{aB} is selected.

Step 6: Compute C_{p1} as in (9.63).

$$C_{p1} = \frac{1}{\omega_0 \left[\beta \left(\omega_0 \right) - X_{aB} \right]} - C_{off2} \geq 0$$

In this step, check if C_{p1} is positive. If not, set $C_{p1} = 0$, then re-design switch S2 by setting

$$C_{off2} = \frac{1}{\omega_0 \left[\beta \left(\omega_0 \right) - X_{aB} \right]}$$

$$C_{off2a} = \frac{C_{off2}}{2 \times \pi \times f_{0a} \times R}$$

The actual channel resistor is given by

$$R_{on2a} = \frac{672 \times 10^{-15}}{C_{off2a}}$$

and its normalized value is

$$R_{on2} = \frac{R_{on2a}}{R}$$

Step 7: Compute R_{bB} and X_{bB} as in (9.64).

$$R_{bB} = \frac{R_{aB}}{R_{aB}^2 + X_{aB}^2}$$

and if R_{aB} is small enough, then

$$X_{bB} = -\frac{X_{aB}}{R_{aB}^2 + X_{aB}^2} \cong -\frac{1}{X_{aB}}$$

Step 8a: Compute C_{p2} as in (9.65c).

$$C_{p2,I} = \left[\frac{1}{\omega_0 R_{on4}} \right] \left[\sqrt{\frac{R_{on4} - R_{bB}}{R_{bB}}} \right] \geq 0$$

Step 8b: Check the result using (9.65d).

$$C_{p2,II} = C_1 - C_{off4} = \mu_A - C_{off4} \geq 0$$

Step 8c: Decide if C_{p2} is acceptable. If not, it is always preferable to select $C_{p2} = 0$, which in turn yields

$$C_{off4} = C_1 = \frac{1}{\omega_0}\mu_A = \tan(\theta_A) ; \quad \omega_0 = 1$$

In other words, we re-design switch S4. In this case, off-state actual switch capacitor is given by

$$C_{off4a} = \frac{C_1}{2 \times \pi \times f_{0a} \times R}$$

In this case, actual channel resistor of S4 is specified by

$$R_{on4a} = \frac{672 \times 10^{-15}}{C_{off4a}}$$

and its normalized value is

$$R_{on4} = \frac{R_{on4a}}{R_a}$$

Step 9: Compute L_{p2} as in (9.65f).

$$L_{p2} = \frac{X_{bB}}{\omega_0 (1 + \omega_0 X_{bB} C_{off3})} > 0$$

Note that positive L_{p2} demands positive X_{bB}, which is satisfied by selecting negative X_{aB}.

Step 10: Computation of actual element values L_{p1a}, L_{p2a}, C_{p1a}, and C_{p2a}.

$$L_{p1a} = \frac{L_{p1}R}{2 \times \pi \times f_{0a}}$$

$$L_{p2a} = \frac{L_{p2}R}{2 \times \pi \times f_{0a}}$$

$$C_{p1a} = \frac{C_{p1}}{2 \times \pi \times f_{0a} \times R}$$

$$C_{p2a} = \frac{C_{p2}}{2 \times \pi \times f_{0a} \times R}$$

This step completes Algorithm 4.

Now, let us run an example to show the utilization of the above formulas.

Example 9.7A: Let $\theta_B = 5°$ and $\theta_A = 35°$, which yields $\triangle\theta = 40°$.
Let $C_{offa-opt} = 90\ fF$; $R_a = 100\ \Omega$; $\omega_0 = 1$ and $f_{0a} = 10\ GHz$.

a. Compute the normalized element values of "$3S - DPS$".
b. Determine the actual element values "$3S - DPS$".
c. Plot the phase and loss characteristics of "$3S - DPS$".

Solution

Algorithm 4 is systematically programmed under the $MatLab$ Program

"*Main_SSS_DPS_Example7A.m*".

Part (a) and Part (b) cover the total of 10 steps of Algorithm 4 as follows

Step 1: Set all the switch capacitors to selected optimum capacitor value $C_{offa-opt}$.

$$C_{off1a} = C_{off2a} = C_{off3a} = C_{off4a} = 90 \times 10^{-15}\ Farad$$

Step 2a: Compute the actual on-mode channel resistor for n-MOS switches.

$$R_{ona} = R_{on1a} = R_{on2a} = R_{on3a} = R_{on4a} = 672 \times \frac{10^{-15}}{C_{offa-opt}} = 7.4667\ \Omega$$

Step 2b: Compute the on-mode normalized channel resistors.

$$R_{on} = R_{on1} = R_{on2} = R_{on3} = R_{on4} = 0.0747$$

Step 3: Compute the major design parameters of State-A: μ_A, L_1, C_1, and L_{p1} as follows.

$$\mu_A = tan\left(\frac{\theta_A}{2}\right) = tan\left(\frac{35}{2}\right) = 0.3153$$

$$L_1 = C_1 = \left(\frac{1}{\omega_0}\right)\mu_A = 0.3153$$

$$L_{p1} = \frac{L_1}{1 + \omega_0^2 L_1 C_{off1}} = 0.2676$$

Step 4: Compute R_{ab} and $\beta(\omega_0)$ for $\omega_0 = 1$.

$$R_{aB}(\omega_0) = \frac{\omega_0^2 R_{on1} L_{p1}^2}{R_{on1}^2 + \omega_0^2 L_{p1}^2} = 0.0693$$

$$\beta(\omega_0) = \frac{\omega_0 R_{on1}^2 L_{p1}}{R_{on1}^2 + \omega_0^2 L_{p1}^2} = 0.0193$$

Step 5a: Compute γ_B

$$\gamma_B = tan(\theta_B) = tan(5°) = 0.0875$$

Step 5b: Compute discriminant $\triangle$

$$\triangle = 1 - \gamma_B^2 \left(R_{aB}^2 - 1\right) = 1.0076$$

Step 5c: Compute $X_{aB,1}$

$$X_{aB,1} = \frac{1 + \sqrt{\triangle}}{\gamma_B} = 22.9036 > 0$$

Step 5d: Compute $X_{aB,2}$

$$X_{aB,2} = \frac{1 - \sqrt{\triangle}}{\gamma_B} = -0.0435 < 0$$

Step 5e: Select negative value for $X_{aB} = X_{aB,2} = -0.0435$ ($KFLAG = -1$ Case) since θ_B is positive.

Step 6: Compute C_{p1} as in (9.63)

$$C_{p1} = \frac{1}{\omega_0 \left[\beta(\omega_0) - X_{aB}\right]} - C_{off2} = 15.3628 > 0$$

It is noted that, since C_{p1} is positive, there is no need to re-design S2.

Step 7a: Compute R_{bB}

$$R_{bB} = \frac{R_{aB}}{R_{aB}^2 + X_{aB}^2} = 10.3596 > 0$$

Step 7b: Compute X_{bB}

$$X_{bB} = -\frac{X_{aB}}{R_{aB}^2 + X_{aB}^2} = 6.4981 > 0$$

Step 8a: Compute $C_{p2,I}$

$$C_{p2,I} = \left[\frac{1}{\omega_0 R_{on4}}\right]\left[\sqrt{\frac{R_{on4} - R_{bB}}{R_{bB}}}\right] = 0.0000 + 13.3445i \geq 0$$

Notice that $C_{p2,I}$ is a complex number since $R_{on4} - R_{bB} = 0.0747 - 6.498 = -6.4233$ is negative.

Step 8b: Compute $C_{p2,II}$

$$C_{p2,II} = C_1 - C_{off4} = \mu_A - C_{off4} = 0.2406 > 0$$

Remarks: $C_{p2,II} = 0.2406$ is an acceptable value but it is different from that of $C_{p2,I}$. Therefore, at the end of the performance analysis, we expect to observe a discrepancy from the target phase shift $\triangle\theta = 45°$.

At this point, we have the freedom to select $C_{p2} = 0$. In this case, S4 must be re-designed for

$$C_{off4} = \mu_A = 0.3153$$

Or actual value of C_{off4} is given by

$$C_{off4a} = \frac{C_1}{2 \times \pi \times f_{0a} \times R} = 5.0181e - 14 \; Farad = 50.181 \; fF$$

On-mode channel resistor is given by

$$R_{on4a} = \frac{672 \times 10^{-15}}{C_{off4a}} = 13.3914 \; \Omega$$

Step 9: Compute L_{p2} at $\omega_0 = 1$.

$$L_{p2} = \frac{X_{bB}}{\omega_0 (1 + \omega_0 X_{bB} C_{off3})} = 1.3901 > 0$$

Table 9.7   MatLab codes to analyze the electrical performance of the designed 3S-DPS phase shifter

```
w1=0; w2=2; N=10001; dw=(w2-w1)/(N-1);

w=w1;

for i=1:N

    W(i)=w;

[FI21A,GainA,VSWRA]=SSS_DPS_State_A(w,Ron2,Ron3,Lp1,Lp2,Cp1,
Cp2,Coff1,Coff4);

[FI21B,GainB,VSWRB]=SSS_DPS_State_B(w,Ron1,Ron4,Lp1,Lp2,
Cp1,Cp2,Coff2,Coff3);

Phase_A(i)=FI21A; Phase_B(i)=FI21B;

GainA_dB(i)=GainA; GainB_dB(i)=GainB;

Phase_Shift(i)=FI21B-FI21A;

w=w+dw;

end

Plot_3S_DPS(W,Phase_A,Phase_B,Phase_Shift,GainA_dB,GainB_dB)
```

Step 10: Computation of the actual element values.

$$L_{p1a} = \frac{L_{p1}R}{2 \times \pi \times f_{0a}} = 4.2588e-10\ Henry = 0.42588\ nH$$

$$L_{p2a} = \frac{L_{p2}R}{2 \times \pi \times f_{0a}} = 2.2124e-09\ Henry = 2.2124\ nH$$

$$C_{p1a} = \frac{C_{p1}}{2 \times \pi \times f_{0a} \times R} = 2.4451e-12\ Farad = 2.445\ pF$$

$$C_{p2a} = \frac{C_{p2}}{2 \times \pi \times f_{0a} \times R} = 0$$

Part (c) is also performed under "$Main\_SSS\_DPS\_Example\ 7A.m$" by platting phase and gain characteristics of 3S-DPS as shown in Table 9.7.

Thus, results of the electrical performance analysis of the designed $45°$ phase shifting unit are depicted in Figure 9.20.

In Figure 9.20a, phase shift performance is shown.

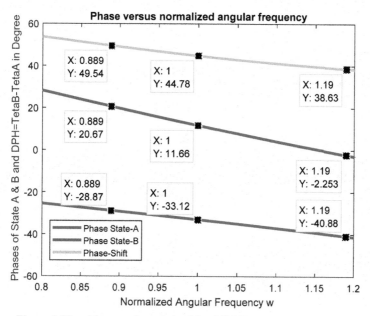

Figure 9.20a Phase performance of the 45° SSS-DPS over wide band.

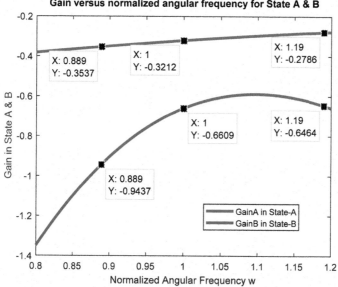

Figure 9.20b Gain performance of the 45° SSS-DPS over wide band.

Loss performance of the phase shifter is plotted in Figure 9.20b.

As it is seen from Figure 9.20a, the phase-shift $\triangle\theta$ varies from $38°.63'$ to $49°.54'$ or it is simply

$$\triangle\theta = 44°.085' \mp 5°.45'$$

over normalized angular frequencies $\omega_1 = 0.889$-to $\omega_2 = 1.19$ (i.e., from 8.89 GHz to 11.9 GHz).

Our target was to hit $\triangle\theta(\omega_0) = 40°$ at $\omega_0 = 1$; instead, we have $44.78°$. This discrepancy is due to approximations made in the course of solutions. However, by shifting center frequency ω_0 beyond 1, we may get a better result as in the following example.

Example 9.7B: Repeat Example 9.7A for $\omega_0 = 1.1$. In other words, keep all the inputs as it is, and change only ω_0 as $\omega_0 = 1.1$.

Solution

Step 1 & Step 2 are as same as Step 1 & Step 2 of Example 9.7A.

Step 3: $\mu_a = tan\left(\frac{35}{2}\right) = 0.3153$ as before; However, $L_1 = C_1 = \frac{1}{\omega_0}\mu_A = 0.2866$;

$$L_{p1} = \frac{L_1}{1+\omega_0^2 L_1 C_{off1}} = 0.2396 \text{ (different from Example 9.7A)}$$

Step 4: $R_{aB} = \frac{\omega_0^2 R_{on1} L_{p1}^2}{R_{on1}^2 + \omega_0^2 L_{p1}^2} = 0.0691$; $\beta = \frac{\omega_0 R_{on1}^2 L_{p1}}{R_{on1}^2 + \omega_0^2 L_{p1}^2} = 0.0196$ (different from Example 9.7A)

Step 5 is as same as Step 5 of Example 9.7A (we choose the negative value for $X_{aB,2} = -0.0435$).

Step 6: $C_{p1} = \frac{1}{\omega_0[\beta(\omega_0)-X_{aB}]} - C_{off2} = 13.8573$ (different from Example 9.7A)

Step 7a: $R_{bB} = \frac{R_{aB}}{R_{aB}^2 + X_{aB}^2} = 10.3694$ (different from Example 9.7A)

Step 7b: $X_{bB} = -\frac{X_{aB}}{R_{aB}^2 + X_{aB}^2} = 6.5187$ (different from Example 9.7A)

Step 8: $C_{p2} = 0$ selected as in Example 7A. $C_{off4} = \mu_A = 0.3153$ as before.

Step 9: $L_{p2} = \frac{X_{bB}}{\omega_0(1+\omega_0 X_{bB} C_{off3})} = 1.2896$ (different from Example 9.7A)

Step 10: Actual Element Values (different from Example 9.7A)

$$L_{p1a} = 0.38139 \, nH$$
$$L_{p2a} = 2.0524 \, nH$$
$$C_{p1a} = 2.2055 \, pF$$
$$C_{p2a} = 0 \, pF$$

Part (c): Performance of the new 3S-DPS unit designed for $\theta_A = 35°$, $\theta_B = 5°$, $f_{0a} = 10 \, GHz$, $\omega_0 = 1.1$ (i.e., $\triangle\theta = 40°$) is depicted in Figure 9.21. In this design, desired phase shift $\triangle\theta_0 = \triangle\theta \, (\omega_0 = 1.1) = 40°$ is realized as $\triangle\theta_r = 40.5°$. The relative error is given by

$$\delta_{\theta 0} = \frac{\triangle\theta_r - \triangle\theta_0}{\triangle\theta_0} = \frac{40.5 - 40}{40} = \frac{0.5}{40} = 1.25\%$$

which is pretty good. Center frequency is located at $f_{0a} = 10 \, GHz$, which is associated with $\omega_0 = 1.1$.

From $f_1 = 0.9838 * f_{0a} = 9.838 \, GHz$ to $f_2 = 1.249 * f_{0a} = 12.49 \, GHz$, the phase variation is given by

$$\triangle\theta = (40.5) \mp 5 \, degree$$

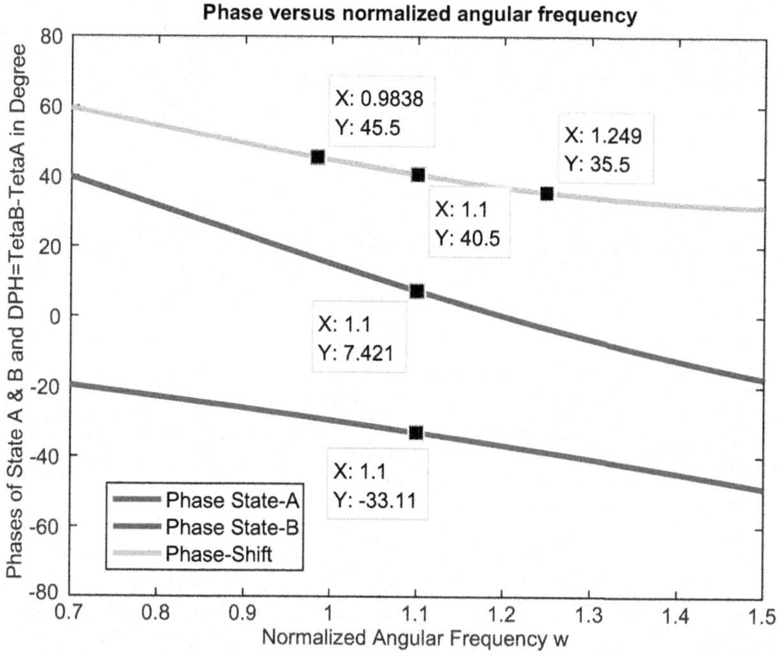

Figure 9.21a Phase performance of the 40° SSS-DPS of Example 7B.

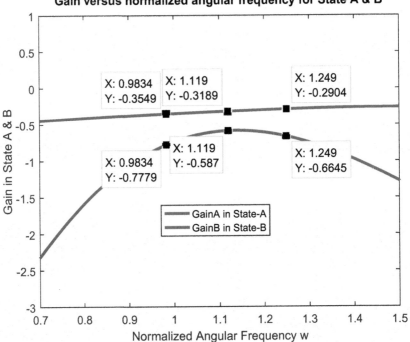

Figure 9.21b   Gain performance of the $40°$ SSS-DPS of Example 9.7B.

The loss characteristics of State-A and State-B of recently designed phase shifter are shown in Figure 9.21b.

A close examination of the above figures reveals that over the frequency band of $\triangle F = 2.66\ GHz = (9.38\ GHz - 12.49\ GHz)$, insertion loss for both states is above 0.8 dB, which is reasonable.

In conclusion, what we see that, phase shifting performance of the 3S–DPS may be optimized by shifting the location of the center frequency ω_0 properly.

As mentioned before, the MatLab programs developed for Algorithm 4, can as well be utilized to determine the component values of 3S-DPS based on the negative value of the phase θ_B of State-B. (i.e., $\theta_B = \varphi_{21B}(\omega_0) < 0$). This way, we may further improve the phase shifting performance of the propose digital phase shifter unit as detailed in the following example.

Example 9.7C: Repeat Example 7A for $\theta_B = -10 < 0$, $\theta_A = 55$, $f_{0a} = 8GHz$, $\omega_0 = 1$ employing the off-mode switch capacitors specified as follows:

$$C_{off1a} = 25 \, fF; C_{off2a} = C_{off3a} = C_{off4a} = 90 \, fF;$$

Solution

Step 1: Normalize the off-mode switch capacitors.

$$C_{off1} = (2\pi f_{01}) \, R_a C_{off1a} = (2\pi \times 8 \times 10^9) \times 100 \times 25 \times 10^{-15} = 0.1257,$$

$$C_{off2} = C_{off3} = C_{off4} = (2\pi \times 8 \times 10^9) \times 100 \times 90 \times 10^{-15} = 0.4524$$

Step 2a: Compute the actual on-mode switch channel resistors.

$$R_{on1a} = \frac{672(fF \times ohm)}{C_{off1a}(fF)} = \frac{672(fF \times ohm)}{25 \, (fF)} = 26.8800\Omega$$

$$R_{on2a} = R_{on3a} = R_{on4a} = \frac{672(fF \times ohm)}{90 \, (fF)} \quad 7.4667\Omega$$

Step 2b: Compute the normalized on-mode switch channel resistors

$$R_{on1} = \frac{26.88}{100} = 0.2688; \quad R_{on2} = R_{on3} = R_{on4} = \frac{7.466}{100} = 0.0747$$

Step 3: Compute the major design parameters for State-A.

$$\mu_A = \tan(\theta_A) = 0.5206$$

$$L_1 = C_1 = 0.5206$$

$$L_{p1} = L_{p1} = \frac{L_1}{1 + \omega_0^2 L_1 C_{off1}} = 0.4886$$

Step 4: Compute R_{aB} and β as in (9.54) and (9.56), respectively.

$$R_{aB} = \frac{\omega_0^2 R_{on1} L_{p1}^2}{R_{on1}^2 + \omega_0^2 L_{p1}^2} = 0.2063; \quad \beta = \frac{\omega_0 R_{on1}^2 L_{p1}}{R_{on1}^2 + \omega_0^2 L_{p1}^2} = 0.1135$$

Step 5a: Compute the major design parameter $\gamma_B = \tan(\theta_B)$ of State-B.

$$\gamma_B = -0.1763$$

Step 5b: Compute the Discriminant $\triangle$ of the second-order equation given by (9.61).

$$\triangle = 1 - \gamma_B^2 \left(R_{aB}^2 - 1 \right) = 1.0298$$

Step 5c: Compute $X_{aB,1}$.

$$X_{aB,1} = \frac{1 + \sqrt{1 - \gamma_B^2 \left(R_{aB}^2 - 1 \right)}}{\gamma_B} = -11.4264$$

Step 5d: Compute $X_{aB,2}$.

$$X_{aB,2} = \frac{1 - \sqrt{1 - \gamma_B^2 \left(R_{aB}^2 - 1 \right)}}{\gamma_B} = 0.0838$$

Step 5e: Select the positive value for X_{aB} since θ_B is negative.

$$X_{aB} = 0.0838$$

Step 6: Compute C_{p1}

$$C_{p1} = \frac{1}{\omega_0 \left[\beta \left(\omega_0 \right) - X_{aB} \right]} - C_{off2} = 33.1842$$

Step 7a: Compute RbB

$$RbB = \frac{R_{aB}}{R_{aB}^2 + X_{aB}^2} = 4.1602$$

Step 7b: Compute

$$XbB \cong -\frac{1}{X_{aB}} = -11.9345$$

Step 8a: Compute $C_{p2,I}$ as in (9.65c).
It is a complex quantity. Therefore, it is useless.

Step 8b: Compute cross arm capacitor $C_{p2,II}$ as in (9.65d).

$$C_{p2,II} = C_1 - C_{off4} = \mu_A - C_{off4} = 0.0682 \; that \; is \; okay$$

Choose positive value for $C_{p2} = 0.0682$.

Note: We may as well set $C_{p2} = 0$.

Step 9: Compute the cross-arm inductor L_{p2}.

$$L_{p2} = \frac{XbB}{\omega_0 \left(1 + \omega_0 XbB C_{off3} \right)} = 2.7130$$

Step 10: Compute actual component values

$$L_{p1a} = \frac{R_a \times L_{p1}}{2\pi \times f_{0a}} = \frac{100 \times 0.4886}{2\pi \times 8 \times 10^9} = 0.97205 \; nH;$$

$$L_{p2a} = \frac{R_a \times L_{p1}}{2\pi \times f_{0a}} = \frac{100 \times 2.7130}{2\pi \times 8 \times 10^9} = 5.3973 \; nH;$$

$$C_{p1a} = \frac{C_{p1}}{2 \times \pi \times f_{0a} \times R_a} = \frac{33.1842}{2\pi \times 8 \times 10^9 \times 100} = 6.6018 \; pF;$$

$$C_{p2a} = \frac{C_{p2}}{2 \times \pi \times f_{0a} \times R_a} = \frac{0.0682}{2\pi \times 8 \times 10^9 \times 100} = 13.568 \; fF$$

Remark: Ideally, symmetrical lattice structures used in phase shifter designs are lossless. In this regard, phase of the transfer scattering parameter is determined directly from the reactive parts of the series or cross arm impedances of the symmetric lattice. In practice, to make computations simpler, target phases are determined directly from the reactive parts of the impedances, which in turn results in the component values of the phase shifter. Inclusion of the real parts of the impedances, mostly effects the insertion loss characteristics of the phase shifter under consideration. Therefore, we have experienced that, in Step 7b, in determining X_{bB} from X_{aB}, the form $X_{bB} \cong -\frac{1}{X_{aB}}$ yields better phase-shift performance than that of the form $X_{bB} = -\frac{R_{aB}^2}{R_{aB}^2 + X_{aB}^2}$. Anyhow, these forms converge to each other as R_{aB} approaches zero.

Part (c): Plot the phase-shifting performance of 3S–DPS under consideration.

Phase characteristics of the phase shifting states are depicted in Figure 9.22a.

From Figure 9.22a, the relative phase error is read as

$$\delta_{\theta 0} = \frac{\triangle\theta_r - \triangle\theta_0}{\triangle\theta_0} = \frac{44.6 - 45}{45} = \frac{0.6}{45} = 1.33\%$$

which is pretty good. At the input, the center frequency is located at $f_{0a} = 8 \; GHz$, which corresponds to normalized angular frequency $\omega_0 = 1$.

From $f_1 = 0.75 * f_{0a} = 6 \; GHz$ to $f_2 = 1.5 * f_{0a} = 12 \; GHz$, the phase variation is given by

$$\triangle\theta = (45.24) \mp 1.18 \; degree$$

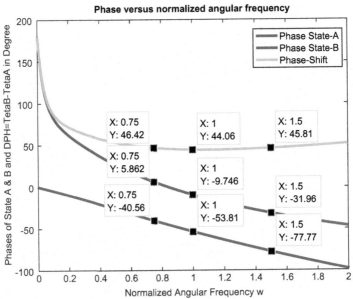

Figure 9.22a   Phase performance of the 40° SSS-DPS of Example 9.7B.

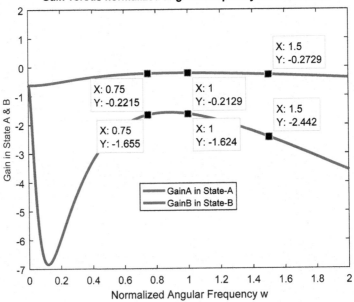

Figure 9.22b   Gain performance of the 40° SSS-DPS of Example 9.7B.

which is very good. The loss characteristics of State-A and State-B are shown in Figure 9.22b.

A close examination of the above figure reveals that over the frequency band of $\triangle F = 0.75 * 87 - 1.5 * 8\ GHz = 6\ GHz - 12\ GHz = 6\ GHz$, insertion loss for both states is above –2.5 dB, which may be acceptable for many applications.

In conclusion, we have exhibited that the phase shifting performance of the 3S-DPS for uneven phase distribution between the states results in wider frequency band with less phase fluctuations over the same topology, which performs even phase distributions between the states. In return, the new circuit yields relatively higher insertion loss in State-B.

9.14 Investigation of Unequal Phase Distribution between the States with Negative Phases: An Alternative Approach

In this section, we initiate the design of an 3S-DPS starting from the cross-arm impedance z_{bB} in State-B.

Let us consider S_{21B} for an all-pass lattice structure as specified by (9.7) and (9.11b).

$$S_{21B} = \frac{z_{bB} - 1}{z_{bB} + 1} \tag{9.66a}$$

In (9.66a), Z_{bB} is the normalized cross-arm impedance of the $3S - DPS$ in $State - B$ and it is specified in terms of its real and imaginary parts as follows

$$z_{bB} = R_{bB} + jX_{bB} \tag{9.66b}$$

where R_{bB} is the real part and X_{bB} is the imaginary part of Z_{bB}.

Hence,

$$S_{21B} = \frac{R_{bB} + jX_{bB} - 1}{R_{bB} + jX_{bB} + 1} = (-1) \times \frac{(1 - R_{bB}) - jX_{bB}}{(1 + R_{bB}) + jX_{bB}}$$
$$= \rho_{21B}(\omega)e^{j\varphi_{21B}(\omega)} \tag{9.66c}$$

with

$$\varphi_{21B}(\omega) = -\theta_B = \mp 180 - tan^{-1}\left(\frac{X_{bB}}{1 - R_{bB}}\right) - tan^{-1}\left(\frac{X_{bB}}{1 + R_{bB}}\right)$$
$$\tag{9.66d}$$

In (9.66d), the phase θ_B is a positive quantity and it varies from $+0°$ to $+180°$.

It is noted that, in the above expression "-1" is represented by $e^{\mp j180°}$ (i.e. $-1 = e^{\mp j180°}$).

If $R_{bB} \ll 1$, then

$$- \theta_B \cong \mp 180 - 2tan^{-1}(X_{bB}) \qquad (9.66e)$$

or

$$tan^{-1}(X_{bB}) \cong \frac{\theta_B}{2} \mp 90° \leq 0 \qquad (9.66f)$$

or

$$X_{bB} \cong tan\left(\frac{\theta_B}{2} \mp 90°\right) = -\frac{1}{tan\left(\frac{\theta_B}{2}\right)} = -\frac{1}{\mu_B} \leq 0 \qquad (9.66g)$$

where

$$\mu_B = tan\left(\frac{\theta_B}{2}\right)$$

On the other hand, by (9.65b)

$$X_{bB}(\omega) = \omega\left[\frac{L_{p2}}{1 - \omega^2 L_{p2} C_{off3}} - \alpha(\omega)\right] \qquad (9.66h)$$

where

$$\alpha(\omega) = \frac{R_{on4}^2 C_{p2}}{1 + (\omega R_{on4} C_{p2})^2} > 0 \qquad (9.66i)$$

In (9.66i), on-mode channel resistance R_{on4} can be considered small enough so that $\alpha(\omega)$ is neglected as compared to $\frac{L_{p2}}{1-\omega^2 L_{p2} C_{off3}}$. Hence, $X_{bB}(\omega)$ is approximated as

$$X_{bB}(\omega) \cong \omega\left[\frac{L_{p2}}{1 - \omega^2 L_{p2} C_{off3}}\right]$$

or at $\omega = \omega_0$

$$L_{p2} = \frac{[X_{bB}/\omega_0]}{\left(1 + \omega_0^2 [X_{bB}/\omega_0] C_{off3}\right)} > 0 \qquad (9.66i)$$

Now, let us derive the expression for C_{p1}. At this point, we consider the all-pass impedance condition

$$z_{aB} = \frac{1}{z_{bB}}$$

or in the open form,

$$R_{aB} + jX_{aB} = \frac{1}{R_{bB} + jX_{bB}}$$

If $R_{aB} \ll |X_{aB}|$ and $R_{bB} \ll |X_{bB}|$, then all-pass condition is approximated as

$$X_{aB} = -\frac{1}{X_{bB}} > 0$$

On the other hand, by (9.55)

$$X_{aB} = \left(\frac{\omega R_{on1}^2 L_{p1}}{R_{on1}^2 + \omega^2 L_{p1}^2} - \frac{1}{\omega (C_{p1} + C_{off2})} \right) > 0$$

Hence, at $\omega = \omega_0$ total capacitor $C_T = C_{p1} + C_{off2}$ is found as

$$C_{p1} + C_{off2} = \left[\frac{1}{\omega_0} \right] \frac{1}{\left(\frac{\omega_0 R_{on1}^2 L_{p1}}{R_{on1}^2 + L_{p1}^2} - X_{aB} \right)} = \left[\frac{1}{\omega_0} \right] \frac{1}{(\eta - X_{aB})} > 0$$

$$(9.66j)$$

where

$$\eta = \frac{\omega_0 R_{on1}^2 L_{p1}}{R_{on1}^2 + L_{p1}^2}$$

(9.66j) is satisfied if

$$\eta = \frac{\omega_0 R_{on1}^2 L_{p1}}{R_{on1}^2 + L_{p1}^2} > X_{aB} = -\frac{1}{X_{bB}} = -\frac{1}{tan\left(\frac{\theta_B}{2} \mp 90° \right)}$$

$$= tan\left(\frac{\theta_B}{2} \right) > 0 \qquad (9.66k)$$

or

$$\eta = \frac{\omega_0 R_{on1}^2 L_{p1}}{R_{on1}^2 + L_{p1}^2} > tan\left(\frac{\theta_B}{2} \right) = \mu_B \qquad (9.66l)$$

It is interesting to note that, in (9.66*l*), L_{p1} is specified by (9.45) such that

$$L_{p1} = \frac{L_1}{1 + \omega_0^2 L_1 C_{off1}}$$

where

$$L_1 = tan\left(\frac{\theta_A}{2}\right) = \mu_A$$

Then,

$$L_{p1} = \frac{tan\left(\frac{\theta_A}{2}\right)}{1 + \omega_0^2 C_{off1} tan\left(\frac{\theta_A}{2}\right)} = \frac{\mu_A}{1 + \omega_0^2 \mu_A C_{off1}}$$

$$= function \ of \ \{\omega_0, \theta_A, C_{off1}\} \tag{9.66m}$$

Thus, (9.66*l*) is expressed as

$$\eta = \frac{\omega_0 R_{on1}^2 L_{p1}}{R_{on1}^2 + L_{p1}^2} = function \ of \ \{\omega_0, \theta_A, C_{off1}, R_{on1}\}$$

$$= function \ of \ \{\omega_0, \theta_A, C_{off1}\} \ > tan\left(\frac{\theta_B}{2}\right) = \mu_B \tag{9.66n}$$

where

$$R_{on1} = \frac{R_{on1a}}{R_a}, \quad R_{on1a} = \frac{672 \times 10^{-15}}{C_{off1a}(Farad)}$$

Hence, once ω_0, θ_A and θ_B are selected, switch S1 must be designed in such a way that (9.66n) must be satisfied.

Finally, the cross-arm capacitors C_{p2} is determined as in (9.65d).

$$C_{p2} = \mu_A - C_{off4} \geq 0$$

If the above equation is not satisfied (i.e., if C_{p2} negative), then we should lower the value of C_{off4} until we end up with positive C_{p2}. Perhaps, the designer may prefer to set $C_{p2} = 0$. In this case, switch S4 is re-designed to yield

$$C_{off4} = \mu_A$$

In order to implement the above design equations, we propose the following algorithm.

Algorithm 5. Design Restriction's Check to Construct 3S-DPS with Negative Phases

In this algorithm, the quantity η is developed step by step if (9.66n) is satisfied.

Inputs:

R_a: Normalization resistance, which is usually selected as
$R_a = 100 \ \Omega$

f_{0a}: Actual Central Frequency (ACF) in Hertz to design 3S-DPS.

ω_0: Normalized Angular Centre Frequency (NACF). It is usually selected either at $\omega_0 = 1$ or in the neighborhood of $\omega_0 = 1$ such as $\omega_0 = 0.7 < 1$ or $\omega_0 = 1.2 > 1$ etc.

C_{off1a}: Desired actual value of off-mode switch capacitor for S1.

C_{off2a}: Desired actual value of off-mode switch capacitor for S2.

C_{off3a}: Desired actual value of off-mode switch capacitor for S3.

θ_A: Desired value of the positive phase shift for State-A such that $\varphi_{21A}(\omega_0) = -\theta_A < 0$. It is noted that θ_A varies between 0° and 180°.

θ_B: Desired value of the positive phase shift for State-B such that $\varphi_{21B}(\omega_0) = -\theta_B < 0$. It is noted that θ_B varies between 0° and 180°.

Step 1: Compute major design parameters

$$\mu_A = \tan\left(\frac{\theta_A}{2}\right)$$

and

$$\mu_B = \tan\left(\frac{\theta_B}{2}\right).$$

Step 2: Compute the normalized values of the "off-mode switch capacitors" C_{off1}, C_{off2}, C_{off3}, C_{off4} and the "on-mode switch channel resistors" R_{on1}, R_{on2}, R_{on3}, and R_{on4} for $NMOS$ switches S1, S2, S3, and S4, respectively.

$$R_{on1a} = \frac{672 \times 10^{-15}}{C_{off1a \ (in \ farad)}}$$

$$R_{on2a} = \frac{672 \times 10^{-15}}{C_{off1a \ (in \ farad)}}$$

$$R_{on3a} = \frac{672 \times 10^{-15}}{C_{off3a \ (in \ farad)}}$$

$$R_{on4a} = \frac{672 \times 10^{-15}}{C_{off4a \ (in \ farad)}}$$

$$R_{on1} = \frac{R_{on1a}}{R_a}$$

$$R_{on2} = \frac{R_{on2a}}{R_a}$$

$$R_{on3} = \frac{R_{on3a}}{R_a}$$

$$R_{on4} = \frac{R_{on4a}}{R_a}$$

$$C_{off1} = 2\pi f_{0a} R_a C_{off1a}$$

$$C_{off2} = 2\pi f_{0a} R_a C_{off2a}$$

$$C_{off3} = 2\pi f_{0a} R_a C_{off3a}$$

$$C_{off4} = 2\pi f_{0a} R_a C_{off4a}$$

Step 3: Compute Series arm inductor L_{p1}

$$L_{p1} = \frac{\mu_A}{1 + \omega_0^2 \mu_A C_{off1}}$$

Step 4: Compute η

$$\eta = \frac{\omega_0 R_{on1}^2 L_{p1}}{R_{on1}^2 + L_{p1}^2}$$

Step 5: Check if η is bigger than μ_B. If yes, then GOTO Step 6. If no, GOTO input-step to either change ω_0 or change C_{off1a}.

Step 6: Compute the imaginary part X_{bB} of the cross-arm impedance z_{bB} as in (9.66g) and compute the imaginary part X_{aB} of the series-arm impedance z_{aB} as in (9.66k).

$$X_{bB} \cong -\frac{1}{\mu_B} < 0$$

$$X_{aB} = -\frac{1}{X_{bB}} > 0$$

Step 7: Computation of series arm components L_{p2} and C_{p1}

Step 7a: Compute the cross-arm inductor L_{p2} as in (9.66i)

$$L_{p2} = \frac{X_{bB}}{(1 + \omega_0 X_{bB} C_{off3})} > 0$$

Check if L_{p2} is positive. If not, re-design switch S3 by reducing C_{off3a}.

Step 7b: Compute the value of the series-arm capacitors C_{p1}

$$C_{p1} = \left[\frac{1}{\omega_0}\right] \left[\frac{1}{\eta - X_{aB}}\right] - C_{off2} > 0$$

Check if C_{p1} is positive. If not, re-design switch S2. The best value of C_{off2} is found when we set $C_{p1} = 0$, which minimizes the loss of switch S2. In this case, C_{off2} is given by

$$C_{off2} = \left[\frac{1}{\omega_0}\right] \left[\frac{1}{\eta - X_{aB}}\right] > 0$$

Step 8: Compute the value of the cross-arm capacitors C_{p2} as in (9.65d)

$$C_{p2} = \mu_A - C_{off4} \geq 0$$

If C_{p2} is found negative, then set it to zero and re-design switch S4 as

$$C_{off4} = \mu_A$$
$$C_{off4a} = \frac{C_{off4}}{2\pi f_{0a} R_a}$$
$$R_{on4a} = \frac{672 \times 10^{-15}}{C_{off4a}}$$
$$R_{on4} = \frac{R_{on4a}}{R_a}$$

Step 9: Compute all the actual component values of 3S-DPS and analyze the performance.

This step completes the algorithm.

Now, let us run an example to implement all the steps of the above algorithm.

Example 9.8:

Let $\theta_A = 55° = -\varphi_{21A}$.

Let $\theta_B = 10° = -\varphi_{21B}$, which yields a phase-shift $\triangle\theta = |\theta_A - \theta_B| = 45°$ between State-A and State-B.

At the beginning, let us select $C_{off1a} = C_{off2a} = C_{off3a} = C_{off4a} = 90 \ fF$.

Let $f_{0a} = 8 \ GHz$ and $\omega_0 = 1$.

Implement Algorithm 5 step by step.

Solution

Step 1: Compute the major design parameters.

$$\mu_A = tan\left(\frac{55}{2}\right) = 0.5206$$

$$\mu_B = tan\left(\frac{10}{2}\right) = 0.0875$$

Step 2: Compute the actual and normalized on-mode channel resistors and normalized off-mode capacitors for switches S1, S2, S3, and S4.

$$R_{ona(i)} = \frac{672 \times 10^{-15}}{90 \times 10^{-15}} = 7.4667 \ \Omega; \quad i = 1, 2, 3, 4$$

$$R_{on(i)} = \frac{R_{on(i)a}}{R_a} = \frac{7.4667}{100} = 0.0747; \quad i = 1, 2, 3, 4$$

$$C_{off(i)} = (2\pi f_{0a} R_a) C_{off(i)a} = (2\pi \times 8 \times 10^9) \times 100 \times 90 \times 10^{-15}$$
$$= 0.4524; \quad i = 1, 2, 3, 4$$

Step 3: Compute the series-arm inductors L_{p1}.

$$L_{p1} = \frac{\mu_A}{1 + \omega_0^2 \mu_A C_{off1}} = \frac{0.5206}{1 + 0.5206 \times 0.4524} = 0.4213$$

Step 4: Compute η

$$\eta = \frac{\omega_0 R_{on1}^2 L_{p1}}{R_{on1}^2 + L_{p1}^2} = \frac{(0.0747)^2 \times (0.4886)}{(0.0747)^2 + (0.4886)^2} = 0.0128$$

Step 5: Check if $\eta = 0.0128 > \mu_B = 0.0875$.

The answer is NO!
Therefore, either C_{off1a} is reduced or ω_0 is increased.

Let us decrease C_{off1a} down to 25 fF. Then, repeat Steps 2–5.
Thus, we have the following results

$$R_{on1a} = 26.88 \ \Omega; R_{on1} = 0.2688$$

$$C_{off1a} = 25 \ fF;; C_{off1} = 0.1257$$

$$L_{p1} = 0.4886$$

$$?$$

$$Check \; if \; \eta = 0.1135 > \mu_B = 0.0875$$

The answer is YES!

Therefore, we can continue with follow-up steps.

Step 6: Compute $X_{bB} \cong -\frac{1}{\mu_B}$ and $X_{aB} = -\frac{1}{X_{bB}} > 0$

$$X_{bB} \cong -\frac{1}{\mu_B} = -\frac{1}{0.0875} = -11.4301$$

$$X_{aB} = -\frac{1}{X_{bB}} = 0.0875 > 0$$

Step 7: Compute L_{p2} and C_{p1}

Step 7a: Compute L_{p2}

$$L_{p2} = \frac{X_{bB}}{(1 + \omega_0 X_{bB} C_{off3})} = \frac{-11.4301}{1 - 11.4301 \times 0.4524} = 2.7405 \overset{?}{>} 0$$

Yes, L_{p2} is positive. We do not have to re-design switch S3.

Step 7b: Compute the normalized value of cross-arm capacitors C_{p1}.

$$C_{p1} = \left[\frac{1}{\omega_0}\right]\left[\frac{1}{\eta - X_{aB}}\right] - C_{off2} = \frac{1}{0.1135 - 0.0875} = 37.9627 > 0$$

Step 8: Compute the normalized value of the cross-arm capacitors C_{p2}

$$C_{p2} = \mu_A - C_{off4} = 0.5206 - 0.4524 = 0.0682 \geq 0$$

Step 9a: Compute the actual elements of 3S-DPS

$$L_{p1a} = R_a \frac{L_{p1}}{2\pi f_{0a}} = 9.7205e - 10 = 0.97205 \; nH$$

$$L_{p2a} = R_a \frac{L_{p2}}{2\pi f_{0a}} = 5.4520e - 09 = 5.452 \; nH$$

$$C_{p1a} = \frac{C_{p1}}{2\pi f_{0a} R_a} = 7.5524e - 12 = 7.55 \; pF$$

$$C_{p2a} = \frac{C_{p2}}{2\pi f_{0a} R_a} = 1.3564e - 14 = 13.65 \; fF$$

Switch S1:

$$C_{off1a} = 25 \; fF; R_{on1a} = \; 26.88 \; \Omega$$

Switch S2:

$$C_{off2a} = 90 \; fF; R_{on2a} = 7.4667 \; \Omega$$

Switch S3:

$$C_{off3a} = 90 \; fF; R_{3a} = 7.4667 \; \Omega$$

Switch S4:

$$C_{off4a} = 90 \; fF; R_{4a} = 7.4667 \; \Omega$$

Step 9b: Electrical Performance of the designed 3S-DPS

Algorithm 5 is programmed as *"Main_SSS_DPS_Example8.m"* under *MatLab* environment and it is listed in Program List 9.18. Performance analysis is completed in this program and it is depicted in Figure 9.22.

In Figure 9.23a, phase shifting performance between the states is depicted. At the first place, we should keep in mind that design equations derived in this section are approximate and developed at the normalized central frequency ω_0. Therefore, we expect some discrepancies between the given data and the resulting performance. For example, for State B, at $\omega_0 = 1$, phase $\varphi_{21B}(\omega_0)$ is fixed as $\theta_B = 10°$. However, actual phase is found as $\varphi_{21B}(\omega_0) = -10.2$, which introduces a relative phase error $\delta_{21\theta_B}$ as

$$\delta_{21\theta_B} = \left| \frac{(-\theta_B) - \varphi_{21B}(\omega_0)}{\theta_B} \right| = \left| \frac{-10 + 10.2}{10} \right| = 0.02 = 2\%$$

In State-A, the relative phase error is given by

$$\delta_{21\theta_A} = \left| \frac{(-\theta_A) - \varphi_{21A}(\omega_0)}{\theta_A} \right| = \left| \frac{-55 + 53.85}{55} \right| = 0.0209 = 2.09 \; \%$$

At the center frequency $\omega_0 = 1$, relative phase shift error $\delta_{21\theta}$ is given by

$$\delta_{21\theta} = \left| \frac{\Delta\theta - \Delta\theta\,(\omega_0)}{\Delta\theta} \right| = \left| \frac{45 - 43.67}{45} \right| = 0.0296 = 2.96\%$$

Furthermore, over an octave frequency band, more specifically from 6 GHz ($\omega_1 = 0.75$) to 12 GHz ($\omega_2 = 1.5$), the phase shift $\Delta\theta\,(\omega)$ between the states is almost flat. Let us open this statement as follows.

Referring to Figure 9.22a, let maximum phase-shift deviation over an octave bandwidth of 6–12 GHz be $\Delta\theta_{max}$. Then,

$$\Delta\theta_{max} = 45.99$$

Over the same bandwidth, let the minimum phase-shift deviation be $\Delta\theta_{min}$. Then,

$$\Delta\theta_{min} = 43.67$$

Similarly, let the average phase-shift be designated by $\Delta\theta_{av}$. Then,

$$\Delta\theta_{av} = \frac{\Delta\theta_{max} + \Delta\theta_{min}}{2} = 44.83$$

Let the phase-shift error be referred as $\epsilon_{\Delta\theta}$ such that

$$\epsilon_{\Delta\theta} = \Delta\theta_{max} - \Delta\theta_{av} = \Delta\theta_{av} - \Delta\theta_{min} = 1.16$$

Then, over one octave bandwidth of 6–12 GHz, the phase-shift variation is given by

$$\Delta\theta\,(\omega) = \theta_{av} \mp \epsilon_{\Delta\theta} = 44.83 \mp 1.16$$

For the $45°$ phase-shift, loss characteristic of the proposed 3S-DPS configuration is depicted in Figure 9.23b. A close examination of this figure reveals that State-B loss of the proposed phase shifter is much higher than that of State-A (maximum loss of 2.441 dB of State-B versus maximum loss of 0.8674 of State-A). Anyhow, the maximum loss of State-B is still less than 3dB, which is acceptable.

The 3S-DPS phase shifting cells can be manufactured as MMIC perhaps using silicon or Si-Ge-based 0.180 nm VLSI technology up to X or Ku Band, respectively. In the next chapter, 3-Bit 45, 90, and $180°$ digital phase shifter designs and their VLSI implementation processes are introduced.

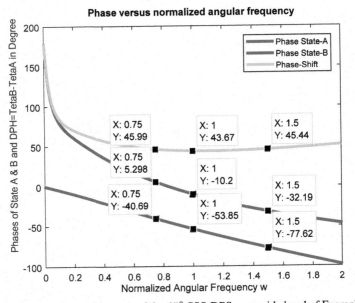

Figure 9.23a   Phase performance of the 45° SSS-DPS over wide band of Example 9.8.

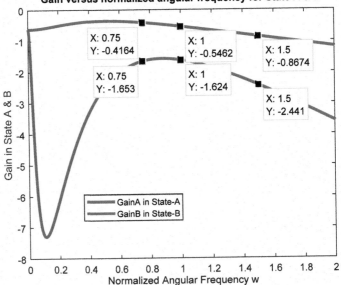

Figure 9.23b   Gain performance of the 45° SSS-DPS over wide band of Example 9.8.

9.15 On-Chip Inductor Design

On-chip passive components performance over desired frequency band is the most important factor, in order to successfully realize 3S-DPS structure within SI substrate. Especially, design of wide band inductor having high-quality factor requires special design considerations [8–12].

There are mainly four types of on-chip inductor topology, which are square, hexagonal, octagonal, and circular, which are given in Figure 9.24.

Among all the types of inductors, circular shape inductor type is more efficient and exhibiting better performance. However, circular shape inductor is the least used type of inductor because of its limitations of the availability of the process fabrications requirements. Most process restricts to the routing angles to 45° so that implementing on-chip spiral shape inductor within these processes is impossible. For this reason, closer to the spiral inductor shape and performance other than the others, hexagonal shape inductor is used within the design of 3S-DPS.

Although simulators analyze inductor drawing using electromagnetic field distribution computations, in order to deeply understand the inductor effect on the 3S-DPS performance, modeling concepts of the on-chip inductors should be addressed first. In this work, a generic lumped pi model developed by Yue and Wong is given [13] In Figure 9.25, inductor cross section on a silicon substrate together with the parasitic components is given.

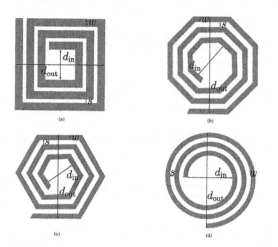

Figure 9.24 On chip inductor types: (a) square, (b) hexagonal, (c) octagonal, (d) circular.

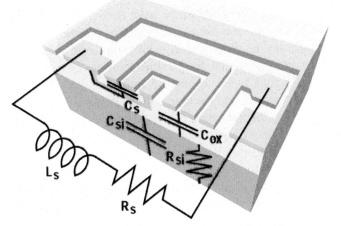

Figure 9.25   Cross section of inductor together with model parameters.

Ls is the total inductance of the designed inductor.

Rs is the series resistance of the inductor arises from metal resistivity together with skin and proximity effects.

Cs is the capacitive coupling between the spiral and metal connection between inner end of inductor to external circuitry.

Cox is the area capacitance between the inductor metals and silicon substrate.

Rsi: Coupling resistance with the silicon substrate, which shows the ohmic losses of the substrate.

Csi is the coupling capacitance of the silicon substrate.

The lumped physical model of the given cross section is shown in Figure 9.26.

There are several inductance expressions already reported in the literature. Among these, Mohan's method [14] is used to roughly calculate the inductance value since his calculations give good approximation of inductance value and can easily be extended to other geometries. In his method, inductance can be calculated as follows:

$$L = \frac{\mu_0 N^2 D_{avg} \alpha_1}{2} \left(ln\frac{\alpha_2}{\rho} + \alpha_3\rho + \alpha_4\rho^2 \right)$$

where μ_0 is permeability of vacuum, N is inductor number of turns, D_{avg}, fill factor, and ρ and α coefficients (for octave-shaped inductors), which are

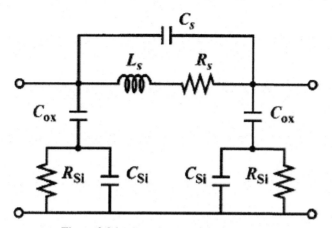

Figure 9.26 Generic model of an inductor.

given in below equations.

$$D_{avg} = \frac{D_{out} + D_{in}}{2}$$

$$\rho = \frac{D_{out} - D_{in}}{D_{out} + D_{in}}$$

Where D_{out} and D_{in} are shown in Figure 9.24. For octave-shaped inductors, α coefficients are given below.

$$\alpha_1 = 1.07, \alpha_2 = 2.29, \alpha_3 = 0, \alpha_4 = 0.19$$

Although these, α coefficients are given for octave shape, using these equations for hexagonal architecture has a good enough approximation for selecting the inductor structure.

The main parasitic element that affects the quality factor of inductor is series resistance Rs. It is mainly arises from metal resistivity of the inductor turning metals and this component of the series resistance R_{s0} can be approximated as

$$R_{s0} = \frac{\rho l_T}{wt}$$

where ρ is the resistivity of the wire, l_T is the total inductor length, w is inductor metal width, and t is physical thickness of the metal.

The other components related to Rs element are skin and proximity effects. These effects especially need to be taken account in higher frequencies. The frequency-related alternating currents within the inductor metal wires generate electric field, which in turn induce its own magnetic field around the wire. This time-varying magnetic field generates an electric field. According to Lenz law, resulting electric field within the conductor behaves as a counteracting element to the overall magnetic field of the inductor. This counteracting induced field within the conductor force the flowing current to the edge of the metal wires. As a result, the overall magnetic field of the inductor can only penetrate up to a depth, which is called the skin depth and this effect is called as skin effect.

The skin depth can be given as:

$$\delta = \sqrt{\frac{2\rho}{\mu\omega}}$$

where ρ is resistivity of the material, μ is the magnetic permeability, and ω is the angular frequency.

As can be seen from the equation, skin depth is frequency related and for high frequencies skin depth reduces, which in turn affects the total resistivity. The effect of the skin depth together with the metal resistivity effect can be given within the below equation.

$$R_s = \frac{\rho l_T}{wt_{eff}}$$

where ρ is the resistivity of the wire, l_T is the total inductor length, w is inductor metal width, and t_{eff} is given as [8, 10, 12].

$$t_{eff} = \delta\left(1 - e^{-t/\delta}\right)$$

where t is the physical thickness of the wire and δ is the skin depth.

The substrate coupling resistance Rsi is the silicon conductivity related to majority carrier concentration. Large Rsi degrades the performance of inductor, and for this case, a large area inductor should be used. This resistance simple estimation can be given as

$$Rsi = \frac{2}{l_T * w * G_{sub}}$$

where l_T is total inductor length, w is width of metal traces of inductor, and G_{sub} is conductance per unit area for the silicon substrate.

Series capacitance Cs is the coupling capacitance between the input and output ports of the inductor. This capacitance occurs directly between the inductor and shows another path to flow input to output for the incoming signal. For large values of Cs degrades inductor performance especially for high frequency operations. The components that form Cs are coupling capacitances between metal traces of inductor and overlap capacitance between the inductor and underpass metal wire to connect the inner end of inductor spiral to external components.

Appendix 9: MatLab Program for Chapter 9

In this appendix, we present the MatLab programs specifically developed for the chapter under consideration.

Program List 9.1. Main Program Main_SS-DPS_Example1.m

```
% Main Program Main_SS-DPS_Example1.m
% January 20, 2018, Vanikoy
% Developed by BS Yarman
% This program computes the element values of SS-DPS for D_Teta=45
  degree
% at w0=1.
% Inputs:
%    Del_Teta_w0
% Output:
%    Del_Teta_w
% -----------------------------------------------------------------
clc; close all;clear
Del_Teta_w0=input('Del_Teta_w0 in degree=')
N=1001;
w1=1e-8;w2=2; DW=(w2-w1)/(N-1);
w=w1;
% Component values:
% -----------State-A:-----------------------------------------
Lp1=tand(Del_Teta_w0/4)
Cp2=Lp1
% -----------State-B:-----------------------------------------
Cp1=1/tand(Del_Teta_w0/4)
Lp2=Cp1
% -----------------------------------------------------------------
%
for i=1:N
    W(i)=w;
    Teta_B(i)=2*atand(1/(w*Cp1));
    Teta_A(i)=-2*atand(w*Lp1);
    Del_Teta_w=2*(atand(1/(w*Cp1))+atand(w*Lp1));
    Del_Teta(i)=Del_Teta_w;
    w=w+DW;
end
figure
plot(W,Del_Teta)
```

```
title ('Phase Variation of SS-DPS')
xlabel('Normalized Angular Frequency w')
ylabel('Teta-B-Teta-A)')
% --------------------------------------------------------------------
figure
plot(W,Teta_B, W,Teta_A)
legend('Teta-B','Teta-A')
title('Teta-B and Teta-A versus W')
xlabel('Normalized Angular Frequency')
ylabel('Teta-B, Teta-A')
```

Program List 9.2. Main Program Main_SS_DPS_Example2.m

```
% Main Program Main_SS_DPS_Example2.m
% January 21, 2018, Vanikoy
% Developed by BS Yarman
% This program computes the normalized element values of SS-DPS
% for D_Teta=45,90,135 and 180 and plots the results
% --------------------------------------------------------------------
clc; close all;clear
w0=1;
Del_Teta_1=45;Del_Teta_2=90; Del_Teta_3=180; Del_Teta_4=135;
[ Lp145,Cp145,Lp245,Cp245,W,Del_Teta_45 ] = SS_DPS_Net_Phase_Shift(w0,
    Del_Teta_1);
[ Lp190,Cp190,Lp290,Cp2,W,Del_Teta_90 ] = SS_DPS_Net_Phase_Shift(w0,
    Del_Teta_2);
[ Lp1180,Cp1180,Lp2180,Cp2180,W,Del_Teta_180 ] =
    SS_DPS_Net_Phase_Shift(w0,Del_Teta_3);
[ Lp1135,Cp1135,Lp2135,Cp2135,W,Del_Teta_135 ] =
    SS_DPS_Net_Phase_Shift(w0,Del_Teta_4);
figure
plot(W,Del_Teta_45,W,Del_Teta_90,W,Del_Teta_135,W,Del_Teta_180);
legend('SS-DPS 45','SS-DPS 90','SS-DPS135','SS-DPS 180')
%--------------------------------------------------------------------
z(1)=abs((45-50.29)/45)*100; y(1)=50.29;
z(2)=abs((90-97.15)/90)*100; y(2)=97.15;
z(3)=abs((135-139.8)/135)*100;y(3)=139.8;
z(4)=abs((180-180)/180)*100; y(4)=180;
Teta(1)=45;Teta(2)=90;Teta(3)=135;Teta(4)=180;
x=Teta;
%--------------------------------------------------------------------
figure
plot(Teta,z)
title('Phase Perturbation eps-teta versus Teta over an octave
                                           0.6<w<1.2')
xlabel('Teta in Degree')
ylabel('Perturbation in Percent (%)')
%--------------------------------------------------------------------
% 3D Plot of Perturbation
figure
plot3(x,y,z)
```

Program List 9.3. Main Program:Main_SS_DPS_Example3.m

```
% Main Program: Main_SS_DPS_Example3.m
% January 25, 2018, Vanikoy
% Developed by BS Yarman
% This program computes the scattering parameters for Example 3.
%
% -----------------------------------------------------------------
clc; close all;clear
% Basic Design
Teta1=45; foa=10e9;R=100
mu1=tand(Teta1/2);
L1=mu1;C1=L1;
Teta2=135;
mu2=tand(Teta2/2);
C2=1/mu2;
L2=C2;
%----------------
% Actual Element Values
R=100; f0a=10e9;
[ L1a ] = Actual_Inductor(L1,R,f0a);
[ L2a ] = Actual_Inductor(L2,R,f0a);
[ C1a ] = Actual_Capacitor(C1,R,f0a);
[ C2a ] = Actual_Capacitor(C2,R,f0a);
Lagging_Section=[L1a C1a]
Leading_Section=[L2a C2a]
```

Program List 9.4. Main Program:Main_SS_DPS_Example4.m

```
% Main Program: Main_SS_DPS_Example4.m
clc; clear; close all
% This program developed by BS Yarman, January 30, 2018, Vanikoy
% -----------------------------------------------------------------
% Inputs:
% Enter target phase shift:
Teta=45;
% For the 0.18u process, select the optimum OFF mode capacitor Coff1
    for S1
Coff1a=90e-15;
% Select normalization resistor R
R=100;
% Specify the actual center frequency
f0a=10e9;
% Enter vector K=[k2 k4]; where k2 is the multiplier for Coff2a and
% k4 is the multiplier for Coff4a
K=[0.9 0.6];
k2=K(1); k4=K(2);
% -----------------------------------------------------------------
% This program implements the practical design algorithm.
% Part I: Basic Symmetrical Lagging & Leading Section (SLLS) Design:
[ L1,C1, L1a,C1a,L2,C2,L2a,C2a ] = Basic_SLLS(Teta,f0a,R);
% -----------------------------------------------------------------
% Part II: Design of NMOS switches for 0.18u process technology
```

```
%
% Design of S1:
% Select optimim choice for Coff1a:
[ Coff1, Ron1, Ron1a ] = NMOS_Switch_Design(Coff1a,R,f0a);
% Design of S2:
Coff2a_max=C2a; Coff2a=k2*Coff2a_max;
[ Coff2, Ron2, Ron2a ] = NMOS_Switch_Design(Coff2a,R,f0a);
% Design Design of S3:
Coff3a=Coff1a;
[ Coff3, Ron3, Ron3a ] = NMOS_Switch_Design(Coff3a,R,f0a);
% Design of S4:
Coff4a_max=C1a; Coff4a=k4*Coff4a_max;
[ Coff4, Ron4, Ron4a ] = NMOS_Switch_Design(Coff4a,R,f0a);
% Part III: Computation of the elements of SSS-DPS
Lp1=L1/(1+L1*Coff1);
Cp1=C2-Coff2;
%
Lp2=L2/(1+L2*Coff3);
Cp2=C1-k4*Coff4;
% Part IV: Computations of actual elements
[ Lp1a ] = Actual_Inductor(Lp1,R,f0a);
[ Lp2a ] = Actual_Inductor(Lp2,R,f0a);
%
[ Cp1a ] = Actual_Capacitor(Cp1,R,f0a);
[ Cp2a ] = Actual_Capacitor(Cp2,R,f0a);
Normalized_Component_Values= [Lp1 Cp1 Lp2 Cp2]
Actual_Component_Values=[Lp1a Cp1a Lp2a Cp2a]
```

Program List 9.5. function Basic_SLLS

```
function [ L1,C1, L1a,C1a,L2,C2,L2a,C2a ] = Basic_SLLS(Teta,f0a,R)
% This function designs a basic symmetrical Lagging & Leading Lattice
    Section
%    Developed by BS Yarman January 30, 2018
% Basic SLLS Design
%    Inputs:
%        Teta: Target Phase Shift in Degree
%        f0a:  Actual center frequency of the design
%        R:    Normalization resistor (It may be chosen as 100 ohms)
%    Output:
%        L1,C1, L2, C2:  Normalized element values of Basic SLLS
%        L1a,C1a,L2a,C2a:Actal element values of Basic SLLS
mu=tand(Teta/4);
% Normalized element values
L1=mu;C1=L1;
C2=1/L1;L2=C2;
%----------------
% Actual Element Values
[ L1a ] = Actual_Inductor(L1,R,f0a);
[ L2a ] = Actual_Inductor(L2,R,f0a);
[ C1a ] = Actual_Capacitor(C1,R,f0a);
[ C2a ] = Actual_Capacitor(C2,R,f0a);
End
```

Program List 9.6. function NMOS_Switch_Design

```
function [ Coffn, Ron, Rona ] = NMOS_Switch_Design(Coffa,R,f0a)
% This function designs an NMOS switch with normalized element values
% Develeoped by BS Yarman, January 29, 2018
%    Inputs:
%        Coffa: optimum value of the OFF State Capacitor of NMOS
%        in Farad
%        R: Normalization Resistor (For symmetrical Lattice it is
%        100 ohms)
%        f0a: Actual Center Frequency
%    Outputs:
%        Coffn: Normalized OFF State Capacitor
%        Ron: Normalized ON State resistor
%        Rona: Actual ON state resistor.
% Step 1: Normalization
Coffn=2*pi*f0a*R*Coffa;
% Computation of the normalized value of the channel resistor Ron1
% when NMOS is ON state
Rona=672*1e-15/Coffa;
Ron=672*1e-15/Coffa/R;

end
```

Program List 9.7. function Actual_Inductor

```
function [ La ] = Actual_Inductor(Ln,R,f0a)
La=Ln*R/2/pi/f0a;

end
```

Program List 9.8. function Actual_Capacitor

```
function [ Ca ] = Actual_Capacitor(Cn,R,f0a)
Ca=Cn/R/2/pi/f0a;

end
```

Program List 9.9. Main Program:Main_SS_DPS_Example5.m

```
% Main Program: Main_SS_DPS_Example5.m
clc; clear;close all
% This program developed by BS Yarman, February 5, 2018, Vanikoy
% ------------------------------------------------------------------
% Inputs:
% Enter target phase shift:
Teta=45;
% For the 0.18u process, select the optimum OFF mode capacitor Coff1
%    for S1
Coff_opt=90e-15;
% Select normalization resistor R
R=100;
```

```
% Specify the actual center frequency
f0a-10e9;
% Enter vector K=[k2 k4]; where k2 is the multiplier for Coff2a and
% k4 is the multiplier for Coff4a
K=[1 1];
k2=K(1); k4=K(2);
% ------------------------------------------------------------------
[ NCV,ACV,Ron,Coffn,Rona,Coffa ] = SSS_DPS_Design(Teta, R, f0a,
   Coff_opt,K)

Ron1=Ron(1); Ron4=Ron(4); Lp1=NCV(1); Cp1=NCV(2); Lp2=NCV(3); Cp2=NCV
   (4);
Coff2=Coffn(2); Coff3=Coffn(3);
% ------------------------------------------------------------------
w1=0; w2=2;N=101; dw=(w2-w1)/(N-1);
w=w1;
for i=1:N
    W(i)=w;
[ FI21B,ILB,VSWR ] = SSS_DPS_State_B(w,Ron1,Ron4,Lp1,Lp2,Cp1,Cp2,
                  Coff2,Coff3);
Phase_B(i)=FI21B;
Insertion_Loss_B(i)=ILB;
w=w+dw;
end
figure
plot(W,Phase_B)
xlabel('Normalized Angular Frequency w')
ylabel(' Phase of State-B in Degree')
title('Phase versus normalized angular frequency for State-B')
figure
plot(W,Insertion_Loss_B)
xlabel('Normalized Angular Frequency w')
ylabel(' Gain of State-B in Degree')
title('Gain versus normalized angular frequency for State-B')
```

Program List 9.10. function SSS_DPS_State_B

```
function [ FI21B,ILB,VSWR ] = SSS_DPS_State_B(w,Ron1,Ron4,Lp1,Lp2,Cp1,
   Cp2,Coff2,Coff3)
% Simple Single Symmetrical Lattice Digital Phase Shifter in State-B
% Developed by BS Yarman, February 4, 2018, Vanikoy
% Computations for State-B: SSS Lattice is a Leading Symmetrical
   Section
% % ----------------------------------------------------------------
%    Inputs:
%        Lp1,Lp2
%        Cp1,Cp2
%        Ron1, Ron4 (Normalized values)
%        Coff2,Coff3 (Normalized values)
%    Output:
%        FI21B: Phase of S21B in degree
%        ILB:   Insertion Loss in dB. ILB=20log10 ( |S21B| ) in dB
% ------------------------------------------------------------------
```

```
DaB=Ron1*Ron1+w*w*Lp1*Lp1;
RaB=w*w*Ron1*Lp1/DaB;
XaB=w*Ron1*Ron1*Lp1/DaB-1/w/(Cp1+Coff2);
zaB=complex(RaB,XaB);
%
DbB=1+(w*Ron4*Cp2)*(w*Ron4*Cp2);
RbB=Ron4/DbB;
XbB=Lp2/(1-w*w*Lp2*Coff3)-w*Ron4*Ron4*Cp2/DbB;
zbB=complex(RbB,XbB);
% ----------------------------------------------------------------
[ FI21B,VSWR,ILB ] = S_Par_SLS(zaB,zbB);

end
```

Program List 9.11. S_Par_SLS

```
function [ FI21B,VSWR,ILB ] = S_Par_SLS(zaB,zbB)
% This function generates the scattering parameters of a symmetrical
% Lattice defined by means of its series and cross arm impedances zaB
    and
% zbB
% This function is developed by BS Yarman, Feb 5, 2018
%   Inputs:
%       Complex series arm impedance zaB
%       Complex cross arm impedance zbB
%   Output:
%       FI21B: Phase of S21B
%       ILB: 20log10( |S21B| ) insertion loss in dB
%       VSWR: Voltage Standing Wave Ratio
% ----------------------------------------------------------------
S11B=(zaB*zbB-1.0)/(zaB*zbB+zaB+zbB+1.0);
S21B=(zbB-zaB)/(zaB*zbB+zaB+zbB+1.0); R21B=real(S21B);X21B=imag(S21B);
ro21B=abs(S21B);
ILB=20*log10(ro21B);
FI21B=atan2d(X21B,R21B);
ro11B=abs(S11B);
VSWR=(1+ro11B)/(1-ro11B);

end
```

Program List 9.12. function SSS_DPS_Component_Values

```
function [ NCV,ACV,RON,COFF, RONA, COFFA  ] = SSS_DPS_Component_Values
    (Teta,R, Coffa,f0a)
% This function generates the estimated component values of an SSS-DPS
% Unit.
% In this function, Cp1 & Cp2 "ComponentValue-Control vector
    K=[k2 k4]"
% is automatically generated from the selected optimum off state
    capacitor
% Coff_opt and from the basic component values of basic symmetrical
% LC sections. That is from C1 and C2.k2=Coff_opt/C2, k4=Coff_opt/C1
    and
```

```
% Cp1=C2-k2'C2=C2(1-k2), Cp2=C1(1-k4).
% In this function the purpose is that "to the optimum value of switch
% capacitors when possible.
% Developed By B.S. Yarman, January 29, 2018
% Inputs:
% Teta: Target phase Shift in degree at w0=1.
% R : Normalization resistor for S-Parameters. It may be R=100 ohm.
% Coff: Optimum OFF state capacitor of NMOS switch in Farad
% fo0 : Actual Center Frequency in Hz.
% K : OFF State Capacitor Control Vector with two enrees K=[k2 k4]
% Outputs:
% (Lp1,Cp1): Series arm inductor and capacitor
% (Lp2,Cp2): Cross arm inductor and capacitor
% NCV: Normalized Component Values as a vector [Lp1, Cp1, Lp2, Cp2]
%
% ------------------------------------------------------------------
% Part I: Define Basic SSS-DPS Cell
[ L1,C1,L2,C2 ] = Basic_SSS_DPS_Design(Teta);
% Compute the normalized value of the actual optimum OFF state
                             capacitor
Coffn=2*pi*f0a*Coffa*R;
%------------------------------------------------------------------
% Part II: Generate the switch parameters for the optimum values of
% OFF state capacitor Coff.
% k2=K(1); k4=K(2);
% ---- Design Switch S1:-------------------------------------------
[ Coff1, Ron1, ~ ] = NMOS_Switch_Design(Coffa,R,f0a);
% ---- Switch S2:--------------------------------------------------
Coff2_max=C2;
% Compute the control number k2 automatically
k2=Coffn/Coff2_max;
Coff2=k2*Coff2_max;
[ Coff2a ] = Actual_Capacitor(Coff2,R,f0a);
[ ~, Ron2, ~ ] = NMOS_Switch_Design(Coff2a,R,f0a);
% ---- Design Switch S3: --------------------------
Coff3=Coff1;
Ron3=Ron1;
% ---- Design Switch S4: ----------
Coff4_max=C1;
k4=Coffn/Coff4_max;
Coff4=k4*Coff4_max;
[ Coff4a ] = Actual_Capacitor(Coff4,R,f0a);
[ ~, Ron4, ~ ] = NMOS_Switch_Design(Coff4a,R,f0a);
%------------------------------------------------------------------
% Part III: Compute the series armcomponents values:
Cp1=C2-Coff2;
% Check S2 if it is okay:
if Cp1<0
% k2=1; % if Cp1 is negative, then set C2=Coff2, k2=1 case.
    Cp1=0;
    Coff2=C2;
    [ Coff2a ] = Actual_Capacitor(Coff2,R,f0a);
    [ Coff2, Ron2, Ron2a ] = NMOS_Switch_Design(Coff2a,R,f0a);
```

```
  end
Lp1=L1/(1+L1*Coff1);
  % Part IV: Compute the cross arm component values
Cp2=C1-Coff4;
  if Cp2<0
      % k4=1; % if Cp2 is negative, then set C1=Coff4, k4=1 case.
     Cp2=0;
     Coff4=C1;
     [ Coff4a ] = Actual_Capacitor(Coff4,R,f0a);
     [ Coff4, Ron4, Ron4a ] = NMOS_Switch_Design(Coff4a,R,f0a);
  end
Lp2=L2/(1+L2*Coff3);
  % Computation of Actual Compenet Values ACV
[ Lp1a ] = Actual_Inductor(Lp1,R,f0a);
[ Cp1a ] = Actual_Capacitor(Cp1,R,f0a);
  %
[ Lp2a ] = Actual_Inductor(Lp2,R,f0a);
[ Cp2a ] = Actual_Capacitor(Cp2,R,f0a);

NCV=[Lp1 Cp1 Lp2 Cp2];
ACV=[Lp1a Cp1a Lp2a Cp2a];
  %-----------------------------------------------------------------
% Generate the ON state resistor vector:
RON=[Ron1 Ron2 Ron3 Ron4]; RONA=R*RON;
COFF=[Coff1 Coff2 Coff3 Coff4]; COFFA= Actual_Capacitor(COFF,R,f0a);

  end
```

Program List 9.13. function UnevenTeta_Basic_SLLS

```
  function [ L1,C1, L1a,C1a,L2,C2,L2a,C2a ] = UnevenTeta_Basic_SLLS(
     Del_Teta,TetaB,f0a,R)
  % This function designs a basic symmetrical Lagging & Leading
     Lattice Section
  % Developed by BS Yarman Feb 15, 2018
  % Basic SLLS Design
  % Inputs:
     % Del_Teta: Target Phase Shift in Degree
     % TetaB: Phase Shift of Leading Symmetrical Section
     % f0a: Actual center frequency of the design
     % R: Normalization resistor (It may be chosen as 100 ohms)
  % Output:
     % L1,C1, L2, C2: Normalized element values of Basic SLLS
     % L1a,C1a,L2a,C2a:Actal element values of Basic SLLS
  %-----------------------------------------------------------------
  % Note: TetaA is the absolute value of the phase of "State-A".
  % Stae-A phase is the lagging state which yields a negative phase.
  % Teta=TetaB+TetaA. Therefore, it is given by
  % TetaA=Teta-TetaB where TetaB is a positive leading phase.
  % TetaA/2 and TetaB/2 must vary between 0 and 90 degree. Therefore
     TetaA &
  % TetaB must be less than 180 degree.
  %-----------------------------------------------------------------
```

```
% Step 1: For Specified Del_Teta and TetaB Determine TetaA
TetaA=Del_Teta-TetaB;
% Step 2: Generate Major Design Parameters (MDP) for the Basic
   Design muA=tand(TetaA/2);
muB=tand(TetaB/2);
% Step 3: Compute the Normalized Component Values (NCV) of the
   Basic Design
L1=muA;C1=L1;
C2=1/muB;L2=C2;
%---------------------------------------------------------------
% Step 3: Compute the Actaul Component Values (ACV) of the Basic
                           Design
[ L1a ] = Actual_Inductor( L1,R,f0a);
[ L2a ] = Actual_Inductor( L2,R,f0a);
[ C1a ] = Actual_Capacitor(C1,R,f0a);
[ C2a ] = Actual_Capacitor(C2,R,f0a);
%---------------------------------------------------------------

   end


Program List 9.14. function UnevenPhase_DPS_Component_Values

function [ Lp1,Cp1,Lp2,Cp2,RON,COFF,RONA, COFFA  ] =
   UnevenPhase_DPS_Component_Values(Teta,TetaB, R, Coffa,f0a)
% This function generates the estimatedcomponent values of an SSS-DPS
% Unit.
% Developed By B.S. Yarman, Feb 7, 2018
% Inputs:
   % Teta: Target phase Shift in degree at w0=1.
   % TetaB: Un-evenly distributed phase of state-B at w=1
   % TetaA: Unevenly distributed phase of State-A. TetaA=Teta-TetaB
   % R: Normalization resistor for S-Parameters. It may be R=100 ohm.
   % Coffa: Actual Optimum value of the "OFF state capacitor" of NMOS
      switch in Farad
   % f0a : Actual Center Frequency in Hz.
% K : OFF State Capacitor Control Vector with two entrees K={k2 k4}
% Note: k2=1, k4=1 corresponds to the ideal values of switch
      capacitor.
% Ideal situation may yield big values. Therefore, k2 and k2
      must be equal or
% less than 1.Cp1=C2-k2*Coff_max>0.k2 may be selected as
      k2=C2/Coff_opt
% Similarly Cp2=C1-k4*Coff4_max,Coff4_max=C1. k4=Coff_opt/Coff4_max
%
% Outputs:
     % (Lp1,Cp1): Series arm inductor and capacitor
     % (Lp2,Cp2): Cross arm inductor and capacitor
%---------------------------------------------------------------
% Part I: Define Basic SSS-DPS Cell
[ L1,C1,~,~,L2,C2,~,~ ]=UnevenTeta_Basic_SLLS(Teta,TetaB,f0a,R);
% Compute the normalized value of the optimum OFF state capacitance
of
% NMOS
```

```
Coffn=2*pi*f0a*Coffa*R;
%-------------------------------------------------------------
% Part II: Generate the switch parameters for the optimumvalues of
% OFF state capacitor Coff.
% k2=K(1); k4=K(2);
% ---- Design Switch S1:------------------------------------
Coff1a=Coffa;
[ Coff1, Ron1, Ron1a ] = NMOS_Switch_Design(Coff1a,R,f0a);
% ---- Design Switch S2:------------------------------------
% Set the maximum OFF state capacitance ofS2:
Coff2_max=C2;
k2=Coffn/Coff2_max;
Coff2=k2*Coff2_max;
[ Coff2a ] = Actual_Capacitor(Coff2,R,f0a);
[ ~, Ron2, Ron2a ] = NMOS_Switch_Design(Coff2a,R,f0a);
% ---- Design Switch S3:------------------------------------
Coff3=Coff1;
Coff3a=Coff1a;
Ron3=Ron1;
Ron3a=Ron1a;
% ---- Design Switch S4:------------------------------------
Coff4_max=C1;
k4=Coffn/Coff4_max;
Coff4=k4*Coff4_max;
[ Coff4a ] = Actual_Capacitor(Coff4,R,f0a);
Ron4=672*1e-15/Coff4a/R;
[ Coff2, Ron4, Ron4a ] = NMOS_Switch_Design(Coff4a,R,f0a);
%-------------------------------------------------------------
% Part III: Compute the series arm components values:
Cp1=C2-Coff2;
 if Cp1<0
    Cp1=0;k2=0;end
Lp1=L1/(1+L1*Coff1);
% Part IV: Compute the cross arm component values
Cp2=C1-Coff4;
 if Cp2<0
    Cp2=0; k4=0; end
Lp2=L2/(1+L2* Coff3);
%-------------------------------------------------------------
% Generate the ON state resistor vector:
RON=[Ron1 Ron2 Ron3 Ron4];
RONA=[Ron1a Ron2a Ron3a Ron4a];
COFF=[Coff1 Coff2 Coff3 Coff4];
COFFA=[Coff1a Coff2a Coff3a Coff4a];
 end

% Main Program: Main_SS_DPS_Example6.m

clc; clear;close all
% This program is developed by BS Yarman, February 7, 2018, Vanikoy
% -------------------------------------------------------------------
% Inputs:
% Enter target phase shift:
```

```
Del_Teta=input('Enter Del_Teta=')
% For the 0.18u process, select the optimum OFF mode capacitor Coff1
for S1
Coff_opt=90e-15;
% Select normalization resistor R
R=100;
% ------------------------------------------------------------------
TetaB=input('Enter TetaB=')
% Specify the actual center frequency
f0a=input('Enter Actual Center Frequency f0a=')
%
% Computational Steps:
% Step 1: Design SSS_DPS for unevenly distributed phase-shift
  Del_Teta,
% TetaB and TetaA. User specify Del_Teta and TetaB. TetaA=Del_TetaB>0
%
[ Lp1,Cp1,Lp2,Cp2,Ron,Coffn,RONA,  COFFA  ] =
   UnevenPhase_DPS_Component_Values(Del_Teta,TetaB, R, Coff_opt,f0a)
%
Ron1=Ron(1); Ron2=Ron(2); Ron3=Ron(3); Ron4=Ron(4);
Coff1=Coffn(1); Coff2=Coffn(2); Coff3=Coffn(3); Coff4=Coffn(4);
% --------------------------------------------------------------
w1=1e-9; w2=2;N=10001; dw=(w2-w1)/(N-1);
w=w1;
for i=1:N
    W(i)=w;
[ FI21A,GainA,VSWRA ] = SSS_DPS_State_A(w,Ron2,Ron3,Lp1,Lp2,Cp1,Cp2,
    Coff1,Coff4);
[ FI21B,GainB,VSWRB ] = SSS_DPS_State_B(w,Ron1,Ron4,Lp1,Lp2,Cp1,Cp2,
    Coff2,Coff3);
Phase_A(i)=FI21A;
Phase_B(i)=FI21B;
GainA_dB(i)=GainA;
GainB_dB(i)=GainB;
Phase_Shift(i)=FI21B-FI21A;
w=w+dw;
end
Plot_3S_DPS(W,Phase_A,Phase_B,Phase_Shift,GainA_dB,GainB_dB)
NCV=[Lp1 Cp1 Lp2 Cp2]
[ Lp1a ] = Actual_Inductor(Lp1,R,f0a);
[ Cp1a ] = Actual_Capacitor(Cp1,R,f0a);
[ Lp2a ] = Actual_Inductor(Lp2,R,f0a);
[ Cp2a ] = Actual_Capacitor(Cp2,R,f0a);
L_3SDPS=[Lp1a Lp2a]
C_3SDPS=[Cp1a Cp2a]
```

Program List 9.15. Main Program:Main_SSS_DPS_Example7.m

```
% Main Program: Main_SSS_DPS_Example7.m
clc; clear;close all
% This program is developed by BS Yarman, February 15, 2018, Vanikoy
% --------------------------------------------------------------------
% Inputs:
```

```
% Enter target phase shift:
Del_Teta=input('Enter a positive phase-shift for Del_Teta=')
% For the 0.18u process, select the optimum OFF mode capacitor
Coff1 for S1 Coff_opt=90e-15;
% Select normalization resistor R
R=100;
%----------------------------------------------------------------
TetaB=input('Enter a positive phase-shift for TetaB=')
% Specify the actual center frequency
w0=input('Enter Center Frequency w0=')
f0a=input('Enter Actual Center Frequency f0a=')
%
% Computational Steps:
% Step 1: Design SSS_DPS for unevenly distributed phase-shift Del_Teta,
% TetaB and TetaA. User specify Del_Teta and TetaB. TetaA=TetaA=
%    Del_Teta-TetaB>0
%
[ Lp1,Cp1,Lp2,Cp2,Ron,Coffn,RONA, COFFA  ] =
    UnevenPhase_DPS_Component_Values(w0, Del_Teta,TetaB, R, Coff_opt,
    f0a)
%
Ron1=Ron(1); Ron2=Ron(2); Ron3=Ron(3); Ron4=Ron(4);
Coff1=Coffn(1); Coff2=Coffn(2); Coff3=Coffn(3); Coff4=Coffn(4);
% ----------------------------------------------------------------
w1=-5; w2=5;N=10001; dw=(w2-w1)/(N-1);
w=w1;
for i=1:N
    W(i)=w;
[ FI21A,GainA,VSWRA ] = SSS_DPS_State_A(w,Ron2,Ron3,Lp1,Lp2,Cp1,Cp2,
    Coff1,Coff4);
[ FI21B,GainB,VSWRB ] = SSS_DPS_State_B(w,Ron1,Ron4,Lp1,Lp2,Cp1,Cp2,
    Coff2,Coff3);
Phase_A(i)=FI21A;
Phase_B(i)=FI21B;
GainA_dB(i)=GainA;
GainB_dB(i)=GainB;
Phase_Shift(i)=FI21B-FI21A;
w=w+dw;
end
Plot_3S_DPS(W,Phase_A,Phase_B,Phase_Shift,GainA_dB,GainB_dB)
NCV=[Lp1 Cp1 Lp2 Cp2]
[ Lp1a ] = Actual_Inductor(Lp1,R,f0a);
[ Cp1a ] = Actual_Capacitor(Cp1,R,f0a);
[ Lp2a ] = Actual_Inductor(Lp2,R,f0a);
[ Cp2a ] = Actual_Capacitor(Cp2,R,f0a);
L_3SDPS=[Lp1a Lp2a]
C_3SDPS=[Cp1a Cp2a]
```

Program List 9.16. Main Program:Main_SSS_DPS_Example7.m

```
% Main Program: Main_SSS_DPS_Example7.m
clc; clear;close all
% This program is developed by BS Yarman,February 15, 2018, Vanikoy
```

```
%------------------------------------------------------------------
% Inputs:
% Enter target phase shift:
Del_Teta=input('Enter a positive phase-shift for Del_Teta=')
% For the 0.18u process, select the optimumOFF mode capacitor
  Coff1 for S1
Coff_opt=90e-15;
% Select normalization resistor R
R=100;
%------------------------------------------------------------------
TetaB=input('Enter a positive phase-shift for TetaB=')
% Specify the actual center frequency
w0=input('Enter Center Frequency w0=')
f0a=input('Enter Actual Center Frequency f0a=')
%
% Computational Steps:
% Step 1: Design SSS_DPS for unevenly distributed phase-shift Del_Teta,
% TetaB and TetaA. User specify Del_Teta and TetaB.
  TetaA=TetaA=Del_Teta-TetaB>0

[ Lp1,Cp1,Lp2,Cp2,Ron,Coffn,RONA,  COFFA  ] =
   UnevenPhase_DPS_Component_Values(w0, Del_Teta,TetaB, R, Coff_opt,
   f0a)
%
Ron1=Ron(1); Ron2=Ron(2); Ron3=Ron(3); Ron4=Ron(4);
Coff1=Coffn(1); Coff2=Coffn(2); Coff3=Coffn(3); Coff4=Coffn(4);
%------------------------------------------------------------------
w1=-5; w2=5;N=10001; dw=(w2-w1)/(N-1);
w=w1;
for i=1:N
    W(i)=w;
[ FI21A,GainA,VSWRA ] = SSS_DPS_State_A(w,Ron2,Ron3,Lp1,Lp2,Cp1,Cp2,
    Coff1,Coff4);
[ FI21B,GainB,VSWRB ] = SSS_DPS_State_B(w,Ron1,Ron4,Lp1,Lp2,Cp1,Cp2,
    Coff2,Coff3);
Phase_A(i)=FI21A;
Phase_B(i)=FI21B;
GainA_dB(i)=GainA;
GainB_dB(i)=GainB;
Phase_Shift(i)=FI21B-FI21A;
w=w+dw;
end
Plot_3S_DPS(W,Phase_A,Phase_B,Phase_Shift,GainA_dB,GainB_dB)
NCV=[Lp1 Cp1 Lp2 Cp2]
[ Lp1a ] = Actual_Inductor(Lp1,R,f0a);
[ Cp1a ] = Actual_Capacitor(Cp1,R,f0a);
[ Lp2a ] = Actual_Inductor(Lp2,R,f0a);
[ Cp2a ] = Actual_Capacitor(Cp2,R,f0a);
L_3SDPS=[Lp1a Lp2a]
C_3SDPS=[Cp1a Cp2a]
```

```
Program List 9.17. Main Program:Main_SSS_DPS_Example6.m
% Main Program: Main_SSS_DPS_Example6.m
clc; clear;close all
% This program is developed by BS Yarman,February 15, 2018, Vanikoy
%-----------------------------------------------------------------
% Inputs:
% Enter target phase shift:
Del_Teta=input('Enter Del_Teta=')
% For the 0.18u process, select the optimum CFF mode capacitor
%  Coff1 for S1
Coff_opt=90e-15;
% Select normalization resistor R
R=100;
%-----------------------------------------------------------------
TetaB=input('Enter TetaB=')
% Specify the normalized and actual center frequency
w0=input('Enter w0=')
f0a=input('Enter Actual Center Frequency f0a=')
%
% Computational Steps:
% Step 1: Design SSS_DPS for unevenly distributed phase-shift Del_Teta,
% TetaB and TetaA. User specify Del_Teta and TetaB. TetaA=Del_TetaB>0
%
[ Lp1,Cp1,Lp2,Cp2,Ron,Coffn,RONA, COFFA  ] =
    UnevenPhase_DPS_Component_Values(w0, Del_Teta,TetaB, R, Coff_opt,
    f0a);
%
Ron1=Ron(1); Ron2=Ron(2); Ron3=Ron(3); Ron4=Ron(4);
Coff1=Coffn(1); Coff2=Coffn(2); Coff3=Coffn(3); Coff4=Coffn(4);
% -----------------------------------------------------------------
w1=-5; w2=5;N=10001; dw=(w2-w1)/(N-1);
w=w1;
for i=1:N
    W(i)=w;
[ FI21A,GainA,VSWRA ] = SSS_DPS_State_A(w,Ron2,Ron3,Lp1,Lp2,Cp1,Cp2,
    Coff1,Coff4);
[ FI21B,GainB,VSWRB ] = SSS_DPS_State_B(w,Ron1,Ron4,Lp1,Lp2,Cp1,Cp2,
    Coff2,Coff3);
Phase_A(i)=FI21A;
Phase_B(i)=FI21B;
GainA_dB(i)=GainA;
GainB_dB(i)=GainB;
Phase_Shift(i)=FI21B-FI21A;
w=w+dw;
end
Plot_3S_DPS(W,Phase_A,Phase_B,Phase_Shift,GainA_dB,GainB_dB)
NCV=[Lp1 Cp1 Lp2 Cp2]
[ Lp1a ] = Actual_Inductor(Lp1,R,f0a);
[ Cp1a ] = Actual_Capacitor(Cp1,R,f0a);
[ Lp2a ] = Actual_Inductor(Lp2,R,f0a);
[ Cp2a ] = Actual_Capacitor(Cp2,R,f0a);
L_3SDPS=[Lp1a Lp2a]
C_3SDPS=[Cp1a Cp2a]
```

Program List 9.18. Main_Negative_FI21B.m

```
% Main_Negative_FI21B.m
% This program generates the componentvalues of a 3S-DPS employing
% lossy switches. Therefore, effect of thelossy switches is
  minimized on
% the component values.
% This program is developed by BS Yarman,on April 10, 2018, Vanikoy,
% Istanbul.
% Algorithm to design 3S-DPS with arbitrary selection of
% State-B Phase TetaB in degree
% Inputs:
% w0: Normalized angular frequency,
% TetaA=FI21A(w0): Phase of State-A at w=w0,in degree
% TetaB=FI21B(w0): Phase of State-B at w=w0,in degree
% C_off1a: Off-Mode Capacitor of Switch 1 (S1,)
% C_off2a: Off-Mode Capacitor of Switch 2 (S2),
% C_off4a: Off-Mode Capacitor of Switch 4 (S4),
% R_a: Actual Normalization Resistor
% Algorithm to design 3S-DPS with arbitrary selection of State-B
  Phase ?_B
% Inputs:
% w0: Normalized angular frequency,
% TetaA=FI21A(w0): Phase of State-A at w=w0,
% TetaB=FI21B(w0): Phase of State-B at w=w0,
% C_off1a: Off-Mode Capacitor of Switch 1 (S1),
% C_off2a: Off-Mode Capacitor of Switch 2 (S2),
% C_off4a: Off-Mode Capacitor of Switch 4 (S4),
% R_a: Actual Normalization Resistor
%-----------------------------------------------------------------
clc, clear, close all
%-----------------------------------------------------------------
TetaB=input('Enter Negative values for TetaB in degree=')
TetaA=input('Enter positive value of TetaA in degree=')
f0a=input('Enter Actual Center Freuency f0a=')
w0=input('Enter Normalized Angular Center Frequency w0=')
Coffa=input('Enter Coffa=')
Coff1a=Coffa;
Coff2a=Coffa;
Coff3a=Coffa;
Coff4a=Coffa;
Ra=100;
%-----------------------------------------------------------------
% Computational Steps:
% Step-1: Normalized the actual capacitances
wa=2*pi*f0a;
Coff1=wa*Ra*Coff1a;
Coff2=wa*Ra*Coff2a;
Coff3=wa*Ra*Coff3a;
Coff4=wa*Ra*Coff4a;
%-----------------------------------------------------------------
% Step-2: Compute the actual ON-State channel resistors and
% their normalized values
```

```
Ron1a=672e-15/Coff1a;
Ron2a=672e-15/Coff2a;
Ron3a=672e-15/Coff3a;
Ron4a=672e-15/Coff4a;
% Normalized on-channel resistors
Ron1=Ron1a/Ra;
Ron2=Ron2a/Ra;
Ron3=Ron3a/Ra;
Ron4=Ron4a/Ra;
%-------------------------------------------------
% Step-3: Design of Reference State-A using Lagging Section
  (L1 and C1).
% Compute the major design parameters:muA,L1, Lp1 and Cp2 as follows.
  muA=tand(TetaA/2);
L1=muA/w0;
C1=L1;
Lp1=L1/(1+w0*w0*L1*Coff1);
%-------------------------------------------------
% Step-4: Compute RaB and beta as follows.
RaBD=Ron1*Ron1+w0*w0*Lp1*Lp1;
RaB=(w0*w0*Ron1*Lp1*Lp1)/RaBD;
beta=(w0*Ron1*Ron1*Lp1)/RaBD;
%-------------------------------------------------
% Step-5: Solve equation (57)to determine XaB
gammaB=tand(TetaB);
% It should be noted that tand(FI21)=tand[FI21(+/-)180)].
  Therefore, solution
% Xab may yield either FI21B or FI21B (+/-)180.
% gammaB=abs(gammaB);
Discriminant=1-gammaB*gammaB*(RaB*RaB-1);
XaB1=(1+sqrt(Discriminant))/(gammaB);
XaB2=(1-sqrt(Discriminant))/(gammaB);
if XaB1<0;XaB=XaB1;end
if XaB2<0;XaB=XaB2;end
%-------------------------------------------------
% Step-6: Compute Cp1 as in (63)
Cp1=1/w0/(beta-XaB)-Coff2;
if Cp1<0; Cp1=0; C2=Coff2;end
% Design Switch 2:
if Cp1==0
Coff2a=Coff2/2/pi/f0a/Ra;
Ron2a=672e-15/Coff2a;
Ron2=Ron2a/Ra;
end
%-------------------------------------------------
% Step 7: Compute RbB and XbB as in (64)
DenRaB=RaB*RaB+XaB*XaB;
RbB=RaB/DenRaB;
XbB=-XaB/DenRaB;
%-------------------------------------------------
% Step 8a: Compute Cp2 as in (65c)
Cp2a=(1/w0/Ron4)*sqrt((Ron4-RbB)/RbB);
% Step 8b: Check if Cp2 is negative
```

```
if (Ron4-RbB)<0
    Cp2a=0;
end
if Cp2a<0
    Cp2a=0;
end
if Cp2a==0
    Coff4=C1;
    Coff4a=Coff4/2/pi/f0a/Ra;
    Ron4a=672e-15/Coff4a;
    Ron4=Ron4a/Ra;
end
%-------------------------------------------------
% Step 8b
Cp2b=C1-Coff4;
if Cp2b<0
    Cp2b=0;
end
if Cp2b==0
    Coff4=C1;
    Coff4a=Coff4/2/pi/f0a/Ra;
    Ron4a=672e-15/Coff4a;
    Ron4=Ron4a/Ra;
end
%-------------------------------------------------
Cp2=(Cp2b+Cp2a)/2;
%-------------------------------------------------
% Step 9: Compute Lp2
Lp2=XbB/(1+w0*XbB*Coff3);
%-------------------------------------------------
% Step 10: Compute the actual component values.
[ Lp1a ] = Actual_Inductor(Lp1,Ra,f0a);
[ Cp1a ] = Actual_Capacitor(Cp1,Ra,f0a);
[ Lp2a ] = Actual_Inductor(Lp2,Ra,f0a);
[ Cp2a ] = Actual_Capacitor(Cp2,Ra,f0a);
%
%-------------------------------------------------
% Step 11: Normalized and actual component values of 3S-DPS in
%   vector form.
NCV=[Lp1 Cp1 Lp2 Cp2]
L_3SDPS=[Lp1a Lp2a]
C_3SDPS=[Cp1a Cp2a]
%-------------------------------------------------
% Step 12: Plot the results
w1=-5; w2=5;N=10001; dw=(w2-w1)/(N-1);
w=w1;
for i=1:N
    W(i)=w;
[ FI21A,GainA,VSWRA ] = SSS_DPS_State_A(w,Ron2,Ron3,Lp1,Lp2,Cp1,Cp2,
    Coff1,Coff4);
[ FI21B,GainB,VSWRB ] = SSS_DPS_State_B(w,Ron1,Ron4,Lp1,Lp2,Cp1,Cp2,
    Coff2,Coff3);
Phase_A(i)=FI21A;
```

```
Phase_B(i)=FI21B;
GainA_dB(i)=GainA;
GainB_dB(i)=GainB;
Phase_Shift(i)=FI21B-FI21A;
w=w+dw;
end
Plot_3S_DPS(W,Phase_A,Phase_B,Phase_Shift,GainA_dB,GainB_dB)
```

Program List 9.19. Main_SSS_DPS_Example7C.m

```
% Main_SSS_DPS_Example7C.m
% This program generates the component values of a 3S-DPS employing
    the
% lossy switches for Example 7. Therefore, effect of the lossy
    switches is
% minimized on the computations.
% This program is developed by BS Yarman, on April 12, 2018, Vanikoy,
% Istanbul.
% Algorithm 4 to design 3S-DPS with arbitrary selection of
% State-B Phase at w0 is designated by TetaB=FI21B(w0)>0 which is
    positive
% in degree
% Inputs:
% w0: Normalized angular frequency,
% TetaA=FI21A(w0)>0: Phase of State-A at w=w0,in degree
% TetaB=FI21B(w0)>0: Phase of State-B at w=w0,in degree
% C_off1a: Off-Mode Actual Capacitor of Switch 1 (S1),
% C_off2a: Off-Mode Actual Capacitor of Switch 2 (S2),
% C_off4a: Off-Mode Actual Capacitor of Switch 4 (S4),
% Ra=100 ohm: Actual Normalization Resistor
% Algorithm-4 to design 3S-DPS with arbitrary selection of State-B
    Phase
% FI21B(w0)=TetaB>0
% ------------------------------------------------------------------
clc, clear, close all
% ------------------------------------------------------------------
TetaB=-10
TetaA=55
f0a=8e9
w0=1.
%
% Note: In this program KFLAG=-1 results in good design. In other
    words,
% XaB is negative. gammaB=tand(TetaB)>0
KFLAG=+1
Coff1a=25e-15;
Coff2a=90e-15;
Coff3a=90e-15;
Coff4a=90e-15;
Ra=100;
% ------------------------------------------------------------------
% Computational Steps:
```

```
% Step-1: Normalized the actual capacitances
wa=2*pi*f0a;
Coff1=wa*Ra*Coff1a;
Coff2=wa*Ra*Coff2a;
Coff3=wa*Ra*Coff3a;
Coff4=wa*Ra*Coff4a;
% -----------------------------------------------------------------
% Step-2a: Compute the actual ON-State channel resistors and
% their normalized values
Ron1a=672e-15/Coff1a;
Ron2a=672e-15/Coff2a;
Ron3a=672e-15/Coff3a;
Ron4a=672e-15/Coff4a;
% Step 2b: Normalized on-channel resistors
Ron1=Ron1a/Ra;
Ron2=Ron2a/Ra;
Ron3=Ron3a/Ra;
Ron4=Ron4a/Ra;
% -----------------------------------------------------------------
% Step-3: Design of Reference State-A using Lagging Section
%         (L1 and C1).
% Compute the major design parameters: muA,L1,C1 and Lp1 as follows.
muA=tand(TetaA/2);
L1=muA/w0;
C1=L1;
Lp1=L1/(1+w0*w0*L1*Coff1);
% -----------------------------------------------------------------
% Step-4: Compute RaB and beta as follows.
RaBD=Ron1*Ron1+w0*w0*Lp1*Lp1;
RaB=(w0*w0*Ron1*Lp1*Lp1)/RaBD;
beta=(w0*Ron1*Ron1*Lp1)/RaBD;
% -----------------------------------------------------------------
% Step-5: Solve equation (9.61)to determine XaB
% Note that TetaB=FI21B(w0)
% Step 5a: Compute gammaB
gammaB=tand(TetaB);
% It should be noted that tand(FI21)=tand[FI21(+/-)180)]. Therefore,
%    solution
% Xab may yield either FI21B or FI21B (+/-)180.
% gammaB=abs(gammaB);
% Step 5b: Discriminant
Discriminant=1-gammaB*gammaB*(RaB*RaB-1);
% Step 5c: Compute XaB1
XaB1=(1+sqrt(Discriminant))/(gammaB);
% Step 5d: Compute Xab2
XaB2=(1-sqrt(Discriminant))/(gammaB);
% -----------------------------------------------------------------
% Step 5e:  Select negative XaB<0
if KFLAG==-1
if XaB1<0;XaB=XaB1;end
if XaB2<0;XaB=XaB2;end
end
% -----------------------------------------------------------------
```

```
% Computations with positive XaB>0
if KFLAG==+1
if XaB1>0;XaB=XaB1;end
ifxsxs XaB2>0;XaB=XaB2;end
end
% -------------------------------------------------------------------
% Step-6: Compute Cp1 as in (9.63)
Cp1=1/w0/(beta-XaB)-Coff2
if Cp1<0; Cp1=0;
C2=Coff2;end
% Design Switch 2:
if Cp1==0
Coff2a=Coff2/2/pi/f0a/Ra;
Ron2a=672e-15/Coff2a;
Ron2=Ron2a/Ra;
end
% -------------------------------------------------------------------
% Step 7: Compute RbB and XbB as in (9.64)
DenRaB=RaB*RaB+XaB*XaB;
% Step 7a: Compute RbB
RbB=RaB/DenRaB;
% Step 7b: Compute XbB
XbB1=-XaB/DenRaB
XbB2=-1/XaB
XbB=input('Enter XbB=')
% -------------------------------------------------------------------
% Step 8a:  Compute Cp2 as in (9.65c)
Cp2_I=(1/w0/Ron4)*sqrt((Ron4-RbB)/RbB)
% Step 8b:  Check if Cp2 is negative
if (Ron4-RbB)<0;Cp2_I=0;end
if Cp2_I<0;Cp2_I=0;end
% -------------------------------------------------------------------
% Step 8b
Cp2_II=C1-Coff4
if
Cp2_II<0;Cp2_II=0;end
% if Cp2b==0
%      Coff4=C1;
%      Coff4a=Coff4/2/pi/f0a/Ra;
%      Ron4a=672e-15/Coff4a;
%      Ron4=Ron4a/Ra;
% end
% -------------------------------------------------------------------
Cp2=input('Enter Cp2=')
if Cp2==0
    Coff4=C1;
    Coff4a=Coff4/2/pi/f0a/Ra;
    Ron4a=672e-15/Coff4a;
    Ron4=Ron4a/Ra;
end
% -------------------------------------------------------------------
% Step 9: Compute Lp2
Lp2=XbB/(1+w0*XbB*Coff3);
```

```
% --------------------------------------------------------------------
% Step 10: Compute the actual component values.
[ Lp1a ] = Actual_Inductor(Lp1,Ra,f0a);
[ Cp1a ] = Actual_Capacitor(Cp1,Ra,f0a);
[ Lp2a ] = Actual_Inductor(Lp2,Ra,f0a);
[ Cp2a ] = Actual_Capacitor(Cp2,Ra,f0a);
%
% --------------------------------------------------------------------
% Step 11: Normalized and actual component values of 3S-DPS in vector
    form.
NCV=[Lp1 Cp1 Lp2 Cp2]
L_3SDPS=[Lp1a Lp2a]
C_3SDPS=[Cp1a Cp2a]
% --------------------------------------------------------------------
% Step 12: Plot the results
w1=0; w2=2;N=10001; dw=(w2-w1)/(N-1);
w=w1;
for i=1:N
    W(i)=w;
[ FI21A,GainA,VSWRA ] = SSS_DPS_State_A(w,Ron2,Ron3,Lp1,Lp2,Cp1,Cp2,
    Coff1,Coff4);
[ FI21B,GainB,VSWRB ] = SSS_DPS_State_B(w,Ron1,Ron4,Lp1,Lp2,Cp1,Cp2,
    Coff2,Coff3);
Phase_A(i)=FI21A;
Phase_B(i)=FI21B;
GainA_dB(i)=GainA;
GainB_dB(i)=GainB;
Phase_Shift(i)=FI21B-FI21A;
w=w+dw;
end
Plot_3S_DPS(W,Phase_A,Phase_B,Phase_Shift,GainA_dB,GainB_dB)
% --------------------------------------------------------------------
```

Program List 9.20. Main_SSS_DPS_Example8.m

```
% Main_SSS_DPS_Example8.m
% This program is written by BS Yarman on April 28, 2018
% Vanikoy, Istanbul
clc,close all
% Inputs:
Ra=100; % Actual Normalization Resistor (ANR)
f0a=8e9, % Actual Center Frequency (ACF) in Hertz.
w0=1.0, % Normalized Angular Center Frequency (NACF)
% ----------------------------------------------------------
Coff1a=25e-15, % Actual OFF-MODE Capacitor of S1 in Farad
Coff2a=90e-15, % Actual OFF-MODE Capacitor of S2 in Farad
Coff3a=90e-15, % Actual OFF-MODE Capacitor of S3 in Farad
Coff4a=90e-15, % Actual OFF-MODE Capacitor of S4 in Farad
% ----------------------------------------------------------
TetaA=55, % Phase of State-A: FI21A=-TetaA
TetaB=10, % Phase of State-B: FI21B=-TetaB
% Del_Teta=FI21B-FI21A=-35-(-55)=-35+80
% ----------------------------------------------------------
```

```
% Step 1: Compute muA and muB:
 muA=tand(TetaA/2)
 muB=tand(TetaB/2)
% ------------------------------------------------------------------
% Step 2: Compute the normalized value of Coff1 and Ron1
% ------------------------------------
Coff1=2*pi*f0a*Ra*Coff1a
Ron1a=672e-15/Coff1a, Ron1=Ron1a/Ra
% ------------------------------------
Coff2=2*pi*f0a*Ra*Coff2a
Ron2a=672e-15/Coff2a, Ron2=Ron2a/Ra
% ------------------------------------
Coff3=2*pi*f0a*Ra*Coff3a
Ron3a=672e-15/Coff1a, Ron3=Ron3a/Ra
% ------------------------------------
Coff4=2*pi*f0a*Ra*Coff4a
Ron4a=672e-15/Coff4a, Ron4=Ron4a/Ra
% ------------------------------------
% Step 3: Compute Lp1
Lp1=muA/(1+w0*w0*muA*Coff1)
% Step 4: Compute eta=w0*Ron1*Ron1*Lp1/(Ron1*Ron1+Lp1*Lp1)
eta=w0*Ron1*Ron1*Lp1/(Ron1*Ron1+Lp1*Lp1)
% Step 5: Check if eta>muB
if eta>muB
    attention='Design Parameters are GOOD'
end
    if eta<muB
    attention='Design Parameters are NO GOOD. Go back to Input-step
    and reduce coff1a or increase w0'
    end

if eta>muB
% Step 6: Compute the imaginary part X_bB of the cross-arm impedance
    Z_bB as in (9.66g)
% and compute  the imaginary part X_aB of the series-arm impedance
    Z_aB as in (9.66k).
XbB=-1/muB
XaB=-1/XbB
% Step 7: Compute the cross-arm inductor Lp2 as in (9.66i)and series-
    arm
% capacitors Cp1:
% Step 7a: Cross-Arm Inductors Lp2:
Lp2=XbB/(1+w0*XbB*Coff3)
% Check if Lp2 is positive. If not re-design S3
if Lp2<0
    Coff3_max=1/w0/abs(XbB)
    Coff3=input('Re-Design S3 Coff3=')
    Coff3a=Coff3/2/pi/f0a/Ra
    Ron3a=672e-15/Coff3a
    Ron3=Ron3a/Ra
end
% Step 7b: Compute Series-Arm capacitors Cp1:
Cp1=1/w0/(eta-muB)-Coff2
```

```
% Check if Cp1 is positive. If not Set Cp1=0 and re-design S2
if Cp1<0
    Cp1=0
    Cp1a=0
    Attention='Cp1 is negative. Therefore S2 is re-designed
    Coff2=1/w0/(eta-XaB)'
    Coff2=1/w0/(eta-XaB)
    Coff2a=Coff2/2/pi/f0a/Ra
    Ron2a=672e-15/Coff2a
    Ron2=Ron2a/Ra
end
% Step 8: Compute the realizable value of the cross-arm capacitor Cp2
Cp2=muA-Coff4
if Cp2<0
    Cp2=0
    Cp2a=0
    Attention='Cp2 is negative. Therefore S4 is re-designed Coff4=muA'
    Coff4_max=muA
    Coff4=input('Enter new normalized value for Coff4=')
    Coff4a=Coff4/2/pi/f0a/Ra
Ron4a=672e-15/Coff4a, Ron4=Ron4a/Ra
end
% -------------------------------------------------------------------
% Step 9: Electric Performance Analysis
w1=0; w2=2;N=10001; dw=(w2-w1)/(N-1);
w=w1;
for i=1:N
    W(i)=w;
    Fa(i)=w*f0a;
[ FI21A,GainA,VSWRA ] = SSS_DPS_State_A(w,Ron2,Ron3,Lp1,Lp2,Cp1,Cp2,
    Coff1,Coff4);
[ FI21B,GainB,VSWRB ] = SSS_DPS_State_B(w,Ron1,Ron4,Lp1,Lp2,Cp1,Cp2,
    Coff2,Coff3);
Phase_A(i)=FI21A;
Phase_B(i)=FI21B;
GainA_dB(i)=GainA;
GainB_dB(i)=GainB;
Phase_Shift(i)=FI21B-FI21A;
w=w+dw;
end
% -------------------------------------------------------------------
Plot_3S_DPS(W,Phase_A,Phase_B,Phase_Shift,GainA_dB,GainB_dB)
figure
plot(Fa,Phase_Shift)
title('DEL-FI=45 at F=8 GHz')
xlabel('Actual Frequencies')
ylabel('Phase-Shift=FI21B-FI21A')
legend('Del-FI=45 Degree')
% -------------------------------------------------------------------
figure
plot(Fa,GainB_dB,Fa,GainA_dB)
title('DEL-FI=45 at F=8 GHz')
legend('GainB','GainA')
```

```
xlabel('Actual Frequencies')
ylabel('GainB and GainA')
% ----------------------------------------------------------------

NCV=[Lp1 Cp1 Lp2 Cp2]
[ Lp1a ]  = Actual_Inductor(Lp1,Ra,f0a);
[ Cp1a ]  = Actual_Capacitor(Cp1,Ra,f0a);
[ Lp2a ]  = Actual_Inductor(Lp2,Ra,f0a);
[ Cp2a ]  = Actual_Capacitor(Cp2,Ra,f0a);
L_3SDPS=[Lp1a Lp2a]
C_3SDPS=[Cp1a Cp2a]
% ----------------------------------------------------------------
end
% ----------------------------------------------------------------
```

References

[1] E. Atasoy, Ercan, F. Piri and B. S. Yarman, "Symmetric lattice 45°, 90° and 180° digital phase shifter at 3–6 GHz for LTE, WIFI, Radar applications," in *IEEE International Symposium on Signals, Circuits and Systems (ISSCS), 2015*, Iashi, Romania, 2015.

[2] E. Atasoy, F. Piri and B. S. Yarman, "3-Bit phase shifter at 3–6GHz band for WiFi, LTE and 5G applications," in *National Conference on Electrical, Electronics and Biomedical Engineering (IEEE-ELECO)*, Bursa, Turkey, December 2016.

[3] C. Avci, E. O. Gunes and B. S. Yarman, "A novel Broadband-Wide Phase Range digital phase shifter topology," in *IEEE International Symposium on Signals, Circuits and Systems (IEEE-ISSCS)*, Iasi, Romania, 2017.

[4] C. Avci, E. O. Gunes and B. S. Yarman, "Design and Implementation of a Novel and Compact 2-Bit Wide Band Digital Phase Shifter," in *IEEE 18th Mediterranean Microwave Symposium (MMS)*, Marsilia, France, October 2018.

[5] C. Avci, E. O. Gunes and B. S. Yarman, "Design of 0–15GHz band 180° digital phase shifting cell topology," in *24th IEEE International Conference on Electronics, Circuits and Systems (ICECS)*, Batumi, Georgia, December 2017.

[6] C. Avci, E. O. Gunes and B. S. Yarman, "Design of wideband 180° digital phase shifter in CMOS process," in *National Conference on Electrical, Electronics and Biomedical Engineering (ELECO)*, Bursa, Turkey, 2016.

[7] TSMC, "Taiwanese Semiconductor Manufacturing Company", (https://www.tsmc.com/english/aboutTSMC/company_profile.htm), Taiwan, 2019.

[8] I. Bahl, Lumped Elements for RF Microwave Circuits, Artech House, 2003.

[9] I. Bahl and P. Bhartia, Microwave Solide State Circuit Design, Hoboken, N.J., USA: Wiley Interscience, 2003.

[10] K. Chang, Handbook of RF Microwave Components and Engineering, New York, USA, Chechester, UK: Wiley, 2003.

[11] F. W. Grover, Inductance Calculations, New York: D. Van Nostrand Company, 1946.

[12] S. Lucyszyn and L. Roberson, "Synthesis Techniques for High Performance Octave Bandwidth 180 degree Analog Phase Shifters," *IEEE Trans on MTT*, Vol. 40, No. 4, pp. 731–740, 1992.

[13] P. Yue and S. S. Wong,, "Physical modeling of spiral inductors on silicon," *IEEE Transactions on Electron Devices*, vol. 47, no. 3, March 2000.

[14] S. S. Mohan, *The Design, Modeling and Optimization Of On-Chip Inductor and Transformer Circuit*, Palo Alto, CA, USA: Doctoral dissertation, Stanford University, 1999.

[15] H. M. Greenhouse, "Design of planar rectangular microelectronic inductors," *IEEE Transactions on Parts, Hybrids, and Packaging*, vol. 10, pp. 101–109, 1974.

[16] MatLab, "MatLab S/W Package," The MathWorks Inc., Natick, Massachusetts, USA, 2018.

10

360° T-Section Digital Phase Shifter

Summary

In Chapters 6–8, we introduced 180° low-pass or high-pass-based, LC-T or LC-π-type symmetrical digital phase shifters, employing solid-state MOS or PIN diodes. Those types of phase shifters can provide phase shifts up to 90° with reasonable element values. Therefore, they are easy to implement for narrow phase range and narrow bandwidth applications. If we go beyond 90° bandwidth, significantly reduces. Moreover, it is not possible to obtain exact 180° phase shift with the design concepts introduced in Chapters 6–8.

The lattice digital phase shifter topology (3S-DPS) presented in Chapter 9 can provide exact 180° phase shift without any problem. It even goes up to 360° over broadband. On the other hand, lattice structure introduced in Chapter 9 includes eight solid-state switches, which may make the physical implementation difficult.

In this chapter, we introduce a different design concept to construct digital phase shifters to cover 360° phase shift range employing a T-Section-based digital phase shifter topology introduced in Chapter 6. This structure is much simpler than that of lattice digital phase shifter. It only includes three switches. However, it provides much less bandwidth than that of lattice digital phase shifter.

It should be mentioned that the content of this chapter is based on our previous work presented in the classical literature [1–3].

10.1 Derivation of Design Equations for a 360° T-Section Digital Phase Shifter

In Figure 10.1(a), a T-Section-based digital phase shifter topology is depicted. This topology is the same as the one shown in Figure 6.3a of Chapter 6. However, the design concept is different as follows.

435

In State-A, all the switching diodes are forward-biased and the phase shifter topology acts like a low-pass-based LC T-Section as shown in Figure 10.1(b). Hence, at a specified operating frequency w_0 and a phase shift $\theta_A = |\varphi_{21A}(w_0)| > 0$, the series arm inductor is given as in (6.2a) such that

$$L_L = \left(\frac{1}{w_0}\right) \tan\left(\frac{\theta_A}{2}\right) > 0 \qquad (10.1a)$$

In the shunt arm, the capacitor C_L is given as in (6.2b);

$$C_L = \frac{2L_L}{1 + w_0^2 L_L^2} > 0 \qquad (10.1b)$$

Similarly, in State-B, all the diodes are reverse-biased. In this state, the phase shift is $\theta_B = \varphi_{21B}(w_0) > 0$ and the topology shown in Figure 10.1(a), imitates the operation of a high-pass LC T-Section as shown in Figure 10.1(c). In this mode of operation, the series arm capacitor C_H is given as in (8.1a) such that

$$C_H = \frac{1}{w_0} \tan\left(90 - \frac{\theta_B}{2}\right) > 0 \qquad (10.1c)$$

and the shunt arm inductor is determined as in (8.1b) as follows

$$L_H = \left(\frac{1}{w_0^2}\right)\left(\frac{1 + w_0^2 C_H^2}{2C}\right) > 0 \qquad (10.1d)$$

Now, let us develop the design equations considering State-A and State-B operations as follows

In State-A (i.e., diodes are forward-biased), the series arm impedance $z_a(jw)$ is given by

$$z_a(jw) = jwL_1 = jwL_L \qquad (10.2a)$$

or

$$L_1 = L_L = \left(\frac{1}{w_0}\right) \tan\left(\frac{|\varphi_{21A}(w_0)|}{2}\right) \qquad (10.2b)$$

In the shunt arm, the admittance $y_a(jw)$ is

$$y_a(jw_0) = jwC_A + \frac{1}{jwL_A} = jwC_A\left(1 - \frac{1}{w^2 L_A C_A}\right)$$

$$= jwC_A\left(\frac{w^2 L_A C_A - 1}{w^2 L_A C_A}\right) \qquad (10.2c)$$

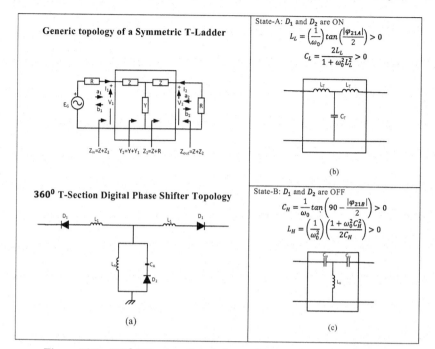

Figure 10.1   360° T-Section digital phase shifter and its switching states.

Then, at ω_0, equivalent shunt arm capacitor C_L of Figure 10.1(b) is given by

$$C_L = \left(\frac{\omega_0^2 L_A C_A - 1}{\omega_0^2 L_A}\right)$$

or

$$\omega_0^2 L_A C_L = \omega_0^2 L_A C_A - 1$$

or

$$\omega_0^2 L_A C_L - \omega_0^2 L_A C_A = -1$$

or

$$\omega_0^2 \left(C_A - C_L\right) L_A = 1$$

$$L_A = \frac{1}{\omega_0^2 \left(C_A - C_L\right)} > 0 \qquad (10.2d)$$

Similar derivations can be completed for State-B as follows.
When all the diodes are reverse-biased (State-B), series arm impedance z_b is given by

$$z_b\left(j\omega\right) = j\omega L_1 + \frac{1}{j\omega C_{D1}} = \frac{1}{j\omega C_{D1}}\left(1 - \omega^2 L_1 C_{D1}\right) \tag{10.2e}$$

At ω_0, the equivalent high-pass T-Section capacitor C_H is simulated as

$$C_H = \frac{C_{D1}}{1 - \omega_0^2 L_1 C_{D1}}$$

or

$$C_H\left(1 - \omega_0^2 L_1 C_{D1}\right) = C_{D1}$$

or

$$\left(1 + \omega_0^2 L_1 C_H\right) C_{D1} = C_H$$

or

$$C_{D1} = \frac{C_H}{1 + \omega_0^2 L_1 C_H} \tag{10.3a}$$

Similarly, shunt admittance of State-B (i.e., all diodes are reverse biased) is given as

$$y_b\left(j\omega\right) = j\omega C_T + \frac{1}{j\omega L_A} = \frac{1}{j\omega L_A}\left(1 - \omega^2 L_A C_T\right) \tag{10.4b}$$

In this case, at ω_0, equivalent shunt inductor L_H is given by

$$L_H = \frac{L_A}{1 - \omega^2 L_A C_T}$$

or

$$L_H\left(1 - \omega^2 L_A C_T\right) = L_A$$

or

$$L_A = \frac{L_H}{1 + \omega_0^2 L_H C_T} \tag{10.3c}$$

Employing (10.2d) and (10.3c), we obtain

$$\frac{L_H}{1 + \omega_0^2 L_H C_T} = \frac{1}{\omega_0^2\left(C_A - C_L\right)}$$

or

$$\left(\omega_0^2 L_H\right) C_A^2 - \left(\omega_0^2 L_H C_L + 1\right) C_A - C_{D2}\left(1 + \omega_0^2 L_H C_L\right) \tag{10.3d}$$

Let

$$A = \omega_0^2 L_H \tag{10.4e}$$

$$B = -\left(\omega_0^2 L_H C_L + 1\right) \tag{10.4f}$$

$$C = -\left(1 + \omega_0^2 L_H C_L\right) C_{D2} = B C_{D2} \tag{10.4g}$$

$$\triangle = B^2 - 4AC = \left(\omega_0^2 L_H C_L + 1\right)^2$$
$$+ 4\left(\omega_0^2 L_H C_L + 1\right) C_{D2} > \left(\omega_0^2 L_H C_L + 1\right) > 0 \tag{10.4h}$$

Then,

$$C_A = \frac{-B + \sqrt{B^2 - 4AC}}{2A} \tag{10.5a}$$

$$C_A = \frac{\left(\omega_0^2 L_H C_L + 1\right) + \sqrt{\left(\omega_0^2 L_H C_L + 1\right)^2 + 4\omega_0^2 L_H C_{D2}\left(\omega_0^2 L_H C_L + 1\right)}}{2\omega_0^2 L_H}$$
$$\tag{10.5b}$$

Based on the above derivation, we can suggest the following design algorithm to construct a 360° T-section digital phase shifter

10.2 Algorithm to Design $360°$ T-Section Digital Phase Shifter

In this section, we propose a design algorithm to construct 360° T-Section Digital Phase Shifter, in short 360°T-DPS, with ideal elements (i.e., lossless components).

Inputs:

θ_A: Desired phase shift at the operating frequency in State-A
θ_B: Desired phase shift at the operating frequency in State-B

It is noted that the designer is free to select phase shifts at each switching states. There is no restriction imposed on the phase shifts over $0 < \{\theta_A \text{ or } \theta_B\} < 180°$.

Remarks:

(a) For State-A, phase $\varphi_{21A}(\omega)$ of the transfer scattering parameter $S_{21A}(j\omega)$ is negative.
(b) For State-B, phase $\varphi_{21B}(\omega)$ of the transfer scattering parameter $S_{21B}(j\omega)$ is positive.
(c) Therefore, the net phase shift between the states is $\triangle\theta = \varphi_{21B}(\omega_0) - \varphi_{21A}(\omega_0) = \theta_B + \theta_A$.

(d) For example, $\triangle\theta = 180°$ is obtained by selecting θ_A between $0°$ and $90°$. Then, θ_B must be selected as $\theta_B = \triangle\theta - \theta_A = 180 - \theta_A$, where $0° < \theta_A \leq 90°$.

f_{0a}: Actual operating frequency in Hz.

ω_0: Normalized angular frequency, which corresponds to f_{0a}.

Remark: For many practical situations, ω_0 is selected as $\omega_0 = 1$. However, the designer is free to pick any value for ω_0.

R: Actual termination resistors for the $360°$ T-Section DPS.

Remark: It is usually selected as $50\ \Omega$.

C_{D2a}: Actual capacitance of the shunt arm diode D_2.

Remark: Most of the applications, it may be preferable to select diodes D_1 and D_2 identical. On the other hand, C_{D1} is computed in Step 3 of this algorithm. Then, one may wish to set $C_{D2} = C_{D1}$ as in Step 6, and omit this input.

Computations Steps:

Step 1: Compute the ideal component values for the low-pass T-Section prototype as in (10.1a & b).

$$L_L = \left(\frac{1}{\omega_0}\right)\tan\left(\frac{\theta_A}{2}\right) > 0$$

$$C_L = \frac{2L_L}{1 + \omega_0^2 L_L^2} > 0$$

Step 2: Compute the ideal component values for the high-pass T-Section prototype as in (10.1c & d).

$$C_H = \frac{1}{\omega_0}\tan\left(90 - \frac{\theta_B}{2}\right) > 0$$

$$L_H = \left(\frac{1}{\omega_0^2}\right)\left(\frac{1 + \omega_0^2 C_H^2}{2C_H}\right) > 0$$

Step 3: Compute the ideal-normalized component values of the series arms.

$$L_1 = L_L = \left(\frac{1}{\omega_0}\right)\tan\left(\frac{\theta_A}{2}\right) > 0$$

$$C_{D1} = \frac{C_H}{1 + \omega_0^2 L_1 C_H}$$

Step 4: Compute the ideal-normalized component values of the shunt arm C_A and L_A and C_T.

$$A = \omega_0^2 L_H$$
$$B = -\left(\omega_0^2 L_H C_L + 1\right)$$
$$C = -\left(1 + \omega_0^2 L_H C_L\right) C_{D2} = B C_{D2}$$
$$\triangle = B^2 - 4AC > 0$$
$$C_A = \frac{-B + \sqrt{B^2 - 4AC}}{2A}$$
$$C_{D2} = C_{D1}$$
$$C_T = \frac{C_A C_{D2}}{C_A + C_{D2}}$$

Step 5: Generate State-A series arms impedance z_a, shunt arm admittance y_a and the scattering parameters.

$$z_a(j\omega) = j\omega L_1$$

$$y_a(j\omega) = j\omega C_A + \frac{1}{j\omega L_A}$$

$$S_{21A} = \frac{2}{z^2 y + 2zy + 2z + y + 2}; \quad \left\{ \begin{array}{c} z = z_a \\ y = y_a \end{array} \right\}$$

$$S_{11A} = \frac{\left(1 - z^2\right) y - 2z}{z^2 y + 2zy + 2z + y + 2}; \quad \left\{ \begin{array}{c} z = z_a \\ y = y_a \end{array} \right\}$$

Step 6: Generate series arms impedance z_b, shunt arm admittance y_b, and the scattering parameters of State-B.

$$z_b(j\omega) = j\omega L_1 + \frac{1}{j\omega C_{D1}}$$

$$y_a(j\omega) = j\omega C_A + \frac{1}{j\omega L_A}$$

$$S_{21B} = \frac{2}{z^2 y + 2zy + 2z + y + 2} \ where \ \left\{ \begin{array}{c} z = z_b \\ y = y_b \end{array} \right\}$$

$$S_{11B} = \frac{\left(1 - z^2\right) y - 2z}{z^2 y + 2zy + 2z + y + 2} \quad where \quad \left\{ \begin{array}{c} z = z_b \\ \\ y = y_b \end{array} \right\}$$

Step 7: Compute the actual element values and plot the performance the 360° T-Section DPS

$$L_{1a} = R \frac{L_1}{(2\pi f_{0a})}$$

$$L_{Aa} = R \frac{L_1}{(2\pi f_{0a})}$$

$$C_{D1a} = \frac{C_{D1}}{(2\pi f_{0a}) R}$$

$$C_{Aa} = \frac{C_A}{(2\pi f_{0a}) R}$$

Let us now, run an example to show the utilization of the above algorithm.

Example 10.1: Let $\theta_A = 90°$. Let $\theta_B = 90°$. Let $f_{0a} = 5\ GHz$. Let $\omega_0 = 1$. Let $R = 50\Omega$. Select $C_{D2} = C_{D1}$.

Employing the algorithm given in Section 10.2, design a 360° T-Section DPS for $\triangle\theta = 180°$ and plot the phase shifting performance.

Solution

Let us follow the computation steps of the T-DPS to construct the desired phase shifter.

Step 1: Ideal component values of Low-pass T-Section

$$L_L = \left(\frac{1}{\omega_0}\right) \tan\left(\frac{\theta_A}{2}\right) = 1$$

$$C_L = \frac{2L_L}{1 + \omega_0^2 L_L^2} = 1$$

Step 2: Ideal component values of High-pass T-Section

$$C_H = \frac{1}{\omega_0} \tan\left(90 - \frac{\theta_B}{2}\right) = 1$$

$$L_H = \left(\frac{1}{\omega_0^2}\right) \left(\frac{1 + \omega_0^2 C_H^2}{2C_H}\right) = 1$$

Step 3: Ideal value of series arm inductor L_1 and reverse bias capacitor C_{D1}

$$L_1 = L_L = 1$$

$$C_{D1} = \frac{C_H}{1 + \omega_0^2 L_1 C_H} = 0.5$$

Step 4: Normalized component values of the shunt arm C_A and L_A and C_T

The above-mentioned components are computed employing the MatLab function

$$[CD1, L1, LA, CA, CT, CAa, LAa, CD1a, L1a, CTa]$$
$$= T360\_DPS(Teta\_A, Teta\_B, f0a, w0, R)$$

Such that

$$C_A = 2.4142$$

$$L_A = 0.7071$$

$$C_T = 0.4142$$

Step 5–7: These steps are completed under the MatLab program "*Main_Example_10_1.m*".

Actual element component values are found as follows:

$$L_{1a} = 1.5915e - 09 \cong 1.6\,nH$$
$$C_{D1a} = 3.1831e - 13 \cong 0.32\,pF$$
$$C_{Aa} = 1.5369e - 12 \cong 1.54\,pF$$
$$L_{Aa} = 1.1254e - 09 \cong 1.125\,nH$$

Phase performance of the 360° T-Section DPS is depicted in Figure 10.2a.

A close examination of Figure 10.2a reveals that 360° T-DPS yields 10% phase perturbation ($180° \mp 18°$) bandwidth of $\triangle\omega = \omega_2 - \omega_1$ such that $\omega_1 = 0.852$ and $\omega_2 = 1.463$.

Gain performance of the phase shifter configuration under consideration is shown in Figure 10.2b.

It is interesting to observe that $-3dB$ (half-power) frequency bandwidth runs from $\omega_1 = 0.803$ up to $\omega_2 = 1.22$ (i.e., $\triangle\omega_{Gain} = 0.4170$). In other words, half-power bandwidth is smaller than that of 10% phase perturbation bandwidth ($\triangle\omega_{Phase} = 1.463 - 0.852 = 0.611$.

In the above example, the phase shift between the states is equally distributed. However, phase range and the gain (or equivalently loss) may

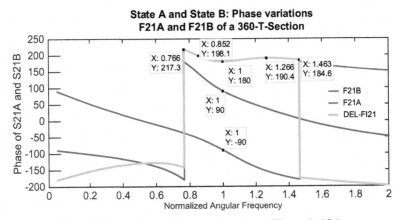

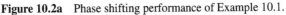

Figure 10.2a   Phase shifting performance of Example 10.1.

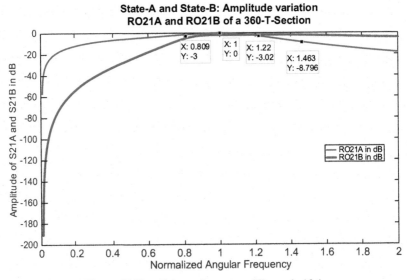

Figure 10.2b   Gain performance of Example 10.1.

be manually optimized by distributing the phase shifts unequally between the states. In the meantime, we may as well play with the operating frequency ω_0 to optimize the phase shift performance of the topology under consideration. In the following example, the phase shift between the states is unevenly distributed to examine the effect of the

10.3 Unequal Distribution of Distribution of the Phase Shift between the States

For a $360°$ T-Section – DPS, design formals are developed in such a way that the designer can distribute the phase shift unevenly between the states. Furthermore, center frequency of operation (i.e., ω_0 may be selected arbitrarily beyond unity as desired. In the following example, we will demonstrate the effect of uneven phase distribution to the performance of the 360 T-Section DPS.

Example 10.2: Let $\omega_0 = 1.5$, let $\theta_A = 120°$ and $\theta_B = 60$. Design a $360°$ T-Section DPS for $f_{0a} = 5e9$ and $R = 50\Omega$.

Answer: Our MatLab program developed for Example 10.1 can be run for the above inputs. Eventually, we end up with the desired DPS design as follows.

In short, normalized component values of the phase shifter are found as

$$L_1 = 1.1547 \text{ and } L_{1a} = 1.8378e - 09 \cong 1.84 \, nH$$
$$C_{D1} = 0.2887 \text{ and } C_{D1a} = 1.8378e - 13 \cong 0.184 \, nF$$
$$L_A = 0.5443 \text{ and } L_{Aa} = 8.6633e - 10 \cong 0.87 \, nH$$
$$C_A = 1.3938 \text{ and } C_{Aa} = 8.8735e - 13 \cong 0.887 \, nF$$
$$C_{D2} = C_{D1} = 0.2887 \text{ and } C_{D2} \cong 0.184 \, nF$$

Phase shift performance of the DPS under consideration is depicted in Figure 10.3a.

10% phase fluctuations of Figure 10.3a occur over the bandwidth of $\triangle\omega = 1.725 - 1.202 = 0.5230$, which is almost half octave.

Gain performance of the uneven phase shift DPS is depicted in Figure 10.3b. It is found that -3 dB bandwidth of Example 10.2 yields $\triangle\omega = 1.675 - 1.257 = 0.4180$, which is narrower than $\triangle\omega = 0.532$.

When these results are compared with those obtained in Example 10.1, they are close enough to each other. Therefore, we conclude that $360°$ T-Section DPS topology does not improve the phase performance if the phase between the states are unevenly distributed.

In the following section, let us investigate the effect of the lossy component on the phase shift performance of the $360°$ T-Section DPS.

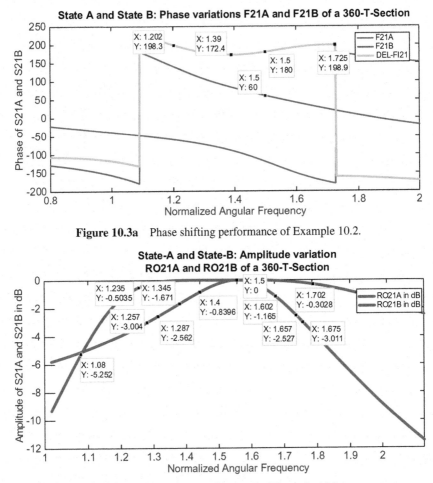

Figure 10.3a   Phase shifting performance of Example 10.2.

Figure 10.3b   Gain performance of Example 10.2.

10.4 Analysis of the Phase Performance of the $360°$ T-Section DPS Topology with Lossy Components

Referring to Figure 10.1, for practical phase shifter designs, one must include the resistive losses of the circuit components, as was introduced in Chapters 6–9.

In this regard, resistive loss r_L in series with an inductor L is approximated as

$$r_L = \frac{\omega_0 L}{Q_L}$$

where Q_L is the quality factor of the inductor.

Similarly, conductive loss G_C associated with a capacitor C is given by

$$G_C = \frac{\omega_0 C}{Q_C}$$

where Q_C is the quality factor of the capacitor C.

For MMIC implementations, Q_L and Q_C vary between 4 and 10. On the other hand, quality factor of discrete components are much higher; it varies between 20 and 100. Moreover, value of the reverse bias MOS capacitor C_{off} depends on the forward-biased channel resistance R_{on} as introduced in previous chapters such that

$$R_{on} C_{off} = Constant$$

The value of $Constant$ depends on the production process of the switches. For example, for a MOS switch produced by TSMC using 180 nm, the technology yields[1]

$$Constant = 672 \times 10^{-15}$$

In this book, we presume that 360° T-DPS is manufactured employing the TSMC 180 Nano-meter technology with identical MOS diodes.

Based on the above discussions, we can generate the lossy impedances for the switching states as follows.

State-A:

In this state, all the diodes are forward-biased. They introduce forward-biased channel resistance R_{F1}. Its normalized value is given by $r_{f1} = r_{f2} = \frac{R_{F1}}{R}$. In this case, the series arm impedance of State-A is determined as

$$z_a(j\omega) = r_a + j\omega L_1$$

where

$$r_a = r_{L1} + r_{f1}$$

$$r_{L1} = \frac{\omega_0 L_1}{Q_{L1}}$$

$$L_1 = \left(\frac{1}{\omega_0}\right) \tan\left(\frac{\theta_A}{2}\right)$$

$$r_{f1} = r_f = \frac{672 \times 10^{-15}}{R \times C_{D1-actual}}$$

---

[1]https://www.tsmc.com/english/default.htm

$$C_{D1} = \frac{C_H}{1 + \omega_0^2 L_1 C_H}; \ C_{D1-actual} = \frac{C_{D1}}{(2\pi f_{0-actual}) \times R}; C_H$$

$$= \left(\frac{1}{\omega_0}\right) \tan\left(90 - \frac{\theta_B}{2}\right)$$

The shunt arm admittance y_a is specified by

$$y_a(j\omega) = \frac{1}{r_{LA} + j\omega L_A} + Y_{TA}$$

where

$$r_{LA} = \frac{\omega_0 L_A}{Q_{LA}}$$

$$Y_{TA} = \frac{1}{Z_{TA}}; \ Z_{TA} = r_{f2} + \frac{1}{G_{CA} + j\omega C_A}$$

$$G_{CA} = \frac{\omega_0 C_A}{Q_{CA}}$$

$$r_{f2} = r_{f1} = r_f = \frac{672 \times 10^{-15}}{R \times C_{D1-actual}}$$

State-B:

In State-B, all the diodes are reverse-biased. Therefore, in the shunt arm, the diode capacitance $C_{D2} = C_{D1}$ since D_1 and D_2 are identical.

In this case, the series arm impedance z_b is given by

$$z_b(j\omega) = (r_{L1} + j\omega L_1) + \frac{1}{(G_{CD1} + j\omega C_{D1})}$$

where

$$G_{CD1} = \frac{\omega_0 C_{D1}}{Q_{CD1}}$$

The shunt arm admittance is specified by

$$y_b(j\omega) = \frac{1}{(r_{LA} + j\omega L_A)} + Y_T(j\omega) = G_B(\omega) + jB(\omega)$$

where

$$Y_T = \frac{1}{Z_T}; Z_T = \frac{1}{G_{CA} + j\omega C_A} + \frac{1}{G_{CD1} + j\omega C_{D1}}$$

Hence, we can propose the following algorithm to design lossy practical 360° T-Section DPS.

10.5 Algorithm: Design of a Lossy $360°$ T-Section DPS

This algorithm designs and generates phase shifting performance of the lossy-$360°$ T-Section Digital Phase Shifter. Here, we presume that DPS is realized using MMIC on a silicon substrate with $0.18\,\mu$m technology of TSMC[2].

Inputs:

θ_A: Phase Shift of State-A
θ_B: Phase Shift of State-B
f_{0a}: Actual center frequency
ω_0: Normalized angular frequency
R: Normalization resistor
Q_L: Quality factor to determine the series loss resistance r_L of a chip inductor L
Q_C: Quality factor to determine the shunt loss conductance G_C of a chip capacitor C

Computational Steps:

Step 1: Compute the ideal element values of the $360°$ T-DPS using our MatLab function T360-DPS.

$$[CD1, L1, LA, CA, CT, CAa, LAa, CD1a, L1a, CTa]$$
$$= T360\_DPS(Teta\_A, Teta\_B, f0a, w0, R)$$

Step 1a: Compute the ideal element values of the 360 degree T-Section Digital Phase Shifter

$$L_L = \left(\frac{1}{\omega_0}\right) \tan\left(\frac{|\theta_A|}{2}\right)$$
$$C_L = \frac{2L_L}{1 + \omega_0^2 L_L^2}$$
$$L_1 = L_L$$

Step 1b: Compute the ideal element values of the high-pass T-Section

$$C_H = \frac{1}{\omega_0} \tan\left(90° - \frac{|\varphi_{21B}(\omega_0)|}{2}\right)$$
$$L_H = \left(\frac{1}{\omega_0^2}\right)\left(\frac{1 + \omega_0^2 L_H^2}{2C_H}\right)$$

---

[2]Taiwanese Semiconductor Manufacturing Company; (https://www.tsmc.com/english/aboutTSMC/company_profile.htm)

Step 1c: Compute C_{D1} and its actual value

$$C_{D1} = \frac{C_H}{1 + \omega_0^2 L_1 C_H}; \ C_{D1-actual} = \frac{C_{D1}}{(2\pi f_{0-actual}) \times R}$$

Step 1d: Compute the ideal-normalized component values of the shunt arm C_A and L_A

$$C_{D2} = C_{D1}$$
$$A = \omega_0^2 L_H$$
$$B = -\left(\omega_0^2 L_H C_L + 1\right)$$
$$C = -\left(1 + \omega_0^2 L_H C_L\right) C_{D2} = B C_{D2}$$
$$\triangle = B^2 - 4AC > 0$$
$$C_A = \frac{-B + \sqrt{B^2 - 4AC}}{2A}$$
$$L_A = \frac{1}{\omega_0^2 \left(C_A - C_L\right)}$$

Step 2: Compute the component losses for both series and shunt arms

Step 2a: Component losses for the series arm

$$r_{L1} = \frac{\omega_0 L_1}{Q_L}$$
$$r_{f1} = r_f = \frac{672 \times 10^{-15}}{R \times C_{D1-actual}}$$
$$r_a = r_{L1} + r_{f1}$$

Step 2b: Component losses for the shunt arm:

$$r_{f2} = r_{f1}$$
$$r_{LA} = \frac{\omega_0 L_A}{Q_L}$$
$$G_{CA} = \frac{\omega_0 C_A}{Q_C}$$
$$G_{CD1} = \frac{\omega_0 C_{D1}}{Q_C}$$

Step 3: Computation of the State-A immittances for both series and shunt arms

Step 3a: Compute the series arm impedance z_a for State-A

$$z_a \left(j\omega\right) = r_a + j\omega L_1$$

Step 3b: Compute the shunt arm admittance y_a for State-A

$$Z_{TA} = r_{f2} + \frac{1}{G_{CA} + j\omega C_A}$$

$$Y_{TA} = \frac{1}{Z_{TA}}$$

$$y_a(j\omega) = \frac{1}{r_{LA} + j\omega L_A} + Y_{TA}$$

Step 4: Computation of the State-B immittances for both series and shunt arms

Step 4a: Computation of series arm impedance in State-B

$$z_b(j\omega) = (r_{L1} + j\omega L_1) + \frac{1}{(G_{CD1} + j\omega C_{D1})}$$

Step 4b: Computation of shunt arm admittance in State-B

$$Z_{TB} = \frac{1}{G_{CA} + j\omega C_A} + \frac{1}{G_{CD1} + j\omega C_{D1}}$$

$$Y_{TB} = \frac{1}{Z_{TB}}$$

$$y_b(j\omega) = \frac{1}{(r_{LA} + j\omega L_A)} + Y_{TB}(j\omega)$$

Step 5: Generate phase and gain performance of LPT-DPS for State-A and State-B

$$S_{21(A,B)} = \frac{2}{z^2 y + 2zy + 2z + y + 2}; \quad \left\{ \begin{array}{l} \text{In State A: } z = z_a, \ y = y_a \\ \text{In State B: } z = z_b, \ y = y_b \end{array} \right\}$$

Step 6: Plot the phase and gain performance

Now, let us run an example to evaluate the performance of a 360° T-DPS constructed with lossy elements.

Example 10.3: Design a 360° T-Section Digital Phase Shifter for the phase range of $\Delta\theta = 180°$. In the course of computations, assume that quality factor for inductors and capacitors are equal and they are given by $Q_L = Q_C = 10$ at $f_{0a} = 5 \ GHz$. We further assume that desired phase shift between the states is evenly distributed at the normalized angular frequency $\omega_0 = 1$. Termination resistor $R = 50 \ \Omega$ selected.

Answer: For this purpose, Algorithm 10.5. is programmed under MatLab program Example 10.3. Inputs to this program are defined as follows.

Inputs:

$$\theta_A = 90°; \theta_B = 90°; f_{0a} = 5 \times 10^9 \ Hz; \omega_0 = 1;$$
$$R = 50 \ \Omega; Q_L = 10; Q_C = 10.$$

Computational Steps:

Step 1: Call function T360_DPS to compute lossless component values of the 360° T-Section digital phase shifter.
    Hence, we end-up with

$$CD1 = 0.5000; L1 = 1; LA = 0.7071; CA = 2.4142;$$

$$CAa = 1.5369e - 12 = 1.54 \ pF; LAa = 1.1254e - 09 = 1.125 \ nH;$$

$$CD1a = 3.1831e - 13 = 0.32 \ pF; L1a = 1.5915e - 09 = 1.6 \ nH.$$

Step 2: Compute the component losses for both series and shunt arms.

Step 2a: Component losses for the series arms as they are normalized.

$$r_{L1} = \frac{\omega_0 L_1}{Q_L} = 0.1$$

$$r_{f1} = r_f = \frac{672 \times 10^{-15}}{R \times C_{D1-actual}} = 0.0422$$

$$r_a = r_{L1} + r_{f1} = 0.1422$$

Step 2b: Component losses for the shunt arm:

$$r_{f2} = r_{f1} = 0.0422$$

$$r_{LA} = \frac{\omega_0 L_A}{Q_L} = 0.0707$$

$$G_{CA} = \frac{\omega_0 C_A}{Q_C} = 0.2414$$

$$G_{CD1} = \frac{\omega_0 C_{D1}}{Q_C} = 0.0500$$

Step 3–6 is programmed under Main_Example_10_3.m.
    Phase shifting performance of the lossy phase shifter is shown in Figure 10.4.

A close examination of Figure 104a reveals that 10% phase range band-width run from $\omega = 0.6$ to $\omega = 1.325$. It looks like that it is more than one octave.

Gain performance of the phase shifter is shown in Figure 10.4b. This figure reveals that gain in both switching states is below -3 dB. Detail loss analysis of the topology under consideration may be completed employing commercially available CAD packages.

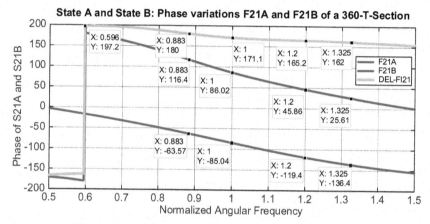

Figure 10.4a   Phase shifting performance of Example 10.3.

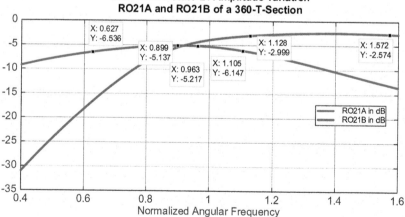

Figure 10.4b   Gain performance of Example 10.3.

10.6 Physical Implementation of 360° T-DPS

The 360° T-DPS may be built by using discrete component technology or MMIC technology. It is expected that discrete component technology offers much higher quality factors for the component values, which in turn improves the electric performance of the phase shifter under consideration. Diode switches may be selected as varactor diodes or PIN diodes. Switching components can be found in the literature such as [4–13] MMIC implementation can be carried out as described in Chapter 9. User may as well make use of the libraries of TSMC foundry of Taiwan [14].

Appendix 10: Program Lists for Chapter 10

Program List 10.1. Main_Example_10_1.m

```
% Main Program: Main_Example_10_1.m
% March 5, 2019
% Developed by BS Yarman, Vanikoy, Istanbul
% This program evaluates the lossy performance of a Highpass T-Section
DPS
% for a specified actual center frequency f0 in Hz.
%-----------------------------------------------------------------------
clc; close all
% Inputs:
Teta_A=input('Design of 360 Degree T Section DPS. Enter Phase of
State-A (Lowpass-T): Teta-A in Degree=')
Teta_B=input('Design of 360 Degree T Section DPS. Enter Phase of
    State-B(Highpass-T): Teta-B in Degree=')
% Inputs
f0a=input('Enter the actual center frequency in Hz f0a =')
w0=input('At f0a, enter the normalized angular frequency w0 =')
R=input('Enter the normaliziation Resistor R =')
%-----------------------------------------------------------------------
% Compute the normalized element values of T-360 Degree-DPS
LL=tand(abs(Teta_A)/2)/w0;
CL=(2*LL)/(1+w0*w0*LL*LL);
%-----------------------------------------------------------------------
%Ideal Highpass T-Section: See Equations (10.2a) and (10.2b)
CH=tand(90-Teta_B/2)/w0;
LH=(1+w0*w0*CH*CH)/2/CH/w0/w0;
[ CD1,L1,LA,CA,CT,CAa,LAa,CD1a,L1a,CTa ] = T360_DPS(Teta_A, Teta_B, f0a
    ,w0,R);
%-----------------------------------------------------------------------
j=sqrt(-1);
w=0;N=2000;w1=0;w2=5;DW=(w2-w1)/N;
FRI(1:(N+1))=zeros;
WA(1:(N+1))=zeros;
DEL_FI21(1:N+1)=zeros;
```

```
for i=1:N+1
WA(i)=w;
%  ------------------
% State A:
%  ------------------

za=j*w*L1;
ya=j*w*CA+1/j/w/LA;
[ S11a,S21a,RO11a,F11a,RO21a,F21a ] = S_Par_T_Section (za,ya);
    F21A(i)=F21a;
    RO21A(i)=20*log10(RO21a);
    F11A(i)=F11a;
    RO11A(i)=20*log10(RO11a);
%  ------------------
% State-B
%  ------------------
zb=j*w*L1+1/j/w/CD1;
yb=j*w*CT+1/j/w/LA;
   [ S11b,S21b,RO11b,F11b,RO21b,F21b ] = S_Par_T_Section (zb,yb);
   F21B(i)=F21b;
   RO21B(i)=20*log10(RO21b);
   F11B(i)=F11b;
   RO11B(i)=20*log10(RO11b);
%  ------------------
 DEL_FI21(i)=F21B(i)-F21A(i);
    w=w+DW;
end
%  ------------------
Plot_State_AB_T360_DPS(WA,F21A,RO21A, F11A,RO11A,F21B,RO21B, F11B,RO11B
    ,DEL_FI21)
```

Program List 10.2. Main_Example_10_2.m

```
% Main Program: Main_Example_10_2.m
% March 3, 2019
% Developed by BS Yarman, Vanikoy, Istanbul
% This program evaluates the lossy performance of a Highpass T-Section
DPS
% for a specified actual center frequency f0 in Hz.
%  ------------------
clc; close all
% Inputs:
Teta_A=input('Design of 360 Degree T Section DPS. Enter Phase of
State-A(Lowpass-T): Teta-A in Degree=')
Teta_B=input('Design of 360 Degree T Section DPS. Enter Phase of
State-B(Highpass-T): Teta-B in Degree=')
% Inputs
f0a=input('Enter the actual center frequency in Hz f0a =')
w0=input('At f0a, enter the normalized angular frequency w0 =')
R=input('Enter the normaliziation Resistor R =')
%  ------------------
% Compute the normalized element values of T-360 Degree-DPS
[ CD1,L1,LA,CA,CT,CAa,LAa,CD1a,L1a,CTa ] = T360_DPS(Teta_A, Teta_B, f0a
    ,w0,R)
```

```
% ------------------------------------------------------------
% ASSUMPTION 1:
% It is assumed that the series loss Ron of a forward biased diode is
% equal to the on channel resistor of an CMOS Switch
% [see Equation (6.1) of Chapter 6].
[ RF1,rf1 ] =Channel_Resistance_of_a_CMOS(CD1,R,f0a);
% ASSUMPTION 2:
% Reverse biased resistive loss of a diode is the "Percent_RVS" amount
of
% its reverse baised impedance at w0
CrDel_D=100;
rr1=1/w0/CD1/CrDel_D;
CD2=CD1;
[ RF2,rf2 ] =Channel_Resistance_of_a_CMOS(CD2,R,f0a);
rr2=1/w0/CD2/CrDel_D;
%
% ------------------------------------------------------------
% Loss Computations for both State-A and State-B:
% Assumption 3: Loss of an inductor is Percent_L amount of its impedance
% value at w0.
% Assumption 4: Connectivity loss of an inductor is "Percent_S"
amount of
% its impedance value at w0.
% Assumption 5: Conductive loss of a Capacitor is "Percent_C" amount
of its
% admittance value at w0.
CrDel_S=100;
CrDEL_L=10;
CrDel_C=100;
% ------------------------------------------------------------
% Resistive loss of the series arms in State-A:
rL1=w0*L1/CrDEL_L;
rLA=w0*LA/CrDEL_L;
rs=w0*L1/CrDel_S;
ra=rf1;
% ------------------------------------------------------------
% Conductive loss of the Shunt arm in State-A:
GCA=w0*CA/CrDel_C;
% GCA=0;GA=0;
GA=GCA+rLA/(rLA*rLA+w0*w0*LA*LA);
% ------------------------------------------------------------
% Resistive loss of the series arms in State-B:
rb=rr1;
% ------------------------------------------------------------
% Conductive loss of the Shunt arm in State-B:
% ------------------------------------------------------------
GB=GCA+rLA/(rLA*rLA+w0*w0*LA*LA)+(w0*w0*rr2*CT*CT)/(1+w0*w0*rr2*rr2*CT*
    CT);
% ------------------------------------------------------------
j=sqrt(-1);
w=0;N=2000;w1=0;w2=2;DW=(w2-w1)/N;
FRI(1:(N+1))=zeros;
WA(1:(N+1))=zeros;
```

```
DEL_FI21(1:N+1)=zeros;
for i=1:N+1
    WA(i)=w;
% -------------------
% State A:
% -------------------
%za=ra+j*w*L1;
za=ra+j*w*L1;
%ya=GA+j*w*CA+1/j/w/LA;
ya=GA+j*w*(CA/(1+w*w*rf2*rf2*CA*CA)-LA/(rLA*rLA+w*w*LA*LA));
[ S11a,S21a,RO11a,F11a,RO21a,F21a ] = S_Par_T_Section (za,ya);
    F21A(i)=F21a;
    RO21A(i)=20*log10(RO21a);
    F11A(i)=F11a;
    RO11A(i)=20*log10(RO11a);
% -------------------
% State-B
zb=rb+j*w*L1+1/j/w/CD1;
yb=GB+j*w*(CT/(1+w*w*rr2*rr2*CT*CT)-LA/(rLA*rLA+w*w*LA*LA));
[ S11b,S21b,RO11b,F11b,RO21b,F21b ] = S_Par_T_Section (zb,yb);
    F21B(i)=F21b;
    RO21B(i)=20*log10(RO21b);
    F11B(i)=F11b;
    RO11B(i)=20*log10(RO11b);
% -------------------
DEL_FI21(i)=F21B(i)-F21A(i);
    w=w+DW;
end
%
Plot_State_AB_T360_DPS(WA,F21A,RO21A, F11A,RO11A,F21B,RO21B, F11B,RO11B
    ,DEL_FI21)
%
ra_actual=ra*R
rb_actual=rb*R
%
GA_actual=GA/R; RA_actual=1/GA_actual
GB_actual=GB/R; RB_actual=1/GB_actual
```

Program List 10.3. Function T360_DPS

```
function [ CD1,L1,LA,CA,CT,CAa,LAa,CD1a,L1a,CTa ] = T360_DPS(Teta_A,
    Teta_B, f0a,w0,R)
% This function generates the element values of an ideal 360 degree
Simple
% T-Section based Digital Phase Shifter
% Inputs:
% Teta_A: Desired phase shift of the Lowpass Based T-Section DPS
% Teta_B: Desired phase shift of the Highpass Based T-Section DPS
% f0a: Actual centre frequency
% w0: Normalized angular frequency. It is selected as w0=1
% R: Port normalization number. It is usually, selected as R=50 ohms
% Outputs:
% CD1: Reverse Biased diode capacitance of the series arms.
```

```
% L1: Series arm inductor
% LA: Shunt arm inductor
%  ------------------------------------------------------
% Ideal Lowpass T-Section: See Equations (10.1a) and (10.1b)
LL=tand(abs(Teta_A)/2)/w0;
CL=(2*LL)/(1+w0*w0*LL*LL);
%  ------------------------------------------------------
% Ideal Highpass T-Section: See Equations (10.2a) and (10.2b)
CH=tand(90-Teta_B/2)/w0;
LH=(1+w0*w0*CH*CH)/2/CH;
%  ------------------------------------------------------
% State-A: Series arm component computations
L1=LL;
CD1=CH/(1+w0*w0*L1*CH);
%  ------------------------------------------------------
% State B: Computation of CA: See Equation (10.4) & (10.5)
A=w0*w0*LH;
B=-(w0*w0*LH*CL+1);
CD2=CD1;
C=B*CD2;
Delta=B*B-4*A*C;
CA=(-B+sqrt(Delta))/2/A;
%  ------------------------------------------------------
LA=1/(CA-CL)/w0/w0;
CT=CA*CD2/(CA+CD2);
%  ------------------------------------------------------
CAa=CA/2/pi/f0a/R;
CD1a=CD1/2/pi/f0a/R;
LAa=R*LA/2/pi/f0a;
L1a=R*L1/2/pi/f0a;
CTa=CT/2/pi/f0a/R;


end
```

Program List 10.4. Main_Example_10_3.m

```
% Main Program: Main_Example_10_3.m
% March 11, 2019
% Developed by BS Yarman, Vanikoy, Istanbul
% This program evaluates the lossy performance of a Highpass T-Section
DPS
% for a specified actual center frequency f0 in Hz.
%  ------------------------------------------------------
clc; close all
% Inputs:
Teta_A=input('Design of 360 Degree T Section DPS. Enter Phase of
State-A(Lowpass-T): Teta-A in Degree=')
Teta_B=input('Design of 360 Degree T Section DPS. Enter Phase of
State-B(Highpass-T): Teta-B in Degree=')
% Inputs
f0a=input('Enter the actual center frequency in Hz f0a =')
w0=input('At f0a, enter the normalized angular frequency w0 =')
```

```
R=input('Enter the normaliziation Resistor R =')
%
% Compute the normalized element values of T-360 Degree-DPS
[ CD1,L1,LA,CA,CT,CAa,LAa,CD1a,L1a,CTa ] = T360_DPS(Teta_A, Teta_B, f0a
    ,w0,R)
%
% ASSUMPTION 1:
% It is assumed that the series loss Ron of a forward biased diode is
% equal to the on channel resistor of an CMOS Switch
% [see Equation (6.1) of Chapter 6].
[ RF1,rf1 ] =Channel_Resistance_of_a_CMOS(CD1,R,f0a);
% ASSUMPTION 2:
% At w0, quality factor for inductors and capacitors
QL=20;
QC=20;

%
CD2=CD1;
[ RF,rf ] =Channel_Resistance_of_a_CMOS(CD1,R,f0a);
%
% Series Arm Losses
rf1=rf;
rL1=(w0*L1)/QL;
ra=rL1+rf1;
% Shunt arm losses
rf2=rf;
rLA=(w0*LA)/QL;
GCA=(w0*CA)/QC;
GCD1=(w0*CD1)/QC;

%
j=sqrt(-1);
w=0;N=2000;w1=0;w2=2;DW=(w2-w1)/N;
FRI(1:(N+1))=zeros;
WA(1:(N+1))=zeros;
DEL_FI21(1:N+1)=zeros;
for i=1:N+1
    WA(i)=w;
%  --------------------
% State A:
%  --------------------
%za=ra+j*w*L1;
za=ra+j*w*L1;
% Computation of shunt arm admittance ya
%  ZTA=r_f2+1/(GCA+jw*CA)
ZTA=rf+1/(GCA+j*w*CA);
YTA=1/ZTA;
ya=1/(rLA+j*w*LA)+YTA;
[ S11a,S21a,RO11a,F11a,RO21a,F21a ] = S_Par_T_Section (za,ya);
    F21A(i)=F21a;
    RO21A(i)=20*log10(RO21a);
    F11A(i)=F11a;
    RO11A(i)=20*log10(RO11a);
%  --------------------
% State-B
```

```
zb=(rL1+j*w*L1)+1/(GCD1+j*w*CD1);
ZTB=1/(GCA+j*w*CA)+1/(GCD1+j*w*CD1);
YTB=1/ZTB;
yb=1/(rLA+j*w*LA)+YTB;
   [ S11b,S21b,RO11b,F11b,RO21b,F21b ] = S_Par_T_Section (zb,yb);
      F21B(i)=F21b;
      RO21B(i)=20*log10(RO21b);
      F11B(i)=F11b;
      RO11B(i)=20*log10(RO11b);

DEL_FI21(i)=F21B(i)-F21A(i);
     w=w+DW;
end

Plot_State_AB_T360_DPS(WA,F21A,RO21A, F11A,RO11A,F21B,RO21B, F11B,RO11B
     ,DEL_FI21)

ra_actual=ra*R
rb_actual=rb*R

GA_actual=GA/R; RA_actual=1/GA_actual
GB_actual=GB/R; RB_actual=1/GB_actual
```

Program List 10.5. Function S_Par_T_Section

```
function [ S11,S21,RO11,F11,RO21,F21 ] = S_Par_T_Section (z,y)
%This function generates the S-Parameters of a T Section
% Phase Shifter from the series arm impedance Z(jw) and
% the shunt arm admittance Y(jw)
%
% Developed by BS Yarman: Feb 20, 2019, Vanikoy, Istanbul
% See Equations (5.9)
%
D=z*z*y+2*z*y+2*z+y+2;
S11=((1-z*z)*y-2*z)/D;
S21=2/D;
R11=real(S11); X11=imag(S11);F11=atan2d(X11,R11);
R21=real(S21); X21=imag(S21);F21=atan2d(X21,R21);
RO11=abs(S11);
RO21=abs(S21);
end
```

Program List 10.6. function Plot_State_AB_T360_DPS

```
function Plot_State_AB_T360_DPS(WA,F21A,RO21A, F11A,RO11A,F21B,RO21B,
     F11B,RO11B,DEL_FI21)
figure
plot(WA,F21A,WA,F21B,WA,DEL_FI21)
title('State A and State B: Phase variations F21A and F21B of a
360-T-Section')
legend('F21A','F21B','DEL-FI21')
xlabel('Normalized Angular Frequency')
ylabel('Phase of S21A and S21B')
```

```
% Amplitude of S21
figure
plot(WA,RO21A,WA,RO21B)
title('State-A and State-B: Amplitude variation RO21A and RO21B of a
360-T-Section')
legend('RO21A in dB','RO21B in dB')
xlabel('Normalized Angular Frequency')
ylabel('Amplitude of S21A and S21B in dB')
% -----------------------------------------------------------------
figure
plot(WA,F11A, WA, F11B)
title('State-A and State-B: Phase variation F11A and F11B of a
360-T-Section')
legend('F11A')
xlabel('Normalized Angular Frequency')
ylabel('Phase of S11A and S11B')
% Amplitude of S11
figure
plot(WA,RO11A, WA, RO11B)
title('State-A and State-B: Amplitude variation RO11A and RO11B of a
360-T-Section')
legend('RO11A in dB','RO11B in dB')
xlabel('Normalized Angular Frequency')
ylabel('Amplitude of S11A and S11B in dB')

end
```

References

[1] Binboga S. Yarman, "Low pass T-section digital phase shifter apparatus,," U.S. Patent: 4630010, Washington DC, USA, December 16, 1986.

[2] B. Yarman, A.Rosen and P.Stabile, "Lowloss EHF Digital Phase Shifters Suitable for Monolithic Implementation," in *IEEE ISCAS,* Montreal, 1984.

[3] B. S. Yarman, "Design of Digital Phase Shifters Suitable for Monolithic Implementations," *Bull. of Technical University of Istanbul,* pp. 185–205, 1985.

[4] Pat Hindle, Editor, "The State of RF and Microwave Switches," *Microwave Journal,* vol. 53, no. 11, p. 20, November 2010.

[5] I. Bahl and B. Prakash, Microwave Solid State Circuit Design, 2. (1. John Wiley & Sons, Ed., 2003.

[6] I. Bahl, Lumped Elements for RF Microwave Circuits, Artech House, 2003.

[7] K. J. Koh and G. M. Rebeiz, "0.13- m CMOS phase shifters for and K-band phased arrays," *IEEE Trans. on MTT,* Vol. 42, No. 11, p. 2535–2546, November 2007.

[8] K. Chang, Handbook of RF Microwave Components and Engineering, New York, USA, Chechester, UK: Wiley, 2003.

[9] D. Kang and S. Hong, "A 4-bit CMOS Phase Shifter Using Distributed Active Switches," *IEEE Trans on MTT,* Vol. 55, No. 7, pp. 1476–1483, 2007.

[10] R. Melik and H. V. Demir, "Implementation OF High Qualityfactor, On-Chip Tuned Microwave Resonators AT 7 GHz," Microwave and Optical Technology Letters, vol. 51, no. 2, pp. 497–501, February 2009 .

[11] H. Fang, T. Xinyi, K. Mouthaan and R. Guinvarch, "Two Octave Digital All-Pass Phase Shifters for Phase Array Applications," *IEEE Radio and Wireless Symposium (RWS),* pp. 169–171, 2013.

[12] L. Yun-Wei, H. Yi-Chieh and C. Chi-Yang, "A Balanced Digital Phase Shifter by a Novel Switching-Mode Topology," *IEEE Trans on MTT,* pp. 2361–2370, 2013.

[13] M. I. L. S.W. Amy, "A Simple Microstrip Phase Shifter," *Journal of Physics,* No. 45, pp. 105–114, 1992.

[14] TSMC, "Taiwanese Semiconductor Manufacturing Company;," (https://www.tsmc.com/english/aboutTSMC/company_profile.htm), Taiwan, 2019.

11

360° PI-Section Digital Phase Shifter

Summary

In this chapter, we extend the previous chapter to design 360° PI-Section digital phase shifter.

The content of this chapter is built on our previous work presented in the literature such as [1–3].

Derivation of design equations for a 360° PI-Section Digital Phase Shifter.

In Figure 11.1a, a PI-Section based digital phase shifter topology is depicted. This topology is as same as the one shown in Figure 7.1 of Chapter 7. However, design concept is different as follows.

In State-A, all the switching diodes are forward-biased and the phase shifter topology acts as a low-pass-based LC PI-Section as shown in Figure 11.1b. Hence, at a specified operating frequency w_0 and a phase shift $\theta_A = |\varphi_{21A}(w_0)| > 0$, the equivalent shunt arm capacitor C_L is given as in (7.1a) such that

$$C_L = \left(\frac{1}{w_0}\right) \tan\left(\frac{\theta_A}{2}\right) > 0 \qquad (11.1a)$$

In the series arm, the equivalent inductor L_L is given as in (7.2b);

$$L_L = L = \frac{2C_L}{1 + w_0^2 C_L^2} > 0 \qquad (11.1b)$$

Similarly, in State-B, all the diodes are reverse-biased. In this state, the phase shift is $\theta_B = \varphi_{21B}(w_0) > 0$ and the topology shown in Figure 11.1a, imitates the operation of a highpass LC PI-Section as shown in Figure 11.1c. In this mode of operation, the shunt arm inductor L_H is given as in (5.34a) such that

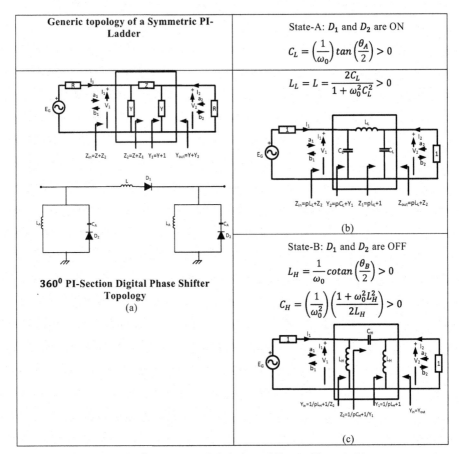

Generic topology of a Symmetric PI-Ladder	State-A: D_1 and D_2 are ON

State-A: D_1 and D_2 are ON

$$C_L = \left(\frac{1}{\omega_0}\right) tan\left(\frac{\theta_A}{2}\right) > 0$$

$$L_L = L = \frac{2C_L}{1 + \omega_0^2 C_L^2} > 0$$

$Z_{in}=Z+Z_2$ $Z_2=Z+Z_1$ $Y_1=Y+1$ $Y_{out}=Y+Y_2$

$Z_{in}=pL_L+Z_2$ $Y_2=pC_L+Y_1$ $Z_1=pL_L+1$ $Z_{out}=pL_L+Z_2$

(b)

360⁰ PI-Section Digital Phase Shifter Topology
(a)

State-B: D_1 and D_2 are OFF

$$L_H = \frac{1}{\omega_0} cotan\left(\frac{\theta_B}{2}\right) > 0$$

$$C_H = \left(\frac{1}{\omega_0^2}\right)\left(\frac{1 + \omega_0^2 L_H^2}{2L_H}\right) > 0$$

$Y_{in}=1/pL_H+1/Z_2$ $Y_1=1/pL_H+1$ $Y_{in}=Y_{out}$
$Z_2=1/pC_H+1/Y_1$

(c)

Figure 11.1   360° PI-Section digital phase shifter and its switching states.

$$L_H = \frac{1}{\omega_0} cotan\left(\frac{\theta_B}{2}\right) > 0 \tag{11.1c}$$

and the series arm capacitor is determined as in (5.34b) as follows

$$C_H = \left(\frac{1}{\omega_0^2}\right)\left(\frac{1 + \omega_0^2 L_H^2}{2L_H}\right) > 0 \tag{11.1d}$$

Now, let us develop the design equations considering State-A and State-B operations as follows:

In State-A (i.e., diodes are forward-biased), the series arm impedance $z_a(j\omega)$ is given by

$$z_a(j\omega) = j\omega L = j\omega L_L \tag{11.2a}$$

In the shunt arm, the admittance $y_a(j\omega)$ is

$$y_a(j\omega_0) = j\omega C_A + \frac{1}{j\omega L_A} = j\omega C_A \left(1 - \frac{1}{\omega^2 L_A C_A}\right) = j\omega C_A \left(\frac{\omega^2 L_A C_A - 1}{\omega^2 L_A C_A}\right)$$

(11.2b)

Then, at ω_0, equivalent shunt arm capacitor C_L of Figure 11.1b is given by

$$C_L = \left(\frac{\omega_0^2 L_A C_A - 1}{\omega_0^2 L_A}\right)$$

or

$$\omega_0^2 L_A C_L = \omega_0^2 L_A C_A - 1$$

or

$$\omega_0^2 L_A C_L - \omega_0^2 L_A C_A = -1$$

or

$$\omega_0^2 (C_A - C_L) L_A = 1$$

$$L_A = \frac{1}{\omega_0^2 (C_A - C_L)} > 0 \qquad (11.2c)$$

Similar derivations can be completed for State-B as follows.

When all the diodes are reverse-biased (State-B), series arm impedance z_b is given by

$$z_b(j\omega) = j\omega L + \frac{1}{j\omega C_{D1}} = \frac{1}{j\omega C_{D1}}\left(1 - \omega^2 L C_{D1}\right) \qquad (11.2d)$$

At ω_0, the equivalent high-pass PI-Section capacitor C_H is simulated as

$$C_H = \frac{C_{D1}}{1 - \omega_0^2 L C_{D1}}$$

or

$$C_H \left(1 - \omega_0^2 L_1 C_{D1}\right) = C_{D1}$$

or

$$\left(1 + \omega_0^2 L_1 C_H\right) C_{D1} = C_H$$

or

$$C_{D1} = \frac{C_H}{1 + \omega_0^2 L C_H} \qquad (11.3a)$$

Similarly, shunt admittance of State-B (i.e., all diodes are reverse biased) is given as

$$y_b\left(j\omega\right) = j\omega C_T + \frac{1}{j\omega L_A} = \frac{1}{j\omega L_A}\left(1 - \omega^2 L_A C_T\right) \tag{11.3b}$$

In this case, at ω_0, equivalent shunt inductor L_H is given by

$$L_H = \frac{L_A}{1 - \omega^2 L_A C_T}$$

or

$$L_H\left(1 - \omega^2 L_A C_T\right) = L_A$$

or

$$L_A = \frac{L_H}{1 + \omega_0^2 L_H C_T} \tag{11.3c}$$

Employing (11.2c) and (11.3d), we obtain

$$\frac{L_H}{1 + \omega_0^2 L_H C_T} = \frac{1}{\omega_0^2\left(C_A - C_L\right)}$$

or

$$\left(\omega_0^2 L_H\right) C_A^2 - \left(\omega_0^2 L_H C_L + 1\right) C_A - C_{D2}\left(1 + \omega_0^2 L_H C_L\right) \tag{11.3d}$$

Let

$$A = \omega_0^2 L_H \tag{11.3e}$$
$$B = -\left(\omega_0^2 L_H C_L + 1\right) \tag{11.3f}$$
$$C = -\left(1 + \omega_0^2 L_H C_L\right) C_{D2} = B C_{D2} \tag{11.3g}$$
$$\Delta = B^2 - 4AC = \left(\omega_0^2 L_H C_L + 1\right)^2 + 4\left(\omega_0^2 L_H C_L + 1\right)$$
$$C_{D2} > \left(\omega_0^2 L_H C_L + 1\right) > 0 \tag{11.3h}$$

Then,

$$C_A = \frac{-B + \sqrt{B^2 - AC}}{2A} \tag{11.4a}$$

$$C_A = \frac{\left(\omega_0^2 L_H C_L + 1\right) + \sqrt{\left(\omega_0^2 L_H C_L + 1\right)^2 + \omega_0^2 L_H C_{D2}\left(\omega_0^2 L_H C_L + 1\right)}}{2\omega_0^2 L_H} \tag{11.4b}$$

Based on the above derivation, we can suggest the following design algorithm to construct a 360° PI-Section digital phase shifter.

11.1 Algorithm to design 360° PI-Section Digital Phase Shifter

In this section, we propose a design algorithm to construct 360° PI-Section Digital Phase Shifter, in short 360° PI-DPS, with ideal elements (i.e., lossless components).

Inputs:

θ_A: Desired phase shift at the operating frequency in State-A
θ_B: Desired phase shift at the operating frequency in State-B

It is noted that the designer is free to select phase shifts at each switching states. There is no restriction is imposed on the phase shifts over $0 < \{\theta_A \text{ or } \theta_B\} < 180°$.

Remarks:

(a) For State-A, phase $\varphi_{21A}(\omega)$ of the transfer scattering parameter $S_{21A}(j\omega)$ is negative.
(b) For State-B, phase $\varphi_{21B}(\omega)$ of the transfer scattering parameter $S_{21B}(j\omega)$ is positive.
(c) Therefore, the net phase shift between the states is $\Delta\theta = \varphi_{21B}(\omega_0) - \varphi_{21A}(\omega_0) = \theta_B + \theta_A$.
(d) For example, $\Delta\theta = 180°$ is obtained by selecting θ_A between 0° and 90°. Then, θ_B must be selected as $\theta_B = \Delta\theta - \theta_A = 180$-$\theta_A$ where $0° < \theta_A \leq 90°$.

f_{0a}: Actual operating frequency in Hz.
ω_0: Normalized angular frequency, which corresponds to f_{0a}.

Remark: For many practical situations, ω_0 is selected as $\omega_0 = 1$. However, the designer is free to pick any value for ω_0.

R: Actual Termination resistors for the 360° PI-Section DPS.

Remark: It is usually selected as 50 Ω.

C_{D2a}: Actual capacitance of the shunt arm diode D_2.

Remark: Most of the applications, it may be preferable to select diodes D_1 and D_2 identical. On the other hand, C_{D1} is computed in Step 3 of this algorithm. Then, one may wish to set $C_{D2} = C_{D1}$ as in Step 6, and omit this input.

Computations Steps:

Step 1: Compute the ideal component values for the lowpass PI-Section prototype as in (11.1a & b).

$$C_L = \left(\frac{1}{\omega_0}\right) \tan\left(\frac{\theta_A}{2}\right) > 0$$

$$L_L = L = \frac{2C_L}{1 + \omega_0^2 C_L^2} > 0$$

Step 2: Compute the ideal component values for the high-pass PI-Section prototype as in (10.1c & d).

$$L_H = \left(\frac{1}{\omega_0}\right) \cotan\left(\frac{\theta_B}{2}\right) > 0$$

$$C_H = \left(\frac{1}{\omega_0^2}\right)\left(\frac{1 + \omega_0^2 L_H^2}{2L_H}\right) > 0$$

Step 3: Compute the ideal-normalized component values of the series arms.

$$L = L_L > 0$$

$$C_{D1} = \frac{C_H}{1 + \omega_0^2 L_1 C_H}$$

Step 4: Compute the ideal-normalized component values of the shunt arm C_A and L_A and C_T.

$$A = \omega_0^2 L_H$$

$$B = -(\omega_0^2 L_H C_L + 1)$$

$$C = -(1 + \omega_0^2 L_H C_L)C_{D2} = BC_{D2}$$

$$\Delta = B^2 - 4AC > 0$$

$$C_A = \frac{-B + \sqrt{B^2 - 4AC}}{2A}$$

$$C_{D2} = C_{D1}$$

$$C_T = \frac{C_A C_{D2}}{C_A + C_{D2}}$$

Step 5: Generate State-A series arms impedance z_a, shunt arm admittance y_a, and the scattering parameters.

$$z_a(j\omega) = j\omega L_1$$

$$y_a \left(j\omega \right) = j\omega C_A + \frac{1}{j\omega L_A}$$

$$S_{21A} = \frac{2}{zy^2 + 2zy + 2y + z + 2}; \quad \left\{ \begin{array}{c} z = z_a \\ y = y_a \end{array} \right\}$$

$$S_{11A} = \frac{z \left(1 - y^2 \right) - 2y}{zy^2 + 2zy + 2y + z + 2}; \quad \left\{ \begin{array}{c} z = z_a \\ y = y_a \end{array} \right\}$$

Step 6: Generate series arms impedance z_b, shunt arm admittance y_b, and the scattering parameters of State-B.

$$z_b \left(j\omega \right) = j\omega L_1 + \frac{1}{j\omega C_{D1}}$$

$$y_a \left(j\omega \right) = j\omega C_A + \frac{1}{j\omega L_A}$$

$$S_{21B} = \frac{2}{zy^2 + 2zy + 2y + z + 2} \text{ where } \left\{ \begin{array}{c} z = z_b \\ y = y_b \end{array} \right\}$$

$$S_{11B} = \frac{z \left(1 - y^2 \right) - 2y}{zy^2 + 2zy + 2y + z + 2} \text{ where } \left\{ \begin{array}{c} z = z_b \\ y = y_b \end{array} \right\}$$

Step 7: Compute the actual element values and plot the performance the $360°$ PI-Section DPS

$$L_a = R \frac{L_1}{\left(2\pi f_{0a} \right)}$$

$$L_{Aa} = R \frac{L_1}{\left(2\pi f_{0a} \right)}$$

$$C_{D1a} = \frac{C_{D1}}{\left(2\pi f_{0a} \right) R}$$

$$C_{Aa} = \frac{C_A}{\left(2\pi f_{0a} \right) R}$$

Let us now, run an example to show the utilization of the above algorithm.

Example 11.1: Let $\theta_A = 90°$. Let $\theta_B = 90°$. Let $f_{0a} = 5\,GHz$. Let $\omega_0 = 1$. Let $R = 50\,\Omega$. Select $C_{D2} = C_{D1}$.

Employing the algorithm given in Section 11.2, design a $360°$ PI-Section DPS for $\Delta\theta = 180°$ and plot the phase shifting performance.

Solution

Let us follow the computation steps of the PI-DPS to construct the desired phase shifter.

Step 1: Ideal component values of low-pass PI-Section

$$C_L = \left(\frac{1}{\omega_0}\right) \tan\left(\frac{\theta_A}{2}\right) = 1$$

$$L_L = L = \frac{2C_L}{1 + \omega_0^2 C_L^2} = 1$$

Step 2: Ideal component values of high-pass T-Section

$$L_H = \left(\frac{1}{\omega_0}\right) \cotan\left(\frac{\theta_B}{2}\right) = 1$$

$$C_H = \left(\frac{1}{\omega_0^2}\right)\left(\frac{1 + \omega_0^2 L_H^2}{2L_H}\right) = 1$$

Step 3: Ideal value of series arm inductor L_1 and reverse bias capacitor C_{D1}

$$L = L_L = 1$$

$$C_{D1} = \frac{C_H}{1 + \omega_0^2 L_1 C_H} = 0.5$$

Step 4: Normalized component values of the shunt arm C_A and L_A and C_T

The above-mentioned components are computed employing the MatLab function

$$[CD1, L1, LA, CA, CT, CAa, LAa, CD1a, L1a, CTa]$$
$$= PI\_360\_DPS(Teta\_A, Teta\_B, f0a, w0, R)$$

Such that

$$C_A = 2.4142$$
$$L_A = 0.7071$$
$$C_T = 0.4142$$

Step 5–7: These steps are completed under the MatLab program "$Main\_Example\_11\_1.m$".

Actual element component values are found as follows:

$$L_a = 1.5915e - 09 \cong 1.6\,nH$$
$$C_{D1a} = 3.1831e - 13 \cong 0.32\,pF$$
$$C_{Aa} = 1.5369e - 12 \cong 1.54\,pF$$
$$L_{Aa} = 1.1254e - 09 \cong 1.125\,nH$$

Phase performance of the 360° PI-Section DPS is depicted in Figure 11.2a.

A close examination of Figure 11.2a reveals that 360° PI-DPS yields 10% phase perturbation ($180° \mp 18°$) bandwidth of $\Delta\omega = \omega_2 - \omega_1$ such that $\omega_1 = 0.735$ and $\omega_2 = 1.172$.

Gain performance of the phase shifter configuration under consideration is shown in Figure 11.2b.

It is interesting to observe that -3dB (half-power) frequency bandwidth runs from $\omega_1 = 0.8175$ up to $\omega_2 = 1.25$ (i.e., $\Delta\omega_{Gain} = 0.4170$). In other words, half-power bandwidth is smaller than that of 10% phase perturbation bandwidth ($\Delta\omega_{Phase} = 1.25 - 0.817 = 0.43$).

In the above example, the phase shift between the states is equally distributed. However, phase range and the gain (or equivalently loss) may be manually optimized by distributing the phase shifts unevenly between the states. In the meantime, we may as well play with the operating frequency ω_0 to optimize the phase shift performance of the topology under consideration.

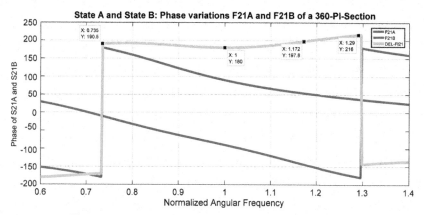

Figure 11.2a Phase shifting performance of Example 11.1.

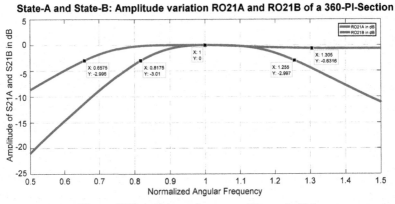

Figure 11.2b   Gain performance of Example 10.1.

In the following example, the phase shift between the states is unevenly distributed to examine the effect of the

11.2 Unequal Distribution of the Phase Shifts between the States

For a 360° PI-Section − DPS, design formals are developed in such a way that the designer can distribute the phase shift unequally between the states. Furthermore, center frequency of operation (i.e., ω_0 may be selected arbitrarily beyond unity as desired). In the following example, we will demonstrate the effect of unequal phase distribution to the performance of the 360° PI-Section DPS.

Example 11.2: Let $\omega_0 = 1.5$, let $\theta_A = 120°$ and $\theta_B = 60$. Design a 360° PI-Section DPS for $f_{0a} = 5e9$ and $R = 50\Omega$.

Solution

Our MatLab program developed for Example 11.1 can be run for the above inputs. Eventually, we end up with the desired DPS design as follows.

In short, normalized component values of the phase shifter is found as

$$L = 1.1547 \text{ and } L_{1a} = 1.8378e - 09 \cong 1.84\,nH$$
$$C_{D1} = 0.2887 \text{ and } C_{D1a} = 1.8378e - 13 \cong 0.184\,nF$$
$$L_A = 0.5443 \text{ and } L_{Aa} = 8.6633e - 10 \cong 0.87\,nH$$

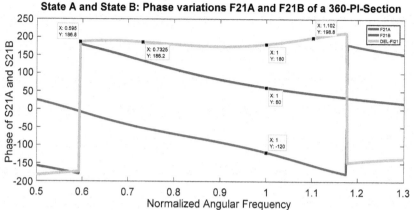

Figure 11.3a Phase shifting performance of Example 11.2.

$$C_A = 1.3938 \text{ and } C_{Aa} = 8.8735e - 13 \cong 0.887 \ nF$$
$$C_{D2} = C_{D1} = 0.2887 \text{ and } C_{D2} \cong 0.184 \ nF$$

Phase shift performance of the DPS under consideration is depicted in Figure 11.3a.

10% phase fluctuations of Figure 11.3a occur over the bandwidth of $\Delta\omega = 1.102 - 0.595 = 0.507$, which is almost half octave.

Gain performance of the uneven phase shift DPS is depicted in Figure 11.3b. It is found that $-3dB$ bandwidth of Example 11.2 yields $\Delta\omega = 1.17 - 0.7 = 0.4680$, which is narrower than $\Delta\omega = 0.507$.

When these results are compared with those obtained in Example 11.1, they are close enough to each other. Therefore, we conclude that 360° PI-Section DPS topology does not much improve the phase performance if the phase between the states are unevenly distributed.

In the following section, let us investigate the effect of the lossy component on the phase shift performance of the 360° PI-Section DPS.

11.3 Analysis of the Phase Performance of the 360° PI-Section DPS Topology with Lossy Components

Referring to Figure 11.1, for practical phase shifter designs, one must include the resistive losses of the circuit components as was introduced in Chapters 6–10.

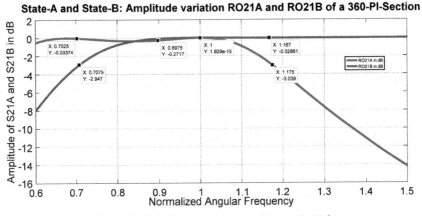

Figure 11.3b   Gain performance of Example 11.2.

In this regard, resistive loss r_L in series with an inductor L is approximated as

$$r_L = \frac{\omega_0 L}{Q_L}$$

where Q_L is the quality factor of the inductor.

Similarly, conductive loss G_C associated with a capacitor C is given by

$$G_C = \frac{\omega_0 C}{Q_C}$$

where Q_C is the quality factor of the capacitor C.

For MMIC implementations, Q_L and Q_C may vary between 4 and 20. On the other hand, quality factor of discrete components is much higher; it varies between 20 and 100. Moreover, the value of the reverse bias MOS capacitor C_{off} depends on the forward-biased channel resistance R_{on} as introduced in previous chapters such that

$$R_{on} C_{off} = Constant$$

The value of "*Constant*" depends on the production process of the switches being used. For example, for an MOS switch produced by TSMC using 180 nm technology yields[1]

$$Constant = 672 \times 10^{-15}$$

---

[1]https://www.tsmc.com/english/default.htm

In this book, we presume that 360° PI-DPS is manufactured employing the TSMC 180 Nano-meter technology with identical MOS diodes.

Based on the above discussions, we can generate the lossy impedances for the switching states as follows.

State-A:

In this state, all the diodes are forward-biased and they introduce forward-biased channel resistance R_{F1} or normalized channel resistance $r_{f1} = r_{f2} = \frac{R_{F1}}{R}$. In this case, the series arm impedance of State-A is given by

$$z_a (j\omega) = r_a + j\omega L$$

where

$$r_a = r_L + r_{f1}$$
$$r_L = \frac{\omega_0 L}{Q_L}$$

where

$$L = L_L = \frac{2C_L}{1 + \omega_0^2 C_L^2}; \quad C_L = \left(\frac{1}{\omega_0}\right) \tan\left(\frac{\theta_A}{2}\right)$$

and

$$r_{f1} = r_{f2} = \frac{672 \times 10^{-15}}{R \times C_{D1-actual}}$$

such that

$$C_{D1} = \frac{C_H}{1 + \omega_0^2 L_1 C_H} \quad \text{where} \quad C_H = \left(\frac{1}{\omega_0^2}\right) \left(\frac{1 + \omega_0^2 L_H^2}{2L_H}\right)$$
$$\text{and} \quad L_H = \left(\frac{1}{\omega_0}\right) \cotan\left(\frac{\theta_B}{2}\right)$$

Then, the actual capacitance value of the reverse-biased switch is given by

$$C_{D1-actual} = \frac{C_{D1}}{(2\pi f_{0-actual}) \times R}$$

The shunt arm admittance y_a is specified by

$$y_a (j\omega) = \frac{1}{r_{LA} + j\omega L_A} + Y_{TA}$$

where

$$r_{LA} = \frac{\omega_0 L_A}{Q_{LA}}$$

$$Y_{TA} = \frac{1}{Z_{TA}}; \quad Z_{TA} = r_{f2} + \frac{1}{G_{CA} + j\omega C_A}$$

$$G_{CA} = \frac{\omega_0 C_A}{Q_{CA}}$$

$$r_{f2} = r_{f1} = \frac{672 \times 10^{-15}}{R \times C_{D1-actual}}$$

State-B:
In State-B, all the diodes are reverse-biased. Therefore, in the shunt arm, the diode capacitance $C_{D2} = C_{D1}$ since D_1 and D_2 are identical.

In this case, the series arm impedance z_b is given by

$$z_b(j\omega) = (r_L + j\omega L) + \frac{1}{(G_{CD1} + j\omega C_{D1})}$$

where

$$G_{CD1} = \frac{\omega_0 C_{D1}}{Q_{CD1}}$$

The shunt arm admittance is specified by

$$y_b(j\omega) = \frac{1}{(r_{LA} + j\omega L_A)} + Y_T(j\omega) = G_B(\omega) + jB(\omega)$$

where

$$Y_T = \frac{1}{Z_T}; Z_T = \frac{1}{G_{CA} + j\omega C_A} + \frac{1}{G_{CD1} + j\omega C_{D1}}$$

Hence, we can propose the following algorithm to design lossy practical 360° PI-Section DPS

11.4 Algorithm: Design of a Lossy 360° PI-Section DPS

This algorithm designs and generates phase shifting performance of the lossy 360° PI-Section Digital Phase Shifter. Here, we presume that DPS is realized using MMIC on a silicon substrate with $0.18\,\mu m$ technology of TSMC[2].

---

[2]Taiwanese Semiconductor Manufacturing Company; (https://www.tsmc.com/english/aboutTSMC/company_profile.htm)

Inputs:

θ_A: Phase Shift of State-A

θ_B: Phase Shift of State-B

f_{0a}: Actual center frequency

ω_0: Normalized angular frequency

R: Normalization resistor

Q_L: Quality factor to determine the series loss resistance r_L of a chip inductor L

Q_C: Quality factor to determine the shunt loss conductance G_C of a chip capacitor C

Computational Steps:

Step 1: Compute the ideal element values of the 360° PI-DPS using our MatLab function PI_360-DPS.

$$[CD1, L, LA, CA, CT, CAa, LAa, CD1a, La, CTa]$$
$$= PI\_360\_DPS(Teta\_A, Teta\_B, f0a, w0, R)$$

Step 1a: Compute the ideal element values of the 360° PI-Section Digital Phase Shifter

$$C_L = \left(\frac{1}{\omega_0}\right) \tan\left(\frac{\theta_A}{2}\right)$$

$$L_L = L = \frac{2C_L}{1 + \omega_0^2 C_L^2}$$

$$L = L_L$$

Step 1b: Compute the ideal element values of the high-pass PI-Section

$$L_H = \left(\frac{1}{\omega_0}\right) \cotan\left(\frac{\theta_B}{2}\right)$$

$$C_H = \left(\frac{1}{\omega_0^2}\right) \left(\frac{1 + \omega_0^2 L_H^2}{2L_H}\right)$$

Step 1c: Compute C_{D1} and its actual value

$$C_{D1} = \frac{C_H}{1 + \omega_0^2 L_1 C_H}; \quad C_{D1-actual} = \frac{C_{D1}}{(2\pi f_{0-actual}) \times R}$$

Step 1d: Compute the ideal normalized component values of the shunt arm C_A and L_A

$$C_{D2} = C_{D1}$$

$$A = \omega_0^2 L_H$$

$$B = -(\omega_0^2 L_H C_L + 1)$$

$$C = -(1 + \omega_0^2 L_H C_L)C_{D2} = BC_{D2}$$

$$\Delta = B^2 - 4AC > 0$$

$$C_A = \frac{-B + \sqrt{B^2 - 4AC}}{2A}$$

$$L_A = \frac{1}{\omega_0^2 (C_A - C_L)}$$

Step 2: Compute the component losses for both series and shunt arms

Step 2a: Component losses for the series arm

$$r_L = \frac{\omega_0 L}{Q_L}$$

$$r_f = \frac{672 \times 10^{-15}}{R \times C_{D1-actual}}$$

$$r_a = r_{L1} + r_{f1}$$

Step 2b: Component losses for the shunt arm

$$r_{f2} = r_f$$

$$r_{LA} = \frac{\omega_0 L_A}{Q_L}$$

$$G_{CA} = \frac{\omega_0 C_A}{Q_C}$$

$$G_{CD1} = \frac{\omega_0 C_{D1}}{Q_C}$$

Step 3: Computation of the State-A immittances for both series and shunt arms

Step 3a: Compute the series arm impedance z_a for State-A

$$z_a(j\omega) = r_a + j\omega L_1$$

Step 3b: Compute the shunt arm admittance y_a for State-A

$$Z_{TA} = r_{f2} + \frac{1}{G_{CA} + j\omega C_A}$$

$$Y_{TA} = \frac{1}{Z_{TA}}$$

$$y_a(j\omega) = \frac{1}{r_{LA} + j\omega L_A} + Y_{TA}$$

Step 4: Computation of the State-B immittances for both series and shunt arms

Step 4a: Computation of series arm impedance in State-B

$$z_b(j\omega) = (r_{L1} + j\omega L_1) + \frac{1}{(G_{CD1} + j\omega C_{D1})}$$

Step 4b: Computation of shunt arm admittance in State-B

$$Z_{TB} = \frac{1}{G_{CA} + j\omega C_A} + \frac{1}{G_{CD1} + j\omega C_{D1}}$$

$$Y_{TB} = \frac{1}{Z_{TB}}$$

$$y_b(j\omega) = \frac{1}{(r_{LA} + j\omega L_A)} + Y_{TB}(j\omega)$$

Step 5: Generate phase and gain performance of 360° PI-DPS for State-A and State-B

$$S_{21(A,B)} = \frac{2}{zy^2 + 2zy + 2y + z + 2}; \quad \left\{ \begin{array}{l} \text{In State A:} \ z = z_a, \ y = y_a \\ \text{In State B:} \ z = z_b, \ y = y_b \end{array} \right\}$$

Step 6: Plot the phase and gain performance

Now, let us run an example to evaluate the performance of an 360° PI-DPS constructed with lossy elements.

Example 11.3: Design a 360° PI-Section Digital Phase Shifter for the phase shift of $\Delta\theta = 180°$. In the course of computations, assume that quality factor for inductors and capacitors are equal and they are given by $Q_L = Q_C = 10$

at $f_{0a} = 5\ GHz$. We further assume that desired phase shift between the states are equally distributed at the normalized angular frequency $\omega_0 = 1$. Termination resistor $R = 50\ \Omega$ selected.

Solution

For this purpose, Algorithm 11.5 is programmed under MatLab program Example 11.3. Inputs to this program are defined as follows.

Inputs:

$$\theta_A = 90°;\ \theta_B = 90°;\ f_{0a} = 5 \times 10^9\ Hz;\ \omega_0 = 1;\ R = 50\ \Omega;$$
$$Q_L = 20;\ Q_C = 20.$$

Computational Steps:

Step 1: Call function PI_360_DPS to compute lossless component values of the 360 degree PI-Section digital phase shifter
   Hence, we end up with

$$CD1 = 0.5000;\ L1 = 1;\ LA = 0.7071;\ CA = 2.4142;$$
$$CAa = 1.5369e - 12 = 1.54\ pF;\ LAa = 1.1254e - 09 = 1.125\ nH;$$
$$CD1a = 3.1831e - 13 = 0.32\ pF;\ L1a = 1.5915e - 09 = 1.6\ nH.$$

Step 2: Compute the component losses for both series and shunt arms
Step 2a: Component losses for the series arms as they are normalized

$$r_L = \frac{\omega_0 L}{Q_L} = 0.1$$

$$r_{f1} = r_{f2} = \frac{672 \times 10^{-15}}{R \times C_{D1-actual}} = 0.0422$$

$$r_a = r_{L1} + r_{f1} = 0.1422$$

Step 2b: Component losses of the shunt arms

$$r_{f2} = r_{f1} = 0.0422$$

$$r_{LA} = \frac{\omega_0 L_A}{Q_L} = 0.0707$$

$$G_{CA} = \frac{\omega_0 C_A}{Q_C} = 0.2414$$

$$G_{CD1} = \frac{\omega_0 C_{D1}}{Q_C} = 0.0500$$

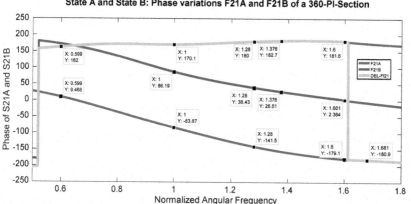

Figure 11.4a Phase shifting performance of Example 11.3.

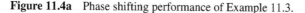

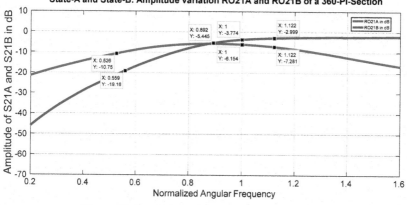

Figure 11.4b Gain performance of Example 11.4.

Steps 3–6 is programmed under Main_Example_11_3.m.

Phase shifting performance of the lossy phase shifter is shown in Figure 11.4a.

A close examination of Figure 11.4a reveals that 10% phase range bandwidth run from $\omega = 0.6$ to $\omega = 1.6$. It looks like that it is more than one octave.

Gain performance of the phase shifter is shown in Figure 11.4b. This figure reveals that gain in both switching states is below $-3dB$. Detail loss analysis of the topology under consideration may be completed employing commercially available CAD packages.

Clearly, 360° PI Section Digital Phase Shifter Cell is highly sensitive to component losses when it is designed for the phase shift of $\Delta\theta = 180°$. For example, in the neighborhood of center frequency f_0 state losses are read as follows

State A: $20log_{10} |S_{21A}(\omega_0)| = -6.154 \, dB$

State B: $20log_{10} |S_{21B}(\omega_0)| = -3.774 \, dB$

In order to understand the effect of the external component losses for series arm inductor L, shunt arm inductors L_A, and shunt arm capacitors C_A, let us increase quality factors Q_L and Q_C to 75, then examine the phase and gain performance of the phase shifter designed in Example 11.1 as in the following example.

Example 11.4: Repeat Example 11.3 for the quality factors $Q_L = \frac{\omega_0 L_{actual}}{r_{L-actual}} = \frac{\omega_0 L_{A-actual}}{r_{L-actual}} = Q_C = \frac{\omega_0 C_{A-actual}}{G_{CA}} = 75$

Solution

For this problem, we run our MatLab program Main_Example_11_3 for the following inputs.

Inputs:

$$\theta_A = \theta_B = 90° \ (i.e. \ \Delta\theta = \varphi_{21B}(\omega_0) - \varphi_{21A}(\omega_0) = \theta_B + \theta_A = 180°$$

$$Q_L = Q_C = 75$$

$$\omega_0 = 1$$

$$f_0 = 5 \, GHz = 5 \times 10^9 \, Hz$$

$$R = 50 \, \Omega$$

Output of the Program:

$$L = 1; L_{actual} = 1.59 \, nH; r_{L-actual} = 0.666 \, \Omega$$

$$C_{D1} = C_{off-Normalized} = 0.5; C_{D1-actual} = C_{off-actual}$$
$$= 318 \, fF; R_{on-actual} = r_{f1} = 2.1 \, \Omega$$

$$L_A = 0.7071; L_{A-actual} = 1.1254 \, nH; r_{LA-actual} = 0.47 \, \Omega$$

$$C_A = 2.4142; C_{A-actual} = 1.5369 \, pF; G_{CA} = 0.0322 \ (normalized)$$

Phase shift performance of the PI-DPS under consideration is depicted in Figure 11.4a.

Based on the 10% phase perturbation, normalized phase bandwidth runs from 0.71 up to 1.48.

Gain performance of the lossy phase shifter is shown in Figure 11.5b.

A close examination of Figure 11.5b indicates that normalized half-power bandwidth of the phase shifter runs from $\omega_1 = 0.84$ to $\omega_2 = 1$. Total bandwidth is approximately 16% with respect to center frequency. This bandwidth is narrow but acceptable.

In conclusion, we say that Example 11.4 provides reasonable solution to design 180° phase shifting cell over the design presented in Example 11.3.

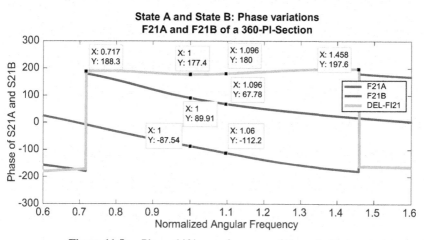

Figure 11.5a   Phase shifting performance of Example 11.4.

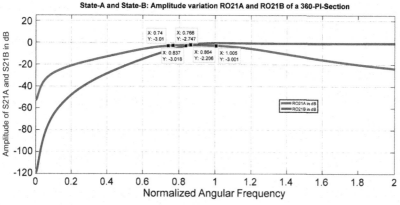

Figure 11.5b   Phase shifting performance of Example 11.4.

The digital phase shifter design under this example can only be realized using discrete components with high-quality factors rather than monolithic chip solution.

MatLab programs of Chapter 11 are presented in the appendix.

11.5 Physical Implementation of 360° PI-DPS

The 360° PI-DPS may be built by using discrete component technology or MMIC technology. It is expected that discrete component technology offers much higher quality factors for the component values, which in turn improves the electric performance of the phase shifter under consideration. Diode switches may be selected as varactor diodes or PIN diodes. Switching components can be found in the literature such as [4–11]. MMIC implementation can be carried out as described in Chapter 9. User may as well make use of the TSMC foundry of Taiwan [12].

Appendix 11: MatLab Programs for Chapter 11

Program List 11.1. Main_Example_11_1.m

```
% Main Program: Main_Example_11.1.m
% March 26, 2019
% Developed by BS Yarman, Vanikoy, Istanbul
% This program evaluates the ideal performance of a Highpass
    PI-Section DPS
% for a specified actual center frequency f0 in Hz.
% ------------------------------------------------------------------
clc; close all
% Inputs:
Teta_A=input('Design of 360 Degree PI Section DPS. Enter Phase of
State-A(Lowpass-PI): Teta-A in Degree=')
Teta_B=input('Design of 360 Degree PI Section DPS. Enter Phase of
State-B(Highpass-PI): Teta-B in Degree=')
% Inputs
f0a=input('Enter the actual center frequency in Hz f0a =')
w0=input('At f0a, enter the normalized angular frequency w0 =')
R=input('Enter the normaliziation Resistor R =')
% ------------------------------------------------------------------
% Compute the normalized element values of PI-360 Degree-DPS
% Ideal Lowpass PI-Section: See Equations (11.1a) and (11.1b)
% CL=(1/w0)*tan(?A/2)>0
CL=(1/w0)*tand(Teta_A/2);
% LL=L=(2C_L)/(1+?_0^2 C_L^2 )>0
LL=2*CL/(1+w0*w0*CL*CL);
L=LL;
```

```
%  -----------------------------------------------------------------
%  Ideal Highpass PI-Section: See Equations (11.2a) and (11.2b)
%  Ideal Highpass T-Section: See Equations (11.2a) and (11.2b)
%  LH=(1/w0)*cotan(?B/2)>0
LH=(1/w0)*cotd(Teta_B/2);
%  CH=(1/(?0^2 ))((1+?0^2 LH^2)/(2LH^2 ))>0
CH=(1/w0/w0)*(1+w0*w0*LH*LH)/2/LH;
%  Compute ideal element values of 360 Degree PI Section Digital Phase
Shifter
[CD1,L,LA,CA,CT,CAa,LAa,CD1a,La,CTa]  =  PI_360_DPS(Teta_A,  Teta_B,  f0a,
    w0,R);
%  -----------------------------------------------------------------

j=sqrt(-1);
w=0;N=2000;w1=0;w2=5;DW=(w2-w1)/N;
FRI(1:(N+1))=zeros;
WA(1:(N+1))=zeros;
DEL_FI21(1:N+1)=zeros;
for  i=1:N+1
     WA(i)=w;
%  -------------------------
%  State A:
%  -------------------------
za=j*w*L;
ya=j*w*CA+1/j/w/LA;
[ S11a,S21a,RO11a,F11a,RO21a,F21a ]  =  S_Par_PI_Section ( za,ya );
     F21A(i)=F21a;
     RO21A(i)=20*log10(RO21a);
     F11A(i)=F11a;
     RO11A(i)=20*log10(RO11a);
%  -------------------------
%  State-B
%  -------------------------
zb=j*w*L+1/j/w/CD1;
yb=j*w*CT+1/j/w/LA;
[ S11b,S21b,RO11b,F11b,RO21b,F21b ]  =  S_Par_PI_Section ( zb,yb );
     F21B(i)=F21b;
     RO21B(i)=20*log10(RO21b);
     F11B(i)=F11b;
     RO11B(i)=20*log10(RO11b);
%  -----------------------------------------------------------------
DEL_FI21(i)=F21B(i)-F21A(i);
     w=w+DW;
end
%  -----------------------------------------------------------------
Plot_State_AB_PI_360_DPS(WA,F21A,RO21A,  F11A,RO11A,F21B,RO21B,  F11B,
    RO11B,DEL_FI21)
```

Program List 11.2. Main_Example_11_2.m

```
%  Main Program: Main_Example_11_2.m
%  March 27, 2019
%  Developed by BS Yarman, Vanikoy, Istanbul
```

```
% This program evaluates the lossy performance of a Highpass PI-Section
    DPS
% for a specified actual center frequency f0 in Hz.
%  -----------------------------------------------------------------
% In this program we used the approximate explicit formulas to
    compute the
% losses for both State-A and State-B
%  -----------------------------------------------------------------
clc; close all
% Inputs:
Teta_A=input('Design of 360 Degree PI Section DPS. Enter Phase of
    State-A(Lowpass-PI): Teta-A in Degree=')
Teta_B=input('Design of 360 Degree PI Section DPS. Enter Phase of
    State-B(Highpass-PI): Teta-B in Degree=')
% Inputs
f0a=input('Enter the actual center frequency in Hz f0a =')
w0=input('At f0a, enter the normalized angular frequency w0 =')
R=input('Enter the normaliziation Resistor R =')
%  -----------------------------------------------------------------
% Compute the normalized element values of PI-360 Degree-DPS
[ CD1,L1,LA,CA,CT,CAa,LAa,CD1a,L1a,CTa ] = PI_360_DPS( Teta_A, Teta_B,
    f0a,w0,R )
%  -----------------------------------------------------------------
% ASSUMPTION 1:
% It is assumed that the series loss Ron of a forward biased diode is
% equal to the on channel resistor of an CMOS Switch
% [see Equation (6.1) of Chapter 6].
[ RF1,rf1 ] =Channel_Resistance_of_a_CMOS( CD1,R,f0a );
% ASSUMPTION 2:
% Reverse biased resistive loss of a diode is the "Percent_RVS"
    amount of
% its reverse baised impedance at w0
CrDel_D=10;
rr1=1/w0/CD1/CrDel_D;
CD2=CD1;
[ RF2,rf2 ] =Channel_Resistance_of_a_CMOS( CD2,R,f0a );
rr2=1/w0/CD2/CrDel_D;
%
%  -----------------------------------------------------------------
% Loss Computations for both State-A and State-B:
% Assumption 3: Loss of an inductor is Percent_L amount of its impedance
% value at w0.
% Assumption 4: Connectivity loss of an inductor is "Percent_S"
    amount of
% its impedance value at w0.
% Assumption 5: Conductive loss of a Capacitor is "Percent_C" amount
    of its
% admittance value at w0.
CrDel_S=10;
CrDEL_L=10;
CrDel_C=10;
%  -----------------------------------------------------------------
% Resistive loss of the series arms in State-A:
```

```
rL1=w0*L1/CrDEL_L;
rLA=w0*LA/CrDEL_L;
rs=w0*L1/CrDel_S;
ra=rf1;
% Conductive loss of the Shunt arm in State-A:
GCA=w0*CA/CrDel_C;
% GCA=0;GA=0;
GA=GCA+rLA/(rLA*rLA+w0*w0*LA*LA);
% Resistive loss of the series arms in State-B:
rb=rr1;
% Conductive loss of the Shunt arm in State-B:
%
GB=GCA+rLA/(rLA*rLA+w0*w0*LA*LA)+(w0*w0*rr2*CT*CT)/(1+w0*w0*rr2*rr2*CT*
    CT);

j=sqrt(-1);
w=0;N=2000;w1=0;w2=2;DW=(w2-w1)/N;
FRI(1:(N+1))=zeros;
WA(1:(N+1))=zeros;
DEL_FI21(1:N+1)=zeros;
for i=1:N+1
    WA(i)=w;
% -------------------
% State A:
% -------------------
%za=ra+j*w*L1;
za=ra+j*w*L1;
%ya=GA+j*w*CA+1/j/w/LA;
ya=GA+j*w*(CA/(1+w*w*rf2*rf2*CA*CA)-LA/(rLA*rLA+w*w*LA*LA));
[ S11a,S21a,RO11a,F11a,RO21a,F21a ] = S_Par_PI_Section (za,ya);
    F21A(i)=F21a;
    RO21A(i)=20*log10(RO21a);
    F11A(i)=F11a;
    RO11A(i)=20*log10(RO11a);
% -------------------
% State-B
zb=rb+j*w*L1+1/j/w/CD1;
yb=GB+j*w*(CT/(1+w*w*rr2*rr2*CT*CT)-LA/(rLA*rLA+w*w*LA*LA));
    [ S11b,S21b,RO11b,F11b,RO21b,F21b ] = S_Par_PI_Section (zb,yb);
    F21B(i)=F21b;
    RO21B(i)=20*log10(RO21b);
    F11B(i)=F11b;
    RO11B(i)=20*log10(RO11b);
% -------------------
DEL_FI21(i)=F21B(i)-F21A(i);
    w=w+DW;
end
%
Plot_State_AB_PI_360_DPS(WA,F21A,RO21A, F11A,RO11A,F21B,RO21B, F11B,
    RO11B,DEL_FI21)
```

```
%  -------------------------------------------------------------------
ra_actual=ra*R
rb_actual=rb*R
%  -------------------------------------------------------------------
GA_actual=GA/R;  RA_actual=1/GA_actual
GB_actual=GB/R;  RB_actual=1/GB_actual
```

Program List 11.3. Main_Example_11_3.m

```
% Main Program: Main_Example_11_3.m
% March 27, 2019
% Developed by BS Yarman, Vanikoy, Istanbul
% This program evaluates the lossy performance of a 360 PI-Section DPS
% for a specified actual center frequency f0 in Hz.
%  -------------------------------------------------------------------
clc; close all
% Inputs:
Teta_A=input('Design of 360 Degree PI Section DPS. Enter Phase of
    State-A(Lowpass-PI): Teta-A in Degree=')
Teta_B=input('Design of 360 Degree PI Section DPS. Enter Phase of
    State-B(Highpass-PI): Teta-B in Degree=')
% Inputs
f0a=input('Enter the actual center frequency in Hz f0a =')
w0=input('At f0a, enter the normalized angular frequency w0 =')
R=input('Enter the normaliziation Resistor R =')
%  -------------------------------------------------------------------
% Compute the normalized element values of PI-360 Degree-DPS
[CD1,L1,LA,CA,CT,CAa,LAa,CD1a,L1a,CTa] = PI_360_DPS(Teta_A, Teta_B, f0a
    ,w0,R)
%  -------------------------------------------------------------------
% ASSUMPTION 1:
% It is assumed that the series loss Ron of a forward biased diode is
% equal to the on channel resistor of an CMOS Switch
% [see Equation (6.1) of Chapter 6].
[RF1,rf1] = Channel_Resistance_of_a_CMOS(CD1,R,f0a);
% ASSUMPTION 2:
% At w0, quality factor for inductors and capacitors
QL=75;
QC=75;
%
CD2=CD1;
[RF,rf] =Channel_Resistance_of_a_CMOS(CD1,R,f0a);
%  -------------------------------------------------------------------
% Series Arm Losses
rf1=rf;
rL1=(w0*L1)/QL;
 ra=rL1+rf1;
% Shunt arm losses
rf2=rf;
rLA=(w0*LA)/QL;
GCA=(w0*CA)/QC;
GCD1=(w0*CD1)/QC;
%  -------------------------------------------------------------------
```

```
j=sqrt(-1);
w=0;N=2000;w1=0;w2=2;DW=(w2-w1)/N;
FRI(1:(N+1))=zeros;
WA(1:(N+1))=zeros;
DEL_FI21(1:N+1)=zeros;
for i=1:N+1
    WA(i)=w;
%  -------------------
% State A:
%  -------------------
%za=ra+j*w*Ll;
za=ra+j*w*Ll;
% Computation of shunt arm admittance ya
% ZTA=rf2+1/(GCA+jwCA)
ZTA=rf+1/(GCA+j*w*CA);
YTA=1/ZTA;
ya=1/(rLA+j*w*LA)+YTA;
[S11a,S21a,RO11a,F11a,RO21a,F21a] = S_Par_PI_Section (za,ya);
    F21A(i)=F21a;
    RO21A(i)=20*log10(RO21a);
    F11A(i)=F11a;
    RO11A(i)=20*log10(RO11a);
%  -------------------
% State-B
zb=(rL1+j*w*L1)+1/(GCD1+j*w*CD1);
ZTB=1/(GCA+j*w*CA)+1/(GCD1+j*w*CD1);
YTB=1/ZTB;
yb=1/(rLA+j*w*LA)+YTB;
[S11b,S21b,RO11b,F11b,RO21b,F21b] = S_Par_PI_Section (zb,yb);
    F21B(i)=F21b;
    RO21B(i)=20*log10(RO21b);
    F11B(i)=F11b;
    RO11B(i)=20*log10(RO11b);
%
    DEL_FI21(i)=F21B(i)-F21A(i);
    w=w+DW;
end
%  -------------------
Plot_State_AB_PI_360_DPS(WA,F21A,RO21A, F11A,RO11A,F21B,RO21B, F11B,
    RO11B,DEL_FI21)
%  -------------------
zb0=(rL1+j*w0*L1)+1/(GCD1+j*w0*CD1);
rb=real(zb0);
%
ra_actual=ra*R
rb_actual=rb*R
%  -------------------
% GA_actual=GA/R; RA_actual=1/GA_actual
% GB_actual=GB/R; RB_actual=1/GB_actual
```

Program List 11.4. function S_Par_PI_Section

```
function [ S11,S21,RO11,F11,RO21,F21 ] = S_Par_PI_Section (z,y)
```

```
%This function generates the S-Parameters of a PI Section
% Phase Shifter from the series arm impedance Z(jw) and
% the shunt arm admittance Y(jw)
% ----------------------------------------------------------
% Developed by BS Yarman: March 26, 2019, Vanikoy, Istanbul
% See Equations (5.9)
% ----------------------------------------------------------
% N=z(1-y^2)-2y
N=z*(1-y*y)-2*y;
% D= zy^2+2zy+2y+z+2
D=z*y*y+2*z*y+2*y+z+2;
S11=N/D;
S21=2/D;
R11=real(S11); X11=imag(S11);F11=atan2d(X11,R11);
R21=real(S21); X21=imag(S21);F21=atan2d(X21,R21);
RO11=abs(S11);
RO21=abs(S21);
end
```

Program List 11.5. function PI_360_DPS

```
function [ CD1,L,LA,CA,CT,CAa,LAa,CD1a,La,CTa ] = PI_360_DPS(Teta_A,
    Teta_B, f0a,w0,R)
% This function generates the element values of an ideal 360 degree
    Simple
% PI-Section based Digital Phase Shifter
% Developed by BS Yarman on March 26, 2019, Vanikoy, Istanbul
% Inputs:
% Teta_A: Desired phase shift of the Lowpass Based PI-Section DPS
% Teta_B: Desired phase shift of the Highpass Based PI-Section DPS
% f0a: Actual centre frequency
% w0: Normalized angular frequency. It is selected as w0=1
% R: Port normalization number. It is usually, selected as R=50 ohms
% Outputs:
% CD1: Reverse Biased diode capacitance of the series arms.
% L: Series arm inductor
% LA: Shunt arm inductor
% CA: Shunt arm capacitor
% ----------------------------------------------------------
% Ideal LowpassPI-Section: See Equations (11.1a) and (11.1b)
% CL=(1/w0)*tan(?A/2)>0
CL=(1/w0)*tand(Teta_A/2);
% LL=L=(2C_L)/(1+?_0^2 C_L^2)>0
LL=2*CL/(1+w0*w0*CL*CL);
% ----------------------------------------------------------
% Ideal Highpass T-Section: See Equations (11.2a) and (11.2b)
% LH=(1/w0)*cotan(?B/2)>0
LH=(1/w0)*cotd(Teta_B/2);
% CH=(1/(?0^2))((1+?0^2 LH^2)/(2LH^2))>0
CH=(1/w0/w0)*(1+w0*w0*LH*LH)/2/LH;
% ----------------------------------------------------------
% State-A: Series arm component computations
```

```
L=LL;
CD1=CH/(1+w0*w0*L*CH);
%  ------------------------------------------------------------
% State B: Computation of CA: See Equation (11.4) % (11.5)
A=w0*w0*LH;
B=-(w0*w0*LH*CL+1);
CD2=CD1;
C=B*CD2;
Delta=B*B-4*A*C;
CA=(-B+sqrt(Delta))/2/A;
%  ------------------------------------------------------------

LA=1/(CA-CL)/w0/w0;
% LA=1/(?0?)/(CA-CL))>0
CT=CA*CD2/(CA+CD2);
%  ------------------------------------------------------------

%
CAa=CA/2/pi/f0a/R;
CD1a=CD1/2/pi/f0a/R;
LAa=R*LA/2/pi/f0a;
La=R*L/2/pi/f0a;
CTa=CT/2/pi/f0a/R;
end
```

Program List 11.6. function Plot_State_AB_PI_360_DPS

```
function Plot_State_AB_PI_360_DPS(WA,F21A,RO21A, F11A,RO11A,F21B,RO21B,
    F11B,RO11B,DEL_FI21)
figure
plot(WA,F21A,WA,F21B,WA,DEL_FI21)
title('State A and State B: Phase variations F21A and F21B of a
    360-PI-Section')
legend('F21A','F21B','DEL-FI21')
xlabel('Normalized Angular Frequency')
ylabel('Phase of S21A and S21B')
% Amplitude of S21
figure
plot(WA,RO21A,WA,RO21B)
title('State-A and State-B: Amplitude variation RO21A and RO21B of a
    360-PI-Section')
legend('RO21A in dB','RO21B in dB')
xlabel('Normalized Angular Frequency')
ylabel('Amplitude of S21A and S21B in dB')
%  ------------------------------------------------------------
figure
plot(WA,F11A, WA, F11B)
title('State-A and State-B: Phase variation F11A and F11B of a
    360-PI-Section')
legend('F11A')
xlabel('Normalized Angular Frequency')
ylabel('Phase of S11A and S11B')
% Amplitude of S11
figure
plot(WA,RO11A, WA, RO11B)
```

```
title('State-A and State-B: Amplitude variation RO11A and RO11B of a
    360-PI-Section')
legend('RO11A in dB','RO11B in dB')
xlabel('Normalized Angular Frequency')
ylabel('Amplitude of S11A and S11B in dB')

end
```

References

[1] B. S. Yarman, "Novel Circuit Configurations to Design Loss Balanced 0–360 degree Digital Phase Shifters," AEU, Vol. 45, No. 2, pp. 96–104, 1991.

[2] B. Yarman, A.Rosen and P.Stabile, "Lowloss EHF Digital Phase Shifters Suitable for Monolithic Implementation," in IEEE ISCAS, Montreal, 1984.

[3] Binboga S. Yarman, "π-section digital phase shifter apparatus," U.S. Patent: 4604593, Washington DC, USA, August 5, 1986.

[4] Pat Hindle, Editor, "The State of RF and Microwave Switches," Microwave Journal, vol. 53, no. 11, p. 20, November 2010.

[5] I. Bahl and B. Prakash, Microwave Solid State Circuit Design, 2. (1. John Wiley & Sons, Ed., 2003).

[6] I. Bahl, Lumped Elements for RF Microwave Circuits, Artech House, 2003.

[7] K. J. Koh and G. M. Rebeiz, "0.13-m CMOS phase shifters for and K-band phased arrays," IEEE Trans. on MTT, Vol. 42, No. 11, p. 2535–2546, November 2007.

[8] K. Chang, Handbook of RF Microwave Components and Engineering, New York, USA, Chechester, UK: Wiley, 2003.

[9] D. Kang and S. Hong, "A 4-bit CMOS Phase Shifter Using Distributed Active Switches," IEEE Trans on MTT, Vol. 55, No. 7, pp. 1476–1483, 2007.

[10] R. Melik and H. V. Demir, "Implementation of High Quality factor, On-Chip Tuned Microwave Resonators at 7 GHz," Microwave and Optical Technology Letters, vol. 51, no. 2, pp. 497–501, February 2009.

[11] H. Fang, T. Xinyi, K. Mouthaan and R. Guinvarch, "Two Octave Digital All-Pass Phase Shifters for Phase Array Applications," IEEE Radio and Wireless Symposium (RWS), pp. 169–171, 2013.

[12] TSMC, "Taiwanese Semiconductor Manufacturing Company," (https://www.tsmc.com/english/aboutTSMC/company_profile.htm), Taiwan, 2019.

12

180° High-pass-based PI-Section Digital Phase Shifter

Summary

A 180° high-pass-based PI Section Digital Phase Shifter (180° HPI-DPS) topology may be regarded as the special form of 360° PI Section DPS of Chapter 11. Just for the sake of completeness of the book, in this chapter, we outline the operation of 180° HP-PI DPS (or equivalently 180° HP-π DPS) topology. The content of this chapter is based on our previous work published patents and papers such as [1–3].

12.1 Derivation of Design Equations for a 180° PI-Section Digital Phase Shifter

As mentioned in summary, a 180° high-pass-based PI Section-DPS is derived directly from Figure 11.1 by setting $\theta_A = 0$, which in turn makes $L_L = L = 0$ and $C_L = 0$. In this case, Figure 11.1 is simplified as shown in Figure 12.1.

In State-A, all the switching diodes are forward-biased. Therefore, at the operating frequency ω_0, the shunt arm immittances constructed with parallel combination of inductor L_A and Capacitor C_A, resonate; performing open circuit operation. In the series arm, ideally, D_1 is shorted.

Thus, in State-A, at the normalized angular frequency ω_0, HPI-DPS topology acts as a "2-pair wire pass", as shown in Figure 12.1(b).

In State-B, all the diodes are reverse-biased. At the operating frequency ω_0, HPI-DPS behaves like a symmetric highpass PI-Section, as depicted in Figure 12.1(c).

Hence, the shunt arm high-pass inductor L_H is given by

$$L_H = \frac{1}{\omega_0}\cotan\left(\frac{\theta_B}{2}\right) > 0 \tag{12.1a}$$

493

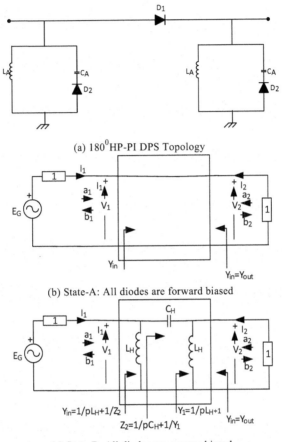

(a) 180^0 HP-PI DPS Topology

(b) State-A: All diodes are forward biased

(c) State-B: All diodes are reverse biased

Figure 12.1   180° HP-PI DPS topology.

and the series arm capacitor is determined as in (5.34b).

$$C_H = \left(\frac{1}{\omega_0^2}\right)\left(\frac{1 + \omega_0^2 L_H^2}{2L_H}\right) > 0 \qquad (12.1b)$$

Now, let us develop the design equations considering State-A and State-B operations.

In State-A (i.e., diodes are forward-biased), the series arm impedance $z_a(j\omega)$ is ideally zero.

$$z_a(j\omega) = 0 \qquad (12.2a)$$

In the shunt arm, the admittance, $y_a\,(j\omega)$ is also zero at ω_0. In other words,

$$y_a\,(j\omega) = j\omega C_A + \frac{1}{j\omega L_A} = j\omega C_A \left(1 - \frac{1}{\omega^2 L_A C_A}\right)$$

$$= j\omega C_A \left(\frac{\omega^2 L_A C_A - 1}{\omega^2 L_A C_A}\right) \tag{12.2b}$$

Then,

$$L_A = \frac{1}{\omega_0^2 C_A} > 0 \tag{12.2c}$$

Similar in State-B all the diodes are reverse-biased (State-B); then, series arm impedance z_b is ideally given by

$$z_b\,(j\omega) = \frac{1}{j\omega C_{D1}} \tag{12.2d}$$

At ω_0, the equivalent highpass PI-Section capacitor C_H is simulated as

$$C_{D1} = C_H \tag{12.3a}$$

Similarly, shunt admittance of State-B (i.e., all diodes are reverse biased) is given by

$$y_b\,(j\omega) = j\omega C_T + \frac{1}{j\omega L_A} = \frac{1}{j\omega L_A}\left(1 - \omega^2 L_A C_T\right) \tag{12.3b}$$

where the capacitor C_T is the series connection of C_A and C_{D2} such that

$$C_T = \frac{C_A C_{D2}}{C_A + C_{D2}}$$

In this case, at ω_0, equivalent shunt inductor L_H is given by

$$L_H = \frac{L_A}{1 - \omega^2 L_A C_T}$$

or

$$L_H \left(1 - \omega^2 L_A C_T\right) = L_A$$

or

$$L_A = \frac{L_H}{1 + \omega_0^2 L_H C_T} \tag{12.3c}$$

Employing (12.2c) and (12.3d), we obtain

$$\frac{L_H}{1 + \omega_0^2 L_H C_T} = \frac{1}{\omega_0^2 C_A}$$

or

$$\left(\omega_0^2 L_H\right) C_A^2 - \left(\omega_0^2 L_H C_L + 1\right) C_A - C_{D2}\left(1 + \omega_0^2 L_H C_L\right) \qquad (12.3d)$$

Let

$$A = \omega_0^2 L_H \qquad\qquad (12.3e)$$

$$B = -1 \qquad\qquad (12.3f)$$

$$C = -C_{D2} \qquad\qquad (12.3g)$$

$$\Delta = B^2 - 4AC = 1 + 4\omega_0^2 L_H C_{D2} > 0 \qquad (12.3h)$$

Then,

$$C_A = \frac{-B + \sqrt{B^2 - 4AC}}{2A} \qquad\qquad (12.4a)$$

$$C_A = \frac{1 + \sqrt{1 + 4\omega_0^2 L_H C_{D2}}}{2\omega_0^2 L_H} \qquad\qquad (12.4b)$$

Based on the above derivation, we can suggest the following design algorithm to construct a 180° HPI-Section digital phase. shifter

12.2 Algorithm to Design 180° PI-Section Digital Phase Shifter

In this section, we propose a design algorithm to construct 180° PI-Section Digital Phase Shifter, in short $360° - PI$-DPS, with ideal elements (i.e., lossless components).

Inputs:

θ_B: Desired phase shift at the operating frequency in State-B.

It is noted that the designer is free to select phase shifts at each switching states. There is no restriction imposed on the phase shifts over $0 < \{\theta_B\} < 180°$.

Remarks:

(a) For State-B, phase $\varphi_{21B}(\omega)$ of the transfer scattering parameter $S_{21B}(j\omega)$ is positive.

(b) Therefore, the net phase shift between the states is $\Delta\theta = \varphi_{21B}(\omega_0) - \varphi_{21A}(\omega_0) = \theta_B$.

(c) Employing 180° HP-PI DPS, it is not possible to obtain $\Delta\theta = 180°$ phase shift. Therefore, $0° < \theta_B \leq 180°$.

f_{0a}: Actual operating frequency in Hz.

ω_0: Normalized angular frequency, which corresponds to f_{0a}.

Remark: For many practical situations, ω_0 is selected as $\omega_0 = 1$. However, the designer is free to pick any value for ω_0.

R: Actual Termination resistors for the 360° PI-Section DPSPI-Section DPS.

Remark: It is usually selected as 50 Ω.

C_{D2a}: Actual capacitance of the shunt arm diode D_2.

Remark: Most of the applications, it may be preferable to select diodes D_1 and D_2 identical. On the other hand, C_{D1} is computed in Step 3 of this algorithm. Then, one may wish to set $C_{D2} = C_{D1}$ as in Step 6, and omit this input.

Computations Steps:

Step 1: Compute the ideal component values for the highpass PI-Section prototype as in (10.1c & d).

$$L_H = \left(\frac{1}{\omega_0}\right) \cot an \left(\frac{\theta_B}{2}\right) > 0$$

$$C_H = \left(\frac{1}{\omega_0^2}\right) \left(\frac{1 + \omega_0^2 L_H^2}{2L_H}\right) > 0$$

Step 2: Compute the ideal-normalized component values of the series arms.

$$C_{D1} = C_H$$

Step 3: Compute the ideal-normalized component values of the shunt arm C_A and L_A and C_T.

$$A = \omega_0^2 L_H$$
$$B = -1$$

$$C = -C_{D2}$$

$$\Delta = B^2 - 4AC > 0$$

$$C_A = \frac{1 + \sqrt{1 + 4\omega_0^2 L_H C_{D2}}}{2\omega_0^2 L_H}$$

$$C_{D2} = C_{D1}$$

$$C_T = \frac{C_A C_{D2}}{C_A + C_{D2}}$$

Step 4: Generate State-A series arms impedance z_a, shunt arm admittance y_a, and the scattering parameters.

$$z_a(j\omega) = 0$$

$$y_a(j\omega) = j\omega C_A + \frac{1}{j\omega L_A}$$

$$S_{21A} = \frac{2}{zy^2 + 2zy + 2y + z + 2}; \quad \left\{ \begin{matrix} z = z_a \\ y = y_a \end{matrix} \right\}$$

$$S_{11A} = \frac{z(1 - y^2) - 2y}{zy^2 + 2zy + 2y + z + 2}; \quad \left\{ \begin{matrix} z = z_a \\ y = y_a \end{matrix} \right\}$$

Step 5: Generate series arms impedance z_b, shunt arm admittance y_b, and the scattering parameters of State-B.

$$z_b(j\omega) = \frac{1}{j\omega C_{D1}}$$

$$y_b(j\omega) = j\omega C_A + \frac{1}{j\omega L_A}$$

$$S_{21B} = \frac{2}{zy^2 + 2zy + 2y + z + 2} \quad \text{where} \quad \left\{ \begin{matrix} z = z_b \\ y = y_b \end{matrix} \right\}$$

$$S_{11B} = \frac{z(1 - y^2) - 2y}{zy^2 + 2zy + 2y + z + 2} \quad \text{where} \quad \left\{ \begin{matrix} z = z_b \\ y = y_b \end{matrix} \right\}$$

Step 6: Compute the actual element values and plot the performance the 180° HP-PI-Section DPS

$$L_a = R\frac{L_1}{(2\pi f_{0a})}$$

$$L_{Aa} = R\frac{L_1}{(2\pi f_{0a})}$$

$$C_{D1a} = \frac{C_{D1}}{(2\pi f_{0a})\,R}$$

$$C_{Aa} = \frac{C_A}{(2\pi f_{0a})\,R}$$

Let us now, run an example to show the utilization of the above algorithm.

Example 12.1: Let $\theta_B = 90°$. Let $f_{0a} = 5\,GHz$. Let $\omega_0 = 1$. Let $R = 50\,\Omega$. Select $C_{D2} = C_{D1}$.

Employing the algorithm given in Section 12.2, design a 180° HP-PI-Section DPS for $\Delta\theta = 90°$ and plot the phase shifting performance.

Solution

Let us follow the computation steps of the HP-PI DPS to construct the desired phase shifter.

Step 1: Ideal component values of high-pass T-Section

$$L_H = \left(\frac{1}{\omega_0}\right)\cotan\left(\frac{\theta_B}{2}\right) = 1$$

$$C_H = \left(\frac{1}{\omega_0^2}\right)\left(\frac{1+\omega_0^2 L_H^2}{2L_H}\right) = 1$$

Step 2: Ideal value of series arm reverse-biased capacitor C_{D1} is

$$C_{D1} = C_H = 1$$

Step 3: Set $\theta_A = 0$ and compute normalized component values of the shunt arm C_A and L_A and C_T by calling out MatLab function PI_360_DPS

$$[CD1, L1, LA, CA, CT, CAa, LAa, CD1a, L1a, CTa]$$
$$= PI\_360\_DPS(Teta\_A,\ Teta\_B,\ f0a, w0, R)$$

Such that

$$C_A = 1.6180$$
$$L_A = 0.6180$$
$$C_T = 0.6180$$

Steps 5–7: These steps are completed under the MatLab program "*Main_Example_12_1.m*".

Actual element component values are found as follows.

$$L_a = 0.98363 \, nH$$
$$C_{D1a} = 0.63662 \, pF$$
$$C_{Aa} = 1.0301 \, pF$$
$$L_{Aa} = 0.98363 \, nH$$

Phase performance of the 180° PI-Section DPS is depicted in Figure 12.2a.

A close examination of Figure 12.2a reveals that 180° HP-PI DPS yields 10% phase perturbation ($90° \mp 9°$) bandwidth of $\Delta\omega = \omega_2 - \omega_1$ such that $\omega_1 = 0.845$ and $\omega_2 = 1.5$.

Gain performance of the phase shifter configuration under consideration is shown in Figure 12.2b.

It is interesting to observe that –3 dB (half-power) frequency bandwidth runs from $\omega_1 = 0.737$ up to $\omega_2 = 1.35$ (i.e., $\Delta\omega_{Gain} = 0.61$). In other words,

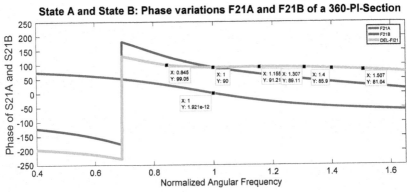

Figure 12.2a   Phase shifting performance of Example 12.1.

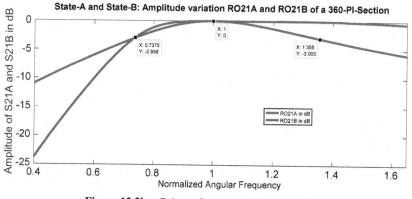

Figure 12.2b Gain performance of Example 10.1.

half-power bandwidth is slightly less than that of 10% phase perturbation bandwidth ($\Delta\omega_{Phase} = 1.5 - 0.845 = 0.65$).

In the following section, let us investigate the effect of the lossy component on the phase shift performance of the 360° PI-Section DPS.

12.3 Analysis of the Phase Performance of the 360° PI-Section DPS Topology with Lossy Components

Referring to Figure 12.1, for practical phase shifter designs, one must include the resistive losses of the circuit components as was introduced in Chapters 6–10.

In this regard, resistive loss r_L in series with an inductor L is approximated as

$$r_L = \frac{\omega_0 L}{Q_L}$$

where Q_L is the quality factor of the inductor.

Similarly, conductive loss G_C associated with a capacitor C is given by

$$G_C = \frac{\omega_0 C}{Q_C}$$

where Q_C is the quality factor of the capacitor C.

For MMIC implementations, Q_L and Q_C may vary between 4 and 20. On the other hand, quality factor of discrete components are much higher; it varies between 20 and 100. Moreover, value of the reverse-biased MOS capacitor C_{off} depends on the forward-biased channel resistance R_{on} as introduced in previous chapters such that

$$R_{on} C_{off} = Constant$$

The value of "*Constant*" depends on the production process of the switches being used. For example, for an MOS switch produced by TSMC using 180 nm technology yields[1]

$$Constant = 672 \times 10^{-15}$$

In this book, we presume that 360° PI-DPS is manufactured employing the TSMC 180 Nano-meter technology with identical MOS diodes.

Based on the above discussions, we can generate the lossy impedances for the switching states as follows.

State-A:

In this state, all the diodes are forward-biased and they introduce forward-biased channel resistance R_{F1} or normalized channel resistance $r_{f1} = r_{f2} = \frac{R_{F1}}{R}$. In this case, the series arm impedance of State-A is given by

$$z_a(j\omega) = r_{f1}$$

where

$$r_a = r_{f1}$$

and

$$r_{f1} = r_{f2} = \frac{672 \times 10^{-15}}{R \times C_{D1-actual}}$$

such that

$$C_{D1} = C_H; \ C_H = \left(\frac{1}{\omega_0^2}\right)\left(\frac{1 + \omega_0^2 L_H^2}{2L_H}\right) \text{ and } L_H = \left(\frac{1}{\omega_0}\right)\cotan\left(\frac{\theta_B}{2}\right)$$

Then, the actual capacitance value of the reverse-biased switch is given by

$$C_{D1-actual} = \frac{C_{D1}}{(2\pi f_{0-actual}) \times R}$$

The shunt arm admittance y_a is specified by

$$y_a(j\omega) = \frac{1}{r_{LA} + j\omega L_A} + Y_{TA}$$

---

[1] https://www.tsmc.com/english/default.htm

where

$$r_{LA} = \frac{\omega_0 L_A}{Q_{LA}}$$

$$Y_{TA} = \frac{1}{Z_{TA}}; \quad Z_{TA} = r_{f2} + \frac{1}{G_{CA} + j\omega C_A}$$

$$G_{CA} = \frac{\omega_0 C_A}{Q_{CA}}$$

$$r_{f2} = r_{f1} = \frac{672 \times 10^{-15}}{R \times C_{D1-actual}}$$

State-B:

In State-B, all the diodes are reverse-biased. Therefore, in the shunt arm, the diode capacitance $C_{D2} = C_{D1}$ since D_1 and D_2 are identical.

In this case, the series arm impedance z_b is given by

$$z_b(j\omega) = \frac{1}{(G_{CD1} + j\omega C_{D1})}$$

where

$$G_{CD1} = \frac{\omega_0 C_{D1}}{Q_{CD1}}$$

The shunt arm admittance is specified by

$$y_b(j\omega) = \frac{1}{(r_{LA} + j\omega L_A)} + Y_T(j\omega) = G_B(\omega) + jB(\omega)$$

where

$$Y_T = \frac{1}{Z_T}; \quad Z_T = \frac{1}{G_{CA} + j\omega C_A} + \frac{1}{G_{CD1} + j\omega C_{D1}}$$

Hence, we can propose the following algorithm to design lossy-practical $360°$ PI-Section DPS

12.4 Algorithm: Design of a Lossy $180°$ HPI Section DPS

The algorithm developed to design lossy $360°$ PI-Section Digital Phase Shifter of Chapter 11 may be utilized by setting $\theta_A = 0$.

Inputs:

$\theta_A = 0$: Phase Shift of State-A, which describes a "two-pair wire pass"

θ_B: Phase Shift of State-B

f_{0a}: Actual center frequency

ω_0: Normalized angular frequency

R: Normalization resistor

Q_L: Quality factor to determine the series loss resistance r_L of a chip inductor L

Q_C: Quality factor to determine the shunt loss conductance G_C of a chip capacitor C

Computational Steps:

Step 1: Compute the ideal element values of the 360° PI-DPS using our MatLab function PI_360-DPS for $\theta_A = 0$

$$[CD1, L, LA, CA, CT, CAa, LAa, CD1a, La, CTa]$$

$$= PI\_360\_DPS(Teta\_A, Teta\_B, f0a, w0, R)$$

Step 1a: Compute the ideal element values of the 360° PI-Section Digital Phase Shifter

$$C_L = \left(\frac{1}{\omega_0}\right) \tan\left(\frac{\theta_A}{2}\right) = 0$$

$$L_L = L = \frac{2C_L}{1 + \omega_0^2 C_L^2} = 0$$

$$L = L_L = 0$$

Step 1b: Compute the ideal element values of the high-pass PI-Section

$$L_H = \left(\frac{1}{\omega_0}\right) \cotan\left(\frac{\theta_B}{2}\right)$$

$$C_H = \left(\frac{1}{\omega_0^2}\right)\left(\frac{1 + \omega_0^2 L_H^2}{2L_H}\right)$$

Step 1c: Compute C_{D1} and its actual value

$$C_{D1} = C_H; \quad C_{D1-actual} = \frac{C_{D1}}{(2\pi f_{0-actual}) \times R}$$

Step 1d: Compute the ideal normalized component values of the shunt arm C_A and L_A by setting $C_L = L_L = 0$

$$C_{D2} = C_{D1}$$

$$C_A = \frac{1 + \sqrt{1 + 4\omega_0^2 L_H C_{D2}}}{2\omega_0^2 L_H}$$

$$L_A = \frac{1}{\omega_0^2 C_A}$$

Step 2: Compute the component losses for both series and shunt arms

Step 2a: Component losses for the series arm

$$r_f = \frac{672 \times 10^{-15}}{R \times C_{D1-actual}}$$

$$r_a = r_f$$

Step 2b: Component losses for the shunt arm

$$r_{f2} = r_f$$

$$r_{LA} = \frac{\omega_0 L_A}{Q_L}$$

$$G_{CA} = \frac{\omega_0 C_A}{Q_C}$$

$$G_{CD1} = \frac{\omega_0 C_{D1}}{Q_C}$$

Step 3: Computation of the State-A immittances for both series and shunt arms

Step 3a: Compute the series arm impedance z_a for State-A

$$z_a(j\omega) = r_a$$

Step 3b: Compute the shunt arm admittance y_a for State-A

$$Z_{TA} = r_{f2} + \frac{1}{G_{CA} + j\omega C_A}$$

$$Y_{TA} = \frac{1}{Z_{TA}}$$

$$y_a(j\omega) = \frac{1}{r_{LA} + j\omega L_A} + Y_{TA}$$

Step 4: Computation of the State-B immittances for both series and shunt arms

Step 4a: Computation of series arm impedance in State-B

$$z_b(j\omega) = \frac{1}{(G_{CD1} + j\omega C_{D1})}$$

Step 4b: Computation of shunt arm admittance in State-B

$$Z_{TB} = \frac{1}{G_{CA} + j\omega C_A} + \frac{1}{G_{CD1} + j\omega C_{D1}}$$

$$Y_{TB} = \frac{1}{Z_{TB}}$$

$$y_b(j\omega) = \frac{1}{(r_{LA} + j\omega L_A)} + Y_{TB}(j\omega)$$

Step 5: Generate phase and gain performance of 180° high-pass PI-DPS for State-A and State-B

$$S_{21(A,B)} = \frac{2}{zy^2 + 2zy + 2y + z + 2}; \quad \left\{ \begin{array}{l} \text{In State A}: z = z_a, \ y = y_a \\ \text{In State B}: \ z = z_b, \ y = y_b \end{array} \right\}$$

Step 6: Plot the phase and gain performance

Now, let us run an example to evaluate the performance of an 180° HPI-DPS constructed with lossy elements.

Example 12.2: Design a 180° high-pass PI Section Digital Phase Shifter for the phase shift of $\Delta\theta = 90°$. In the course of computations, assume that quality factor for inductors and capacitors are equal and they are given by $Q_L = Q_C = 10$ at $f_{0a} = 5\,GHz$. The normalized angular frequency $\omega_0 = 1$. Termination resistor is chosen as $R = 50\,\Omega$.

Solution

For this purpose, Algorithm 12.5 is programmed under MatLab program Example 12.3. This program is exactly the same one developed for Example 11.4 of Chapter 11. However, in this program 180° HPI-DPS operation is obtained by setting $\theta_A = 0°$.

Inputs to this program are defined as follows.

Inputs:

$$\theta_A = 0°; \ \theta_B = 90°; \ f_{0a} = 5 \times 10^9 \ Hz; \ \omega_0 = 1; \ R = 50 \ \Omega;$$

$$Q_L = 10; \ Q_C = 10.$$

Computational Steps:

Step 1: Call function PI_360_DPS with $\theta_A = 0°$ to compute ideal component values of the 180° HPI-Section digital phase shifter.
    Hence, we end up with

$$CD1 = 1; LA = \ 0.6180; CA = \ 1.6180;$$

$$CAa = 1.0301 \ pF; LAa = 0.98363 \ nH;$$

$$CD1a = 0.63662 \ pF$$

Step 2: Compute the component losses for both series and shunt arms

Step 2a: Component losses for the series arms as they are normalized.
$r_{f1} = r_{f2} = \frac{672 \times 10^{-15}}{R \times C_{D1-actual}} = 0.0211$ (normalized with respect to R = 50 Ω)
$r_a = r_{f1} = \ 0.1422$ (normalized with respect to R = 50 Ω)

Step 2b: Component losses for the shunt arms:

$$r_{f2} = r_{f1} = 0.0211$$

$$r_{LA} = \frac{\omega_0 L_A}{Q_L} = 0.0618$$

$$G_{CA} = \frac{\omega_0 C_A}{Q_C} = 0.1618$$

$$G_{CD1} = \frac{\omega_0 C_{D1}}{Q_C} = 0.1000$$

Steps 3–6 are programmed under Main_Example_12_2.m.

Phase shifting performance of the lossy phase shifter is shown in Figure 12.3a.

A close examination of Figure 12.3a reveals that 10% phase range bandwidth run from $\omega = 0.822$ to $\omega = 1.295$.

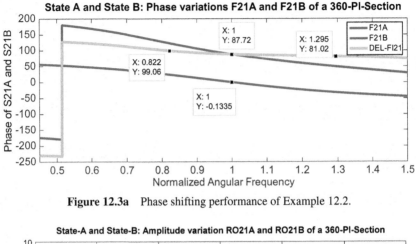

Figure 12.3a   Phase shifting performance of Example 12.2.

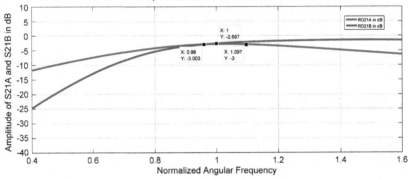

Figure 12.3b   Gain performance of Example 12.2.

Gain performance of the phase shifter is shown in Figure 12.3b. This figure reveals that half-power (–3dB) gain bandwidth runs from $\omega_1 = 0.96$ to $\omega_2 = 1.097$. It looks like that 180° high-pass PI section Digital Phase shifter topology provides narrow bandwidth, which is about 10% in the neighborhood of the operation frequency.

At this point, we should note that in State-B, loss G_{CD1} associated with the diode capacitors ($C_{D1} = C_{D2}$) may be smaller than that of $G_{CD1} = 0.1$. If this is the case, obviously, electrical performance of the 180° HPI-DPS will be better. For example, for many MMIC design processes, G_{CD1} is neglected (i.e., $G_{CD1} = 0$). If this the case, at $\omega_0 = 1$, we found that $\varphi_{21B}(\omega_0) = 89.01$ and the $20\log_{10}|S_{21B}(j\omega_0)| = -1.588$ dB, which is much better than that of –2.69 dB of Figure 12.3b.

12.5 Physical Implementation of $180°$ HPI-DPS

The $180°$ HPI-DPS may be built by using discrete component technology or MMIC technology. It is expected that discrete component technology offers much higher quality factors for the component values, which in turn improves the electric performance of the phase shifter under consideration. Diode switches may be selected as varactor diodes or PIN diodes. Switching components can be found in the literature such as [4–11]. MMIC implementation can be carried out as described in Chapter 9.

Appendix 12: MatLab Programs for Chapter 12

Program List 12.1. Main_Example_12_1.m

```
% Main Program: Main_Example_12_2.m
% March 27, 2019
% Developed by BS Yarman, Vanikoy,Istanbul
% This program evaluates the lossy performance of a 360 PI-Section DPS
% for a specified actual center frequency f0 in Hz.
%-------------------------------------------------------------------
clc; close all
% Inputs:
Teta_A=0;
Teta_B=input('Design of 180 Degree Highpass PI Section DPS. Enter
            Phase of State-B(Highpass-PI): Teta-B in Degree=')
% Inputs
f0a=input('Enter the actual center frequency in Hz f0a =')
w0=input('At f0a, enter the normalized angular frequency w0 =')
R=input('Enter the normaliziation Resistor R =')
%-------------------------------------------------------------------
% Compute the normalized element values of PI-360 Degree-DPS
[CD1,L1,LA,CA,CT,CAa,LAa,CD1a,L1a,CTa] = PI_360_DPS(Teta_A, Teta_B,
f0a,w0,R)
%-------------------------------------------------------------------
% ASSUMPTION 1:
% It is assumed that the series loss Ron of a forward biased diode is
% equal to the on channel resistor of an CMOS Switch
% [see Equation (6.1) of Chapter 6].
[RF1,rf1] =Channel_Resistance_of_a_CMOS(CD1,R,f0a);
% ASSUMPTION 2:
% At w0, quality factor for inductors and capacitors
QL=10;
QC=10;
%-------------------------------------------------------------------
CD2=CD1;
[RF,rf] =Channel_Resistance_of_a_CMOS(CD1,R,f0a);
%-------------------------------------------------------------------
% Series Arm Losses
rf1=rf;
```

```
rL1=(w0*L1)/QL;
ra=rL1+rf1;
% Shunt arm losses
rf2=rf;
rLA=(w0*LA)/QL;
GCA=(w0*CA)/QC;
GCD1=(w0*CD1)/QC;
%----------------------------------------------------------
j=sqrt(-1);
w=0;N=2000;w1=0;w2=2;DW=(w2-w1)/N;
FRI(1:(N+1))=zeros;
WA(1:(N+1))=zeros;
DEL_FI21(1:N+1)=zeros;
for i=1:N+1
    WA(i)=w;
% ---------------------
% State A:
% ---------------------
%za=ra+j*w*L1;
za=ra+j*w*L1;
% Computation of shunt arm admittance ya
% ZTA=rf2+1/(GCA+jwCA)
ZTA=rf+1/(GCA+j*w*CA);
YTA=1/ZTA;
ya=1/(rLA+j*w*LA)+YTA;
[S11a,S21a,RO11a,F11a,RO21a,F21a] = S_Par_PI_Section (za,ya);
    F21A(i)=F21a;
    RO21A(i)=20*log10(RO21a);
    F11A(i)=F11a;
    RO11A(i)=20*log10(RO11a);
% ---------------------
% State-B
zb=(rL1+j*w*L1)+1/(GCD1+j*w*CD1);
ZTB=1/(GCA+j*w*CA)+1/(GCD1+j*w*CD1);
YTB=1/ZTB;
yb=1/(rLA+j*w*LA)+YTB;
[S11b,S21b,RO11b,F11b,RO21b,F21b] = S_Par_PI_Section (zb,yb);
    F21B(i)=F21b;
    RO21B(i)=20*log10(RO21b);
    F11B(i)=F11b;
    RO11B(i)=20*log10(RO11b);
%----------------------------------------------------------
DEL_FI21(i)=F21B(i)-F21A(i);
    w=w+DW;
end
%----------------------------------------------------------
Plot_State_AB_PI_360_DPS(WA,F21A,RO21A, F11A,RO11A,F21B,RO21B, F11B,
    RO11B,DEL_FI21)
%----------------------------------------------------------
zb0=(rL1+j*w0*L1)+1/(GCD1+j*w0*CD1);
rb=real(zb0);
%
ra_actual=ra*R
```

```
rb_actual=rb*R

% GA_actual=GA/R; RA_actual=1/GA_actual
% GB_actual=GB/R; RB_actual=1/GB_actual
```

Program List 12.2. Main_Example_12_2.m

```
% Main Program: Main_Example_12_2.m
% March 27, 2019
% Developed by BS Yarman, Vanikoy, Istanbul
% This program evaluates the lossy performance of a 360 PI-Section DPS
% for a specified actual center frequency f0 in Hz.

clc; close all
% Inputs:
Teta_A=0;
Teta_B=input('Design of 180 DegreeHighpass PI Section DPS. Enter
             Phase of State-B(Highpass-PI): Teta-B in Degree=')
% Inputs
f0a=input('Enter the actual center frequency in Hz f0a =')
w0=input('At f0a, enter the normalized angular frequency w0 =')
R=input('Enter the normaliziation Resistor R=')

% Compute the normalized element values of PI-360 Degree-DPS
[CD1,L1,LA,CA,CT,CAa,LAa,CD1a,L1a,CTa] = PI_360_DPS(Teta_A, Teta_B,
    f0a,w0,R)

% ASSUMPTION 1:
% It is assumed that the series loss Ron of a forward biased diode is
% equal to the on channel resistor of anCMOS Switch
% [see Equation (6.1) of Chapter 6].
[RF1,rf1] =Channel_Resistance_of_a_CMOS(CD1,R,f0a);
% ASSUMPTION 2:
% At w0, quality factor for inductors and capacitors
QL=10;
QC=10;

CD2=CD1;
[RF,rf] =Channel_Resistance_of_a_CMOS(CD1,R,f0a);

% Series Arm Losses
rf1=rf;
rL1=(w0*L1)/QL;
 ra=rL1+rf1;
% Shunt arm losses
rf2=rf;
rLA=(w0*LA)/QL;
GCA=(w0*CA)/QC;
GCD1=(w0*CD1)/QC;

j=sqrt(-1);
w=0;N=2000;w1=0;w2=2;DW=(w2-w1)/N;
```

```
FRI(1:(N+1))=zeros;
WA(1:(N+1))=zeros;
DEL_FI21(1:N+1)=zeros;
for i=1:N+1
    WA(i)=w;
%  -------------------
%  State A:
%  -------------------
%za=ra+j*w*L1;
%za=ra+j*w*L1;
%  Computation of shunt arm admittance ya
%  ZTA=rf2+1/(GCA+jwCA)
ZTA=rf+1/(GCA+j*w*CA);
YTA=1/ZTA;
ya=1/(rLA+j*w*LA)+YTA;
[S11a,S21a,RO11a,F11a,RO21a,F21a] = S_Par_PI_Section (za,ya);
    F21A(i)=F21a;
    RO21A(i)=20*log10(RO21a);
    F11A(i)=F11a;
    RO11A(i)=20*log10(RO11a);
%  -------------------
% State-B
zb=(rL1+j*w*L1)+1/(GCD1+j*w*CD1);
ZTB=1/(GCA+j*w*CA)+1/(GCD1+j*w*CD1);
YTB=1/ZTB;
yb=1/(rLA+j*w*LA)+YTB;
  [ S11b,S21b,RO11b,F11b,RO21b,F21b ] = S_Par_PI_Section (zb,yb);
    F21B(i)=F21b;
    RO21B(i)=20*log10(RO21b);
    F11B(i)=F11b;
    RO11B(i)=20*log10(RO11b);
%  -------------------
DEL_FI21(i)=F21B(i)-F21A(i);
    w=w+DW;
end
%  -------------------
Plot_State_AB_PI_360_DPS(WA,F21A,RO21A, F11A,RO11A,F21B,RO21B,F11B,
    RO11B,DEL_FI21)
%  -------------------
zb0=(rL1+j*w0*L1)+1/(GCD1+j*w0*CD1);
rb=real(zb0);
%
ra_actual=ra*R
rb_actual=rb*R
%  -------------------
% GA_actual=GA/R; RA_actual=1/GA_actual
% GB_actual=GB/R; RB_actual=1/GB_actual
```

Program List 12.3. Function S_Par_PI_Section

```
function [ S11,S21,RO11,F11,RO21,F21 ] = S_Par_PI_Section (z,y)
% This function generates the S-Parameters of a PI Section
```

```
% Phase Shifter from the series arm impedance Z(jw) and
% the shunt arm admittance Y(jw)
%-----------------------------------------------------------
% Developed by BS Yarman: March 26,2019,Vanikoy,Istanbul
% See Equations (5.9)
%-----------------------------------------------------------
% N=z(1-y^2)-2y
N=z*(1-y*y)-2*y;
% D= zy^2+2zy+2y+z+2
D=z*y*y+2*z*y+2*y+z+2;
S11=N/D;
S21=2/D;
R11=real(S11);X11=imag(S11);F11=atan2d(X11,R11);
R21=real(S21);X21=imag(S21);F21=atan2d(X21,R21);
RO11=abs(S11);
RO21=abs(S21);


end
```

Program List 12.4. Function PI_360_DPS

```
function [CD1,L,LA,CA,CT,CAa,LAa,CD1a,La,CTa]
         = PI_360_DPS(Teta_A, Teta_B, f0a,w0,R)
% This function generates the element valuesof an ideal 360 degree
  Simple
% PI-Section based Digital Phase Shifter
% Developed by BS Yarman on March 26,2019,Vanikoy,Istanbul
% Inputs:
           % Teta_A: Desired phase shift of the Lowpass Based
                            PI-Section DPS
           % Teta_B: Desired phase shift of the Highpass Based
                            PI-Section DPS
           % f0a: Actual centre frequency
           % w0: Normalized angular frequency. It is selected
                      as w0=1
           % R: Port normalization number. It is
                        usually, selected as R=50 ohms
% Outputs:
           % CD1: Reverse Biased diode
                      capacitance of the series arms.
           % L: Series arm inductor
           % LA: Shunt arm inductor
           % CA: Shunt arm capacitor
%-----------------------------------------------------------
% Ideal LowpassPI-Section: See Equations(11.1a) and (11.1b)
% CL=(1/w0)*tan(?A/2)>0
CL=(1/w0)*tand(Teta_A/2);
% L_L=L=(2C_L)/(1+?_0^2C_L^2)>0
LL=2*CL/
  (1+w0*w0*CL*CL);
%-----------------------------------------------------------
```

```
% Ideal Highpass T-Section: See Equations(11.2a) and (11.2b)
% LH=(1/w0)*cotan(?B/2)>0
LH=(1/w0)*cotd(Teta_B/2);
% CH=(1/(?0^2))((1+?0^2 LH^2)/(2LH^2))>0
CH=(1/w0/w0)*(1+w0*w0*LH*LH)/2/LH;
%------------------------------------------------------
% State-A: Series arm component computations
L=LL;
CD1=CH/(1+w0*w0*L*CH);
%------------------------------------------------------
% State B: Computation of CA: See Equation (11.4 & 11.5)
A=w0*w0*LH;
B=-(w0*w0*LHCL+1);
CD2=CD1;
C=B*CD2;
Delta=B*B-4*A*C;
CA=(-B+sqrt(Delta))/2/A;
%------------------------------------------------------
LA=1/(CA-CL)/w0/w0;
% LA=1/(?0^2)/(CA-CL))>0
CT=CA*CD2/(CA+CD2);
%------------------------------------------------------
CAa=CA/2/pi/f0a/R;
CD1a=CD1/2/pi/f0a/R;
LAa=R*LA/2/pi/f0a;
La=R*L/2/pi/f0a;
CTa=CT/2/pi/f0a/R;

end
```

References

[1] Binboga S. Yarman, "π-section digital phase shifter apparatus," U.S. Patent: 4604593, Washington DC, USA, August 5, 1986.

[2] B. S. Yarman, "New Circuit Configurations for Designing 0–180° Digital Phase Shifters," *IEE Proceeding H*, Vol. 134, pp. 253–260, 1987.

[3] B. S. Yarman, "Novel Circuit Configurations to Design Loss Balanced 0–360° Digital Phase Shifters," *AEU*, Vol. 45, No. 2 , pp. 96–104, 1991.

[4] Pat Hindle, Editor, "The State of RF and Microwave Switches," *Microwave Journal*, vol. 53, no. 11, p. 20, November 2010.

[5] I. Bahl and B. Prakash, Microwave Solid State Circuit Design, 2. (1. John Wiley & Sons, Ed., 2003).

[6] I. Bahl, Lumped Elements for RF Microwave Circuits, Artech House, 2003.

[7] K. J. Koh and G. M. Rebeiz, "0.13-m CMOS phase shifters for and K-band phased arrays," *IEEE Trans. on MTT*, Vol. 42, No. 11, p. 2535–2546, November 2007.

[8] K. Chang, Handbook of RF Microwave Components and Engineering, New York, USA, Chechester, UK: Wiley, 2003.

[9] D. Kang and S. Hong, "A 4-bit CMOS Phase Shifter Using Distributed Active Switches," *IEEE Trans on MTT*, Vol. 55, No. 7, pp. 1476–1483, 2007.

[10] R. Melik and H. V. Demir, "Implementation of High Quality Factor, on-chip Tuned Microwave Resonators at 7 GHz," *Microwave and Optical Technology Letters*, vol. 51, no. 2, pp. 497–501, February 2009.

[11] H. Fang, T. Xinyi, K. Mouthaan and R. Guinvarch, "Two Octave Digital All-Pass Phase Shifters for Phase Array Applications," *IEEE Radio and Wireless Symposium (RWS)*, pp. 169–171, 2013.

Index

About the Author

Prof. Dr. Binboga Siddik Yarman completed his BSc. in Technical University of Istanbul, Turkey; MSc. Stevens Institute of Technology, Hoboken, NJ, USA, in 1977; and PhD at Cornell University, Ithaca, NY, USA, in 1982.

He is an expert in designing wideband matching networks, broadband RF and microwave amplifiers, and digital phase shifters.

He designed many low-noise small signal and power amplifiers for major government agencies such as US Army, Navy, and Air Force as well as Comsat, INTELSAT, Hughes, TRW, AT&T Bell Labs during his work at RCA David Sarnoff Research Center, Princeton, NJ, USA.

He also designed high-power broadband amplifiers for US security forces during his professional consultancy assignment to MOTOROLA Advance Solutions Labs, in Penang, Malaysia.

He used to teach in many universities as a Professor of Electronic Circuits and Microwave Engineering at Stevens Institute of Technology, Hoboken, NJ, USA; Cornell University, Ithaca, NY, USA; Technical University of Istanbul; Istanbul University; Middle East Technical University of Turkey; Ruhr University of Germany; Wuhan Technological University of China; and Tokyo Institute of Technology of Japan.

Currently, he is the chairman of the Department of Electrical and Electronics Engineering of Istanbul University.

He published many papers in major microwave and circuit and systems transactions such as IEEE CAS and IEEE MTT, IET of UK and JEE of Japan, AEU of Germany, etc. in designing amplifiers, antenna matching networks, and digital phase shifters.

He is the inventor/co-inventor of various "Real Frequency Techniques" to design broadband power transfer networks, RF and microwave amplifiers, and digital phase shifters.

He is the author/co-author of major text/research books such as "Design of Ultra-Wideband Antenna Matching Networks" by Springer, 2008, "Design of Ultra-Wideband Power Transfer Networks" by Wiley 2010, and

"Design of Broadband RF and Microwave Amplifiers" by CRC of Taylor & Franchise 2016.

Beyond many others, he holds four US patents in digital phase shifters.

He has been elected to New York Academy of Science in 1994 and named as "Man of the Year in Science and Technology" by Cambridge Biography Center of UK in 1999.

He is a Fellow IEEE, Fellow of Salzburg, and Fellow of Alexander von Humboldt of Germany.